AF780924

AI for Smart Home Automation

AI for Smart Home Automation

Editor

Daniele Cenni

MDPI • Basel • Beijing • Wuhan • Barcelona • Belgrade • Manchester • Tokyo • Cluj • Tianjin

MDPI

Editor
Daniele Cenni
Università degli Studi di
Firenze
Firenze, Italy

Editorial Office
MDPI
St. Alban-Anlage 66
4052 Basel, Switzerland

This is a reprint of articles from the Special Issue published online in the open access journal *Sensors* (ISSN 1424-8220) (available at: https://www.mdpi.com/journal/sensors/special_issues/home_auto).

For citation purposes, cite each article independently as indicated on the article page online and as indicated below:

LastName, A.A.; LastName, B.B.; LastName, C.C. Article Title. *Journal Name* **Year**, *Volume Number*, Page Range.

ISBN 978-3-0365-8172-9 (Hbk)
ISBN 978-3-0365-8173-6 (PDF)

© 2023 by the authors. Articles in this book are Open Access and distributed under the Creative Commons Attribution (CC BY) license, which allows users to download, copy and build upon published articles, as long as the author and publisher are properly credited, which ensures maximum dissemination and a wider impact of our publications.
The book as a whole is distributed by MDPI under the terms and conditions of the Creative Commons license CC BY-NC-ND.

Contents

About the Editor

Daniele Cenni

Daniele Cenni was a research fellow at the Department of Information Engineering of the University of Florence from 2007 to 2021. He received a Master's degree in Informatics Engineering (Laurea) and a PhD in Telematics and Information Society from the University of Florence. He has worked on many projects funded by the European Commission. He currently works at the "Area for Innovation and Management of Information and Computing Systems" at the University of Florence. His research interests include deep learning, cloud computing, information retrieval, smart cities, and GIS.

Preface

The rapid development of Artificial Intelligence techniques has enabled benefits at various levels in the Information Society. In the field of smart home automation, there has been a proliferation of new application paradigms, enabling more efficient home activities, optimizing the use of devices, and simplifying and automating tasks for users. Various aspects of home life, including home security, control of lights, heating, ventilation, air conditioning, and appliance control, need innovative solutions that can learn from the numerous data made available by home devices. Aspects that benefit from the application of the increasingly complex and specific AI techniques include event-driven scheduling of activities, indoor climate, remote control, energy optimization and monitoring, and the definition of home automation hubs for automated control and management of smart devices through the use of desktop or mobile applications. This Special Issue arises from the need to gather the most recent contributions on some of the aforementioned aspects in the rapidly growing field of Smart Home Automation, which is introducing new and increasingly innovative developments.

Daniele Cenni
Editor

sensors

MDPI

Article

A Secure and Smart Home Automation System with Speech Recognition and Power Measurement Capabilities †

Chandra Irugalbandara [1,2], Abdul Salam Naseem [1], Sasmitha Perera [1], Sithamparanathan Kiruthikan [1] and Velmanickam Logeeshan [1,*]

1 Department of Electrical Engineering, University of Moratuwa, Moratuwa 10400, Sri Lanka; chandra.legendary@gmail.com (C.I.); nazeemthebeta@gmail.com (A.S.N.); sasmithahasanthip@gmail.com (S.P.); kiruthikan011@gmail.com (S.K.)
2 BCS Technology International Pty Ltd., Colombo 00700, Sri Lanka
* Correspondence: logeeshanv@uom.lk

† This paper is an extended version of our paper published in 2022 IEEE World AI IoT Congress (AIIoT), Seattle, WA, USA, 6–9 June 2022 .

Abstract: The advancement in the internet of things (IoT) technologies has made it possible to control and monitor electronic devices at home with just the touch of a button. This has made people lead much more comfortable lifestyles. Elderly people and those with disabilities have especially benefited from voice-assisted home automation systems that allow them to control their devices with simple voice commands. However, the widespread use of cloud-based services in these systems, such as those offered by Google and Amazon, has made them vulnerable to cyber-attacks. To ensure the proper functioning of these systems, a stable internet connection and a secure environment free from cyber-attacks are required. However, the quality of the internet is often low in developing countries, which makes it difficult to access the services these systems offer. Additionally, the lack of localization in voice assistants prevents people from using voice-assisted home automation systems in these countries. To address these challenges, this research proposes an offline home automation system. Since the internet and cloud services are not required for an offline system, it can perform its essential functions, while ensuring protection against cyber-attacks and can provide quick responses. It offers additional features, such as power usage tracking and the optimization of linked devices.

Keywords: home automation systems; speech recognition; natural language understanding; smart plug socket

Citation: Irugalbandara, C.; Naseem, A.S.; Perera, S.; Kiruthikan, S.; Logeeshan, V. A Secure and Smart Home Automation System with Speech Recognition and Power Measurement Capabilities. *Sensors* **2023**, *23*, 5784. https://doi.org/10.3390/s23135784

Academic Editor: Daniele Cenni

Received: 14 February 2023
Revised: 4 March 2023
Accepted: 24 April 2023
Published: 21 June 2023

Copyright: © 2023 by the authors. Licensee MDPI, Basel, Switzerland. This article is an open access article distributed under the terms and conditions of the Creative Commons Attribution (CC BY) license (https://creativecommons.org/licenses/by/4.0/).

1. Introduction

The popularity of home automation systems is increasing in a fast developing world, offering comfort, convenience, and safety to users. These systems, especially those that are voice-activated, have particularly helped elderly people and those with disabilities [1]. These systems are developed with a central controller, that controls appliances, such as power outlets, temperature sensors, lights, security systems, and emergency systems. A major advantage to the user is that the connected devices can be managed and controlled remotely using various devices, such as smartphones, laptops, tablets, desktops, and even voice commands in the latest home automation systems.

The widespread availability of Wi-Fi-enabled devices has exponentially increased the adoption of smart home systems [1]. These systems offer benefits, such as energy savings, ease of use, time conservation, and a better quality of life. As the internet of things continues to advance, home automation system developers are now utilizing cloud-based systems.

The popularity of home automation systems with voice assistance has also risen among consumers. Devices, such as Amazon Echo, Google Home, and Apple HomePod, have become integral to the smart home experience. Many smart appliance manufacturers have incorporated one or more of these voice assistants to increase the sales of their products.

Speech recognition is a crucial component in home automation systems. In the existing automation systems, voice recognition services are performed by third-party cloud services, such as Wit.ai, IBM Watson, Google Cloud Speech, Microsoft Cognitive Services, etc. However, if the internet connection is unstable, the dependency on cloud services can lead to the failure of voice recognition features.

Additionally, these devices must be always connected to the internet to maintain the connection to the cloud, usually through the home Wi-Fi. With increasing cyber-security threats, Wi-Fi connections are becoming increasingly vulnerable, and having a system that can control the household devices attached to it makes it much more dangerous for users. There have been allegations that companies, such as Google and Amazon, have acquired private information from users through their home automation systems, which raises concerns about privacy violations [2]. This results in a trade-off between user privacy and convenience that should not be ignored.

Furthermore, while well-known home automation systems are popular and reliable in some areas of the world, they are highly unreliable in developing nations due to poor internet connectivity. In addition, the lack of localization in voice assistants also restricts the use of home automation systems in developing countries, which makes the home automation industry restricted to specific regions of the world.

The aforementioned limitations of home automation systems are due to the reliance on cloud-based services [3]. This creates a need for an offline-based voice-assisted home automation system. Several studies have proposed different methods to implement an offline-based, voice-assisted, home automation system that does not depend on cloud-based services. However, there are several limitations in their implementation, which are discussed in the literature review section. This research proposes an offline voice-assisted home automation system that mitigates these limitations and has significant improvement in performance. In addition, it requires less memory and lower computational requirements compared to the existing methods. The developed system includes a compatible smart plug in addition to the basic smart home functionality, such as voice assistance, relay control, and energy consumption tracking with improved privacy.

This paper is organized as follows. Section 2 discusses the literature review. Section 3 presents the architecture of the system. Section 4 describes the implementation of the proposed approach. Section 5 presents the results of the evaluation carried out. Finally, Section 6 concludes the outcome of the research.

2. Literature Review

Researchers have proposed various methods to implement offline voice-assisted home automation systems. Errobidart et al. proposed an offline demotic system with voice commands, and is shown in Figure 1 [4]. The system consists of an EasyVR speech recognition module and an Arduino Mega. Since it lacks natural language processing (NLP), a pre-defined set of commands must be assigned to each task. In addition, the cost of the EasyVR shield and the Arduino Mega raises the total cost of the device to a level higher than existing home automation systems on the market [5]. The system was developed with two communication protocols. However, it can only maintain a proper connection up to 50 cm.

G. et al. [6] proposed a method that employs the hidden Markov model toolkit (HTK) to convert speech into text. The overview of the system is shown in Figure 2. It uses a GSM module to transfer signals between the hub and smart plug via SMS. However, the HTK has limitations in terms of shorter time intervals that impacts performance. Natural language processing (NLP) is not utilized in this system either. In addition, the hub does not have an inbuilt microphone.

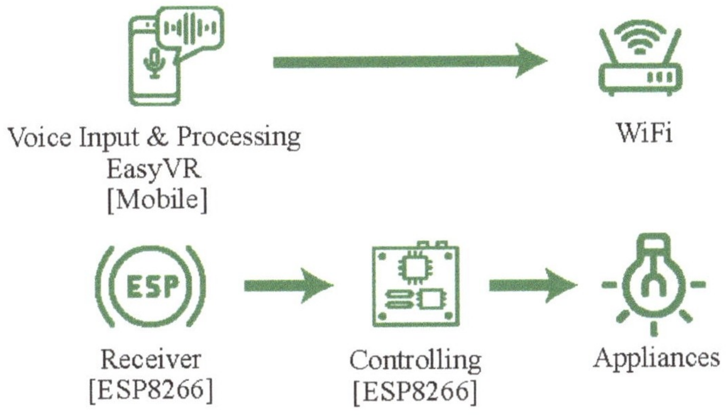

Figure 1. Offline demotic system using voice commands.

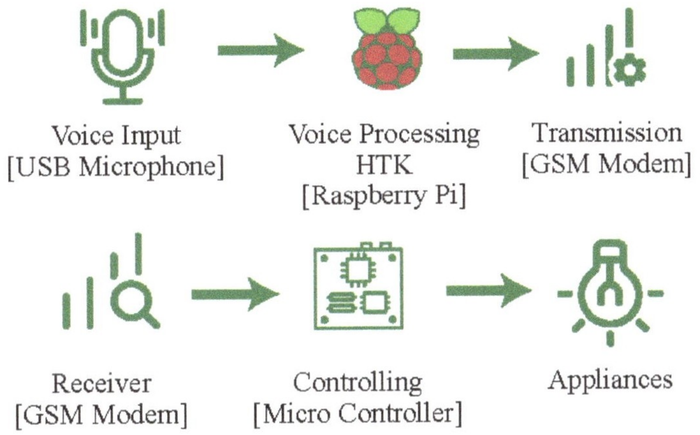

Figure 2. Low-cost home automation system using offline speech recognition.

Elsokah et al., proposed a next-generation home automation system that is based on voice recognition and that uses an Easy VR 2.0 shield in combination with an Arduino microcontroller [7]. The overview of the system is shown in Figure 3. The system communicates between the hub and smart plug through a Wi-Fi module and has the added advantage of incorporating environmental inputs, such as humidity and temperature. However, the number of commands that can be executed is limited due to the use of the Easy VR shield.

Rani et al. proposed a system that utilizes natural language processing (NLP) and uses a mobile phone for voice input and processing, as illustrated in Figure 4 [8]. The commands are then sent to the Arduino, that acts as a controller in a smart plug through Wi-Fi. However, a significant drawback of this system is its reliance on a mobile phone.

Our proposed offline home automation system delivers a significant advancement in the field of smart homes. One of the primary advancements of our system is the significant performance improvements over existing home automation systems. Our system features faster response times, higher accuracy, and efficient utilization of resources. This enables users to control their smart homes quickly, without experiencing frustrating delays or inaccurate responses.

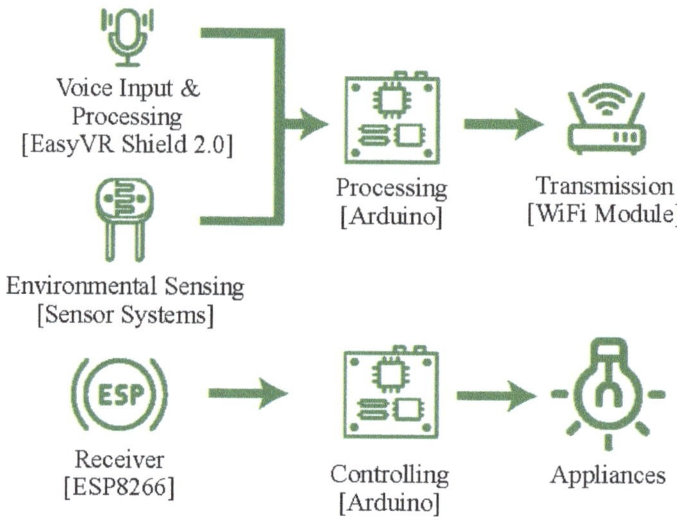

Figure 3. Next generation home automation system based on voice recognition.

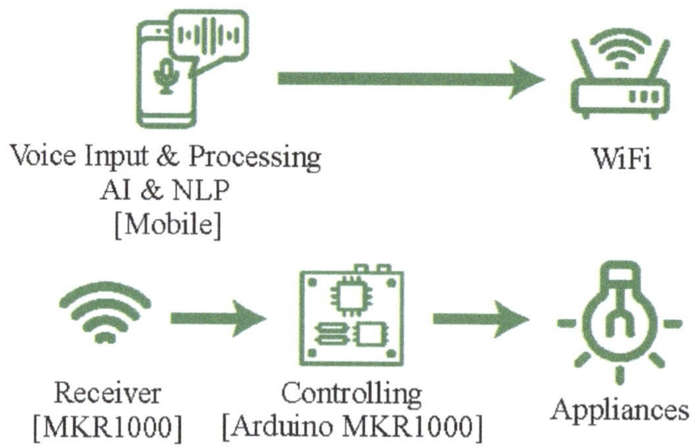

Figure 4. Voice-controlled home automation system using natural language processing and the internet of things.

Another key advantage of our system is the reduction in memory and computational requirements. This results in lower hardware requirements and reduced power consumption than the existing systems, which makes it more accessible and affordable to a wider range of users.

Overall, our offline home automation system represents a significant step forward in the field of smart homes. Due to the improved performance and reduced computational requirements. Our system provides valuable benefits to the users who are looking to simplify and streamline their home automation tasks.

3. System Architecture

The proposed "HomeIO" is a low-cost, offline, and versatile home automation system with built-in speech recognition and intention detection. Its goal is to simplify the system and reduce its cost. This will eliminate the need for high-performance cloud computing which in turn will reduce the privacy and cyber security risks. The system is also flexible, allowing new appliances from other manufacturers to be easily added to the network for safe and secure operation. Additionally, the system includes smart plug sockets with improved distance connectivity and energy measurement capabilities. Figure 5 shows the system architecture consisting of a smart hub and smart plug socket.

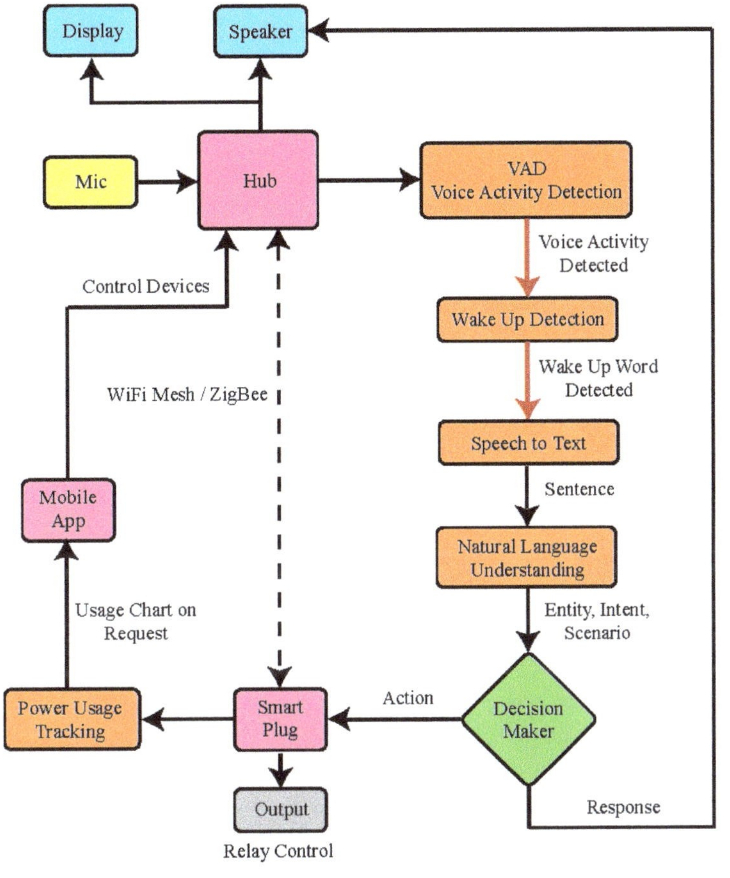

Figure 5. System architecture.

This section is divided into three subsections. The first subsection describes the smart hub component. The second subsection describes the smart plug socket component, and the third subsection describes the communication protocols and mediums used for connectivity.

3.1. Smart Hub

In a home automation system, the smart hub serves as the central control point for intercommunication among components. There can be one or more smart home hubs or none at all. Typically, smart hubs rely on cloud processing to handle the demands of voice assistants, which can result in inoperability if cloud services are unavailable. To overcome this, HomeIO's smart hub features an on-device speech-based user interface, that allows users to communicate with the system without the need for third-party services.

Figure 6 shows an overview of the speech recognition system used in HomeIO. The system consists of four components: a voice activity detection (VAD) model, a wake word detection model, a speech-to-text model, and a natural language understanding model. The VAD model has an algorithm that constantly monitors the surroundings for speech signals and uses VADs to identify speech segments with lower energy consumption. Upon detecting a voiced segment, the wake word detection model is triggered. If the wake word is detected, the speech-to-text model converts the audio signals into sentences. The natural language understanding engine determines the scenario, intent, and entity from the sentence. This is passed on to the control system to make a decision.

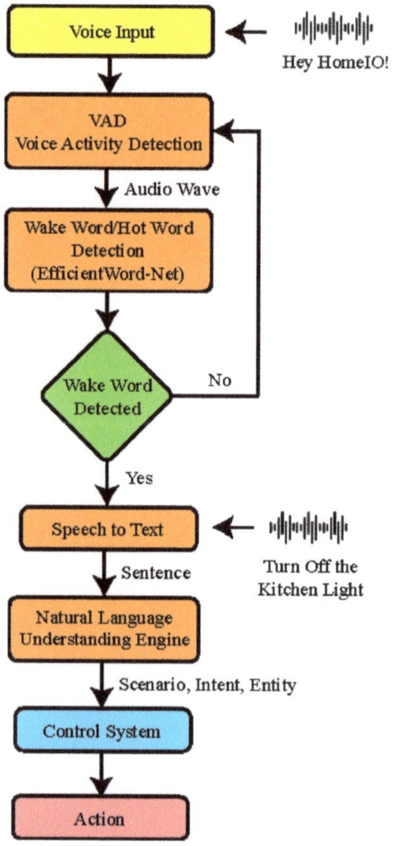

Figure 6. Overview of the speech recognition system.

3.1.1. Voice Activity Detection (VAD)

Voice activity detection (VAD) is an essential component of many speech signal processing programs that separates audio streams into periods of speech activity and periods of silence. It operates continuously when the device is on. Therefore, its algorithm should have a low energy consumption. VAD recognizes when a person is speaking to the device and helps determine when the person has stopped speaking. The VAD algorithm used in this proposed home automation system is based on the open-source Google WebRTC voice activity detector, written in C language for real-time web communications [9]. The system uses Gaussian mixture models (GMMs) to distinguish between voiced and unvoiced speech segments. The VAD only supports 16-bit mono PCM audio with several preset sample rates and frame intervals.

3.1.2. Automatic Speech Recognition (ASR)

Since the speech recognition in HomeIO takes place on the device, the models used must be optimized for small sizes and low computational demands. In order to achieve this, the speech recognition system is divided into two main components: the speech-to-text model and the natural language understanding (NLU) model.

The goal of the speech-to-text (STT) model is to analyse an audio signal, break it down, digitize it into a machine-readable format, and produce the most appropriate text representation. In order to achieve this, the speech-to-text systems rely on two types of models: acoustic models and language models. Our implementation of STT utilizes a simple convolutional neural network (CNN), as shown in Figure 7. The model predicts the letters pronounced by the user, and once there is a silence between words, the predicted letters are passed through a connectionist temporal classification (CTC) beam search decoder with lexicon constraints and a language model to obtain the best estimate of the sentence. the model was trained on Mozilla's common voice English dataset that contains 2886 h of audio data labelled by 79,398 different voices. For the language model, a pre-trained KenLM model is used.

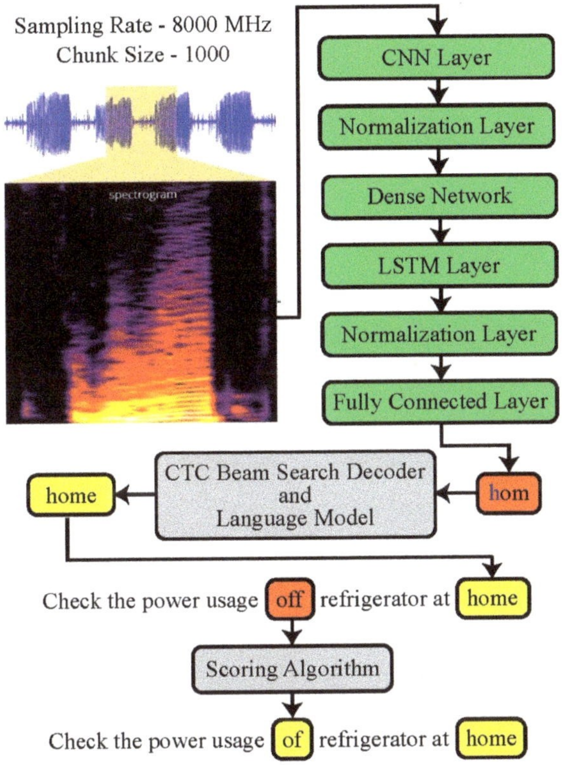

Figure 7. Speech to text model.

The input sentence must be translated into a machine-readable format for the smart hub to understand and execute the instruction. In the past, simple if-else statements were used to check for specific terms in the text, but these methods cannot capture key aspects, such as time, location, and intention. With advancements in NLP and ML, it is now possible to extract meaningful information from a sentence. Our system uses a pre-trained bidirectional encoder representations from transformers (BERT) model. The overview of this model is shown in Figure 8. It extracts the mask from the input sentence, that is then sent to an artificial neural network (ANN), that determines the scenario, intention, and entity from the input sentence [10,11]. An example of this would be the sentence "Turn off the light in the kitchen", where the scenario is "IoT", the intention is "Turn OFF", and the entities are "Kitchen: location" and "lights: device". These variables can then be used in control logic to perform the desired task. The model was trained using open-source data.

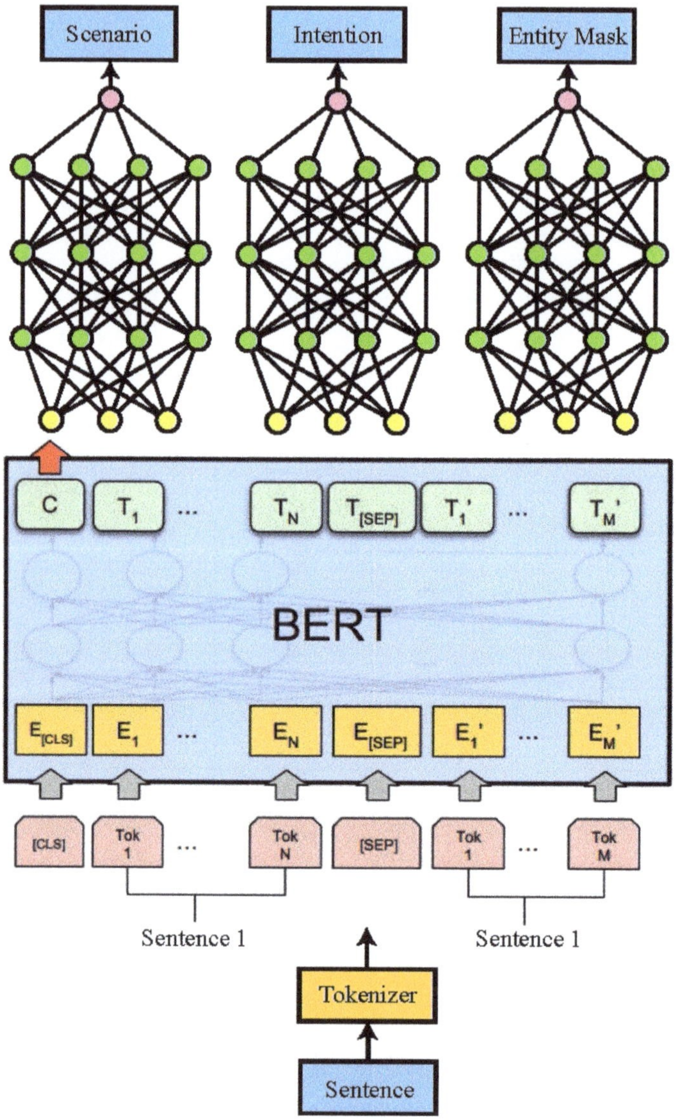

Figure 8. Natural language understanding model.

3.1.3. Device Power and Security Management

The HomeIO system includes a feature to monitor the power usage of the connected devices. This is accomplished through an on-device database that tracks power consumption and temperature readings from sensors on smart devices, and provides real-time updates to the user. This information can be used to set schedules and rules for switching on and switching off of the devices. The mesh network's security is ensured through a two-layered security channel, where a user must know the username and password for the smart hub and the decryption key for each connected device to access it.

3.2. Smart Plug Socket

The smart plugs are designed to be placed between the wall socket and electrical appliances, that enables the user to switch on and switch off the appliance using a smartphone or voice commands via wireless communication. Smart plugs have become a necessity in home automation systems as they help save household energy while maintaining the user's comfort and enhancing their way of life. In comparison, using traditional power sockets requires physical interaction to switch on and switch off the appliances, and they also lack the ability to remotely monitor the appliance's status or measure power usage.

While some smart plugs on the market have energy management capabilities, many brands only offer this feature in a more expensive pro version due to the size and cost of the energy measurement unit. Measuring energy usage is crucial in promoting a more sustainable and environmentally friendly energy use [12].

3.2.1. Power Usage Tracking

According to a study by Ahmed et al. in 2015, a power node was developed to monitor the power usage in smart plugs [13]. It consisted of a voltage sensor, a current sensor, and a Zigbee microcontroller. However, this design has limitations, such as a slow sampling rate, poor communication coverage, and high cost.

The proposed smart plug design includes a feature to monitor the real-time power consumption of an electrical appliance. In the design, HLW8012, a single-phase energy monitoring integrated circuit, is used to gather data, such as RMS current, RMS voltage, and RMS active power. This IC is directly connected to an ESP32 Node MCU, that allows for fast power usage calculations. The use of an HLW8012 power sensor instead of separate current and voltage sensors makes the design more compact and cost-effective.

3.2.2. Relay Operation

The smart plug uses a 5 V electro-mechanical relay to control the connected appliances. The relay is activated by a DC current, that opens or closes the switch contacts. The ESP32 Node MCU receives commands from the user, either through a voice command or a control action using the mobile app, and operates the relay contacts accordingly.

3.3. Connectivity

The connectivity between the hub and devices is a crucial aspect of any home automation system, as it determines its reliability. Communication protocols, such as BLE, Zigbee, Z-Wave, and Wi-Fi are commonly used by home automation systems, each with its own advantages and disadvantages [14]. Factors, such as range, bandwidth, interference resistance, and energy consumption, impact the stability of the connection. The cost-effectiveness, security, and ease of configuration are important considerations for customers [15].

When choosing a communication protocol for HomeIO, which is an offline home automation system that prioritizes security and reliability, factors, such as resistance to external attacks, sufficient bandwidth for real-time data collection from power-monitoring devices, and ease of connection, must be considered. Therefore, a Wi-Fi-based wireless mesh network is the best communication solution for HomeIO.

In order to mitigate the inherent constraints of stand-alone networking systems, such as signal loss as devices move away from the router and interference from electrical equipment, mesh networking was selected. Mesh networks can self-organize and configure dynamically, resulting in reduced installation time. Self-configuration enables dynamic workload distribution. This is useful if several nodes fail simultaneously and results in an improved fault tolerance and maintenance costs. The use of mesh networking in home automation, combined with a suitable communication protocol, creates a reliable, low-maintenance, and fault-tolerant device-to-device or hub-to-hub connection that can work well even in the presence of walls or other electronic devices that may interfere with communication.

Message queuing telemetry transport (MQTT) is a standard IoT messaging protocol developed by the Organization for the Advancement of Structured Information Standards (OASIS). The overview of the protocol is shown in Figure 9. It is a widely used messaging protocol in the IoT field because of its simplicity and security features. It has a low overhead for both coding and network traffic. It uses transport layer security (TLS) encryption and authenticates clients using an open-standard authorization (OAuth) protocol to ensure secure communication between devices [16]. In the HomeIO system, MQTT will be used as the messaging protocol due to its ease of use and compatibility with the programming languages used in the smart hub and smart devices.

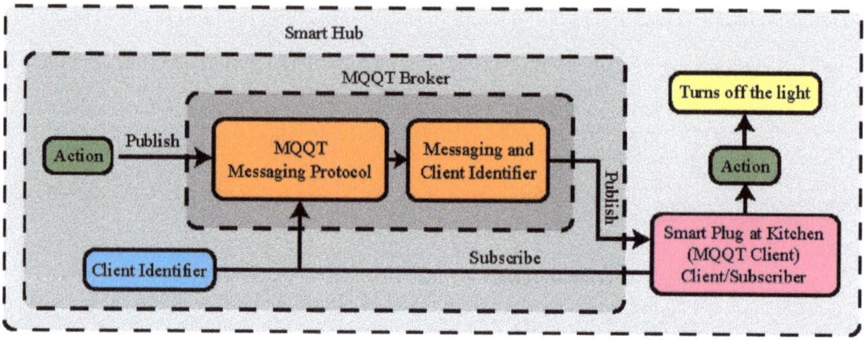

Figure 9. Connection between the smart hub and smart plug socket using MQQT.

4. Implementation

The proposed system consists of two separate devices, a smart hub and a smart plug socket that were created separately and were tested for communication. The prototype includes an offline speech recognition system, power monitoring capability, and the ability to control the smart plug through a relay. The two components are discussed in detail in the following subsections.

4.1. Smart Hub

4.1.1. Hardware Implementation

The smart hub device consists of a Raspberry Pi 4 with 4 gigabytes of memory, a 5 V/2 A power adapter connected to its USB-C port, and a ReSpeaker 2-mic Array Pi HAT as the microphone.

4.1.2. Software Implementation

The configuration of the speech recognition system is optimized for the highest accuracy possible. The voice activity detection model runs continuously while the device is on. All of the other models (wake-word detection, speech-to-text, and natural language processing) are executed in a single script. The firmware components are stored on an SD card, since the Raspberry Pi module does not have an internal memory. The microphone listener event is executed in a separate thread and updates a global circular queue. The VAD takes chunks of audio from the queue using a specific chunk size and determines if the chunk contains speech. If there are 10 consecutive chunks without speech, the process stops and the full command, including the wake-word, is transcribed to text and parsed through the natural language processing for entities, intent, and scenario. These are then sent to the controller to carry out the desired task.

4.2. Smart Plug Socket

For the smart plug socket, an ESP32 NodeMCU microcontroller was chosen to gather sensor data and act as the controller. The necessary codes for the relay operation and energy measurement were written using the Arduino Integrated Development Environment (IDE).

The smart plug socket is powered by a 5 V DC power supply module. To ensure extra safety, a fuse has been included in the relay control circuit design. The 5 V relay has five pins: End 1, End 2, Common (COM), Normally Closed (NC), and Normally Open (NO). End 1 and End 2 pins are used to activate the relay. These are connected to 5 V and the ground, respectively. To control a household appliance, one end should be connected to the common (COM) pin of the relay and the other end should be connected to either the Normally Open (NO) or Normally Closed (NC) pin.

The smart plug socket uses the HLW8012 breakout board for energy measurement. A schematic and a printed circuit board have been designed for the smart plug socket, where all of the modules can fit in a small space. The schematic diagram of the smart plug socket is shown in Figure 10. The smart socket plug is designed to monitor the power usage when it is not connected to the hub, and to transmit the recorded data when it is connected to the hub.

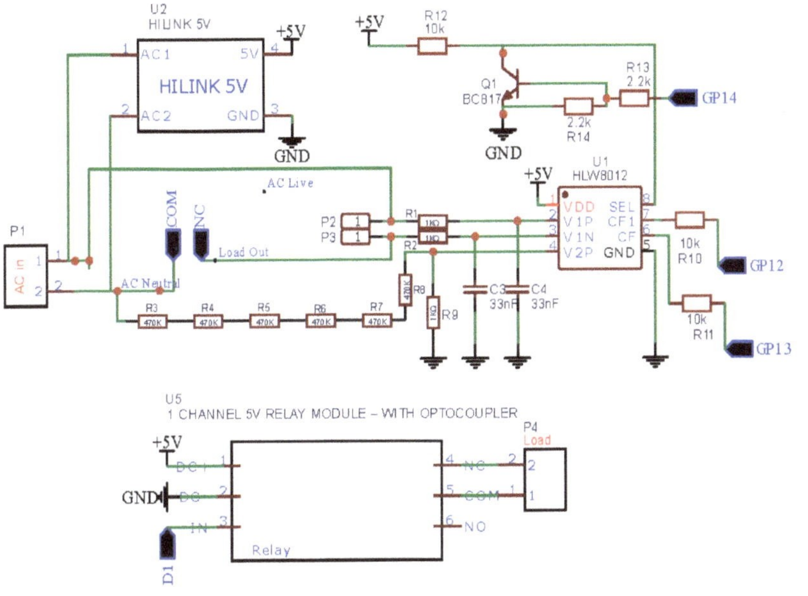

Figure 10. The schematic diagram of the smart plug socket.

4.3. Connectivity

The communication between the smart hub and the smart plug socket is facilitated by Node-RED installed on the Raspberry Pi. The Raspberry Pi functions as the server, while the NodeMCU functions as a client. The MQTT protocol is employed to enable two-way communication with the MQTT broker located in the smart hub, and acts as the server that receives messages from the clients (the smart plug socket) in the network. The mac address of the NodeMCU in the smart plug socket is used as the identifier to establish connections within the Wi-Fi mesh network. To connect a new smart plug socket to the hub, a user needs to scan a QR code or enter the mac address via a mobile app made to connect devices and display the energy consumption. Then, they can assign the device a name (e.g., "kitchen light"), type (e.g., "light"), and location (e.g., "kitchen"), that allows them to control the specific smart plug socket through commands.

4.4. Dashboard

The system includes a dashboard created using ReactJS as the central interface for users to connect with the smart hub and smart plug socket. The interface of the dashboard is shown in Figure 11. Figure 12 shows the measured power usage data of a selected appliance. This dashboard serves as a tool for managing devices and monitoring them, and runs locally on the smart hub. It can be accessed from any device that is connected to the same Wi-Fi mesh network and provides the following user-friendly features to make the operation of the home automation system easier.

1. Ability to customize the connected devices to the hub;
2. Monitor/Change the on-off status of the connected devices;
3. Gives a graphical view of the energy consumption data.

Figure 11. Dashboard interface.

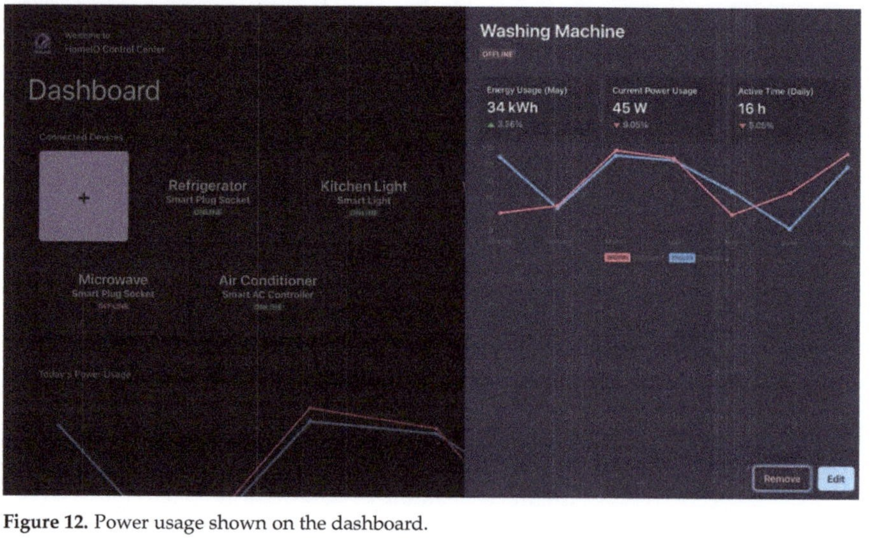

Figure 12. Power usage shown on the dashboard.

This user-friendly dashboard gives the user the ability to easily adjust the number of connected devices. Devices can be quickly added or removed from the network, and the real-time on-off status of each device can be monitored. The user can also turn the devices on or off with a simple toggle operation, which enables remote control and reduces energy waste.

The dashboard plays a crucial role in helping users understand their energy consumption patterns by providing detailed information about the energy consumption of their home for a specified period of time. This can aid in reducing the energy waste and promote energy efficiency.

5. Evaluation and Results

Evaluation of the smart plug was carried out in stages, starting with individual units and then as a complete system. The accuracy of the power measurement in the smart plug ranges from 90–95%. The wireless mesh network was able to efficiently self-organize and continuously transmit data without failure within a range of 8–10 m.

The automatic speech recognition (ASR) system was installed on a Raspberry Pi 4 with 4 GB RAM. The system required 50% of the RAM and less than 40% of the CPU for simultaneous operation of all three models, leading to fast results. The ASR was able to accurately detect English language commands and execute relay operations. Results showed that HomeIO's STT models performed better than other popular STT models in terms of word error rate (WER) and memory usage, as shown in Table 1. Compared to the existing models, our model requires significantly less memory, and has fewer parameters, without a significant drop in accuracy. Memory requirements are reduced by using quantization and pruning. Lazy loading is utilized to reduce the inference time of the neural network, which enables to reduce the computational requirements without increasing the response time. Among the existing models, Quartz Net [17] has the lowest word error rate. However, it has 18.9 Million parameters and requires significantly high memory and computational power. Compared to this model, our model has only 4.3 million parameters, which is nearly 75% less. This saves a lot of memory and computational power, which results in a lower power consumption. Even though the our model requires 180 MB of storage, it can be stored on an SD Card, which is economically feasible. The accuracy of the NLU model in a private dataset was as high as 96%.

Table 1. Performance comparison of the existing STT models and our model.

Models	Model Size (MB)	No of Parameters (Million)	WER
Jasper [18]	1230	201	3.23
Wav2Letter++ [19]	2870	208	3.26
Quartz Net 15 × 5 [17]	81.1	18.9	2.96
CMU-Sphinx (HMM) [20]	70	-	11.4
Deep Speech 2 [21]	1100	47.2	6.71
Our model (Not Optimized, No LM)	55.5	4.3	8.14
Our model (traced, No LM)	18.5	4.3	6.30
Our model (traced, with LM)	180	4.3	4.61

6. Conclusions

The advancement in the internet of things (IoT) has revolutionized home automation systems and has made people's lives more convenient and comfortable. Voice-assisted home automation systems have been especially beneficial for elderly people and those with disabilities. However, the dependency on cloud-based services has made these systems vulnerable to cyber-attacks, particularly in developing countries where the quality of the internet is low. To address these challenges, this research proposes an offline home automation system that operates independent of internet and cloud services. This system ensures protection against cyber-attacks, provides quick responses, and offers additional features, such as power usage tracking and the optimization of linked devices.

There are several potential areas for improvement that we can explore in the future. One potential area for improvement is the accuracy. While our system has already demonstrated impressive results, we believe that there may be room for further improvement. We are currently experimenting with different deep learning model architectures and parameter configurations to find the optimal setup for an automation system. We can also focus on enhancing the user experience of our system. This involves developing a more intuitive user interface. By improving the user experience, we can make our system more appealing to a wider range of users.

Finally, we can focus on enhancing the security of our system. Even though our system does not rely on cloud-based services, it relies on Wi-FI mesh networks to function, which has a certain level of security concern. We can focus on improving the security of our system.

By exploring these potential areas for improvement, we can ensure that our offline home automation system remains competitive and continues to provide value to our users.

Author Contributions: Conceptualization, V.L.; methodology, C.I., A.S.N. and V.L.; software, C.I. and S.P.; validation, A.S.N.; formal analysis, S.K.; investigation, S.P. and S.K.; resources, V.L.; data curation, C.I. and A.S.N.; writing—original draft preparation, V.L. and S.K.; writing—review and editing, V.L. and S.K.; visualization, S.P. and S.K.; supervision, V.L.; project administration, V.L. All authors have read and agreed to the published version of the manuscript.

Funding: This research received no external funding.

Institutional Review Board Statement: Not applicable.

Informed Consent Statement: Not applicable.

Data Availability Statement: Not applicable.

Acknowledgments: We would like to acknowledge BCS Technology International Pty Ltd. for their technical support provided to this research.

Conflicts of Interest: The authors declare no conflict of interest.

Abbreviations

The following abbreviations are used in this manuscript:

IoT	Internet of Things
NLP	Natural Language Processing
HTK	Hidden Markov Model Toolkit
VAD	Voice Activity Detection
GMM	Gaussian Mixture Model
ASR	Automatic Speech Recognition
NLU	Natural Language Understanding
STT	Speech-to-Text
CNN	Convolutional Neural Network
CTC	Connectionist Temporal Classification
BERT	Bidirectional Encoder Representations from Transformers
ANN	Artificial Neural Network
MQTT	Message Queuing Telemetry Transport
OASIS	Organization for the Advancement of Structured Information Standards
TLS	Transport Layer Security

OAuth	Open-standard Authorization
NC	Normally Closed
NO	Normally Open
IDE	Integrated Development Environment

References

1. Katuk, N.; Ku-Mahamud, K.R.; Zakaria, N.H.; Maarof, M.A. Implementation and recent progress in cloud-based smart home automation systems. In Proceedings of the 2018 IEEE Symposium on Computer Applications & Industrial Electronics (ISCAIE 2018), Penang, Malaysia, 28–29 April 2018; pp. 71–77. [CrossRef]
2. Lau, J.; Zimmerman, B.; Schaub, F. Alexa, Are You Listening? *Proc. ACM Hum.-Comput. Interact.* **2018**, 2. [CrossRef]
3. Irugalbandara, I.B.C.; Naseem, A.S.M.; Perera, M.S.H.; Logeeshan, V. HomeIO: Offline Smart Home Automation System with Automatic Speech Recognition and Household Power Usage Tracking. In Proceedings of the 2022 IEEE World AI IoT Congress (AIIoT), Seattle, WA, USA, 6–9 June 2022; pp. 571–577. [CrossRef]
4. Errobidart, J.; Uriz, A.J.; Gonzalez, E.; Gelosi, I.E.; Etcheverry, J.A. Offline domotic system using voice comands. In Proceedings of the 2017 Eight Argentine Symposium and Conference on Embedded Systems (CASE), Buenos Aires, Argentina, 9–11 August 2017; pp. 25–30. [CrossRef]
5. EasyVR 3 Plus Shield for Arduino—COM-15453-SparkFun Electronics. Available online: https://www.sparkfun.com/products/15453 (accessed on 24 July 2022).
6. Prasanna, G.; Ramadass, N. Low Cost Home Automation Using Offline Speech Recognition. *Int. J. Signal Process. Syst.* **2014**, *2*, 96–101. [CrossRef]
7. Elsokah, M.M.; Saleh, H.H.; Ze, A.R. Next generation home automation system based on voice recognition. In Proceedings of the ICEMIS'20: Proceedings of the 6th International Conference on Engineering & MIS 2020, Almaty, Kazakhstan, 14–16 September 2020. [CrossRef]
8. Rani, P.J.; Bakthakumar, J.; Kumaar, B.P.; Kumaar, U.P.; Kuma, S.R. Voice controlled home automation system using natural language processing (NLP) and internet of things (IoT). In Proceedings of the ICONSTEM 2017—2017 Third International Conference on Science Technology Engineering & Management (ICONSTEM), Chennai, India, 23–24 March 2017; Volume 2018-January, pp. 368–373. [CrossRef]
9. WebRTC. Available online: https://webrtc.org/ (accessed on 24 July 2022).
10. Turc, I.; Chang, M.; Lee, K.; Toutanova, K. Well-Read Students Learn Better: On the Importance of Pre-training Compact ModelsAug. *arXiv* **2019**, arXiv:1908.08962.
11. Devlin, J.; Chang, M.; Lee, K.; Toutanova, K. BERT: Pre-training of Deep Bidirectional Transformers for Language Understanding. *arXiv* **2018**, arXiv:1810.04805.
12. Zhen, Y.; Maragatham, T.; Mahapatra, R.P. Design and implementation of smart home energy management systems using green energy. *Arab. J. Geosci.* **2021**, *14*. [CrossRef]
13. Ahmed, M.S.; Mohamed, A.; Homod, R.Z.; Shareef, H.; Sabry, A.H.; Khalid, K.B. Smart plug prototype for monitoring electrical appliances in Home Energy Management System. In Proceedings of the 2015 IEEE Student Conference on Research and Development, SCOReD 2015, Kuala Lumpur, Malaysia, 13–14 December 2015; pp. 32–36. [CrossRef]
14. IoT. Internet of Things (IoT) communication protocols: Review. In Proceedings of the 2017 8th International Conference on Information Technology (ICIT), Amman, Jordan, 17–18 May 2017.
15. Nguyen, K.T.; Laurent, M.; Oualha, N. Survey on secure communication protocols for the Internet of Things. *Ad Hoc Netw.* **2015**, *32*, 17–31. [CrossRef]
16. Sahmi, I.; Abdellaoui, A.; Mazri, T.; Hmina, N. MQTT-PRESENT: Approach to secure internet of things applications using MQTT protocol. *Int. J. Electr. Comput. Eng.* **2021**, *11*, 4577–4586. [CrossRef]
17. Ama, E. *Home Automation with WiFi Innovation*; ResearchGate: Berlin, Germany, 2018; p. 33.
18. Ghori, M.R.; Wan, T.; Anbar, M.; Sodhy, G.C.; Rizwan, A. Review on Security in Bluetooth Low Energy Mesh Network in Correlation with Wireless Mesh Network Security. In Proceedings of the 2019 IEEE Student Conference on Research and Development (SCOReD), Seri Iskandar, Malaysia, 15–17 October 2019; pp. 219–224. [CrossRef]
19. Adomnicai, A.; Fournier, J.J.A.; Masson, L. Hardware security threats against bluetooth mesh networks. In Proceedings of the 2018 IEEE Conference on Communications and Network Security (CNS), Beijing, China, 30 May–1 June 2018. [CrossRef]
20. Clark, S.; Peressini, J.; Crist, T. *A Study of Zigbee Home Automation Integration*; ResearchGate: Berlin, Germany, 2012.
21. Wireless Mesh Networking. 2007. Available online: Https://www.routledge.com/Wireless-Mesh-Networking-Architectures-Protocols-and-Standards/Zhang-Luo-Hu/p/book/9780849373992 (accessed on 24 July 2022).

Disclaimer/Publisher's Note: The statements, opinions and data contained in all publications are solely those of the individual author(s) and contributor(s) and not of MDPI and/or the editor(s). MDPI and/or the editor(s) disclaim responsibility for any injury to people or property resulting from any ideas, methods, instructions or products referred to in the content.

sensors

MDPI

Article

A New NILM System Based on the SFRA Technique and Machine Learning

Simone Mari *, Giovanni Bucci, Fabrizio Ciancetta, Edoardo Fiorucci and Andrea Fioravanti

Dipartimento di Ingegneria Industriale e dell'Informazione e di Economia, Università dell'Aquila, 67100 L'Aquila, Italy; giovanni.bucci@univaq.it (G.B.); fabrizio.ciancetta@univaq.it (F.C.); edoardo.fiorucci@univaq.it (E.F.); andrea.fioravanti@univaq.it (A.F.)
* Correspondence: simone.mari@graduate.univaq.it

Abstract: In traditional nonintrusive load monitoring (NILM) systems, the measurement device is installed upstream of an electrical system to acquire the total aggregate absorbed power and derive the powers absorbed by the individual electrical loads. Knowing the energy consumption related to each load makes the user aware and capable of identifying malfunctioning or less-efficient loads in order to reduce consumption through appropriate corrective actions. To meet the feedback needs of modern home, energy, and assisted environment management systems, the nonintrusive monitoring of the power status (ON or OFF) of a load is often required, regardless of the information associated with its consumption. This parameter is not easy to obtain from common NILM systems. This article proposes an inexpensive and easy-to-install monitoring system capable of providing information on the status of the various loads powered by an electrical system. The proposed technique involves the processing of the traces obtained by a measurement system based on Sweep Frequency Response Analysis (SFRA) through a Support Vector Machine (SVM) algorithm. The overall accuracy of the system in its final configuration is between 94% and 99%, depending on the amount of data used for training. Numerous tests have been conducted on many loads with different characteristics. The positive results obtained are illustrated and commented on.

Keywords: machine learning (ML); nonintrusive load monitoring (NILM); smart home; support vector machine (SVM); sweep frequency response analysis (SFRA)

Citation: Mari, S.; Bucci, G.; Ciancetta, F.; Fiorucci, E.; Fioravanti, A. A New NILM System Based on the SFRA Technique and Machine Learning. *Sensors* **2023**, *23*, 5226. https://doi.org/10.3390/s23115226

Academic Editor: Daniele Cenni

Received: 3 April 2023
Revised: 26 May 2023
Accepted: 29 May 2023
Published: 31 May 2023

Copyright: © 2023 by the authors. Licensee MDPI, Basel, Switzerland. This article is an open access article distributed under the terms and conditions of the Creative Commons Attribution (CC BY) license (https://creativecommons.org/licenses/by/4.0/).

1. Introduction

The goal of energy saving within modern smart homes and energy management systems is pursued by monitoring and controlling household parameters, such as lighting and home temperature [1]. This need has led to a significant increase in attention to nonintrusive load-monitoring (NILM) systems.

Among the energy-monitoring systems, those based on the NILM technique represent one of the most relevant solutions. The total energy consumption of users is monitored and the consumption of each individual load is identified. For this purpose, the measurements of current and voltage are carried out, or often of the current alone; the data collected are then processed with a so-called "disaggregation" algorithm. The main advantages of the nonintrusiveness are the simplicity and cost-effectiveness of installation. Therefore, systems of this type are useful for both consumers and utility companies when analyzing the use and costs of electricity.

In the early 1990s, the first NILM system was proposed [2]. Since then, more advanced algorithms have enabled a significant improvement in energy-unbundling systems.

This is especially true over the past decade, which has seen a significant increase in interest in this topic.

The first NILM systems detected events and classified the various loads using traditional algorithms [2,3]. The most modern, however, use artificial intelligence algorithms, in particular, through machine-learning (ML) techniques. For example, in some studies [4,5], the energy disaggregation problem has been reformulated as an adaptive filtering problem; refs. [6,7] propose model-driven NILM systems and the works proposed in other studies [8–10] are based on hidden Markov chains, while others [11–15] use artificial neural networks. The latter types of algorithms learn from the data provided and can perform certain tasks. Therefore, ML algorithms continue to improve over time by learning from data with minimal human intervention [16].

With regard to the NILM problem, these systems process the active power—and sometimes also the reactive power—absorbed by the monitored system [16]. However, NILM systems have been developed based on transient rather than stationary characteristics or the analysis of other quantities, which differ due to belonging to different domains (time or frequency).

In this sense, the sampling frequency is a fundamental parameter used to define the extractable information. Low-frequency time series were processed to evaluate steady-state characteristics. On the other hand, time series at other frequencies were processed to obtain information about the startup and shutdown transients to be able to discriminate the loads through their dynamic parameters (overshoot, rise time, etc.) or by characterizing appliances based on the pulses produced on the power line. Other attempts have been made by processing the trajectories drawn on the V-I plane.

Today, the division of NILM systems into event-driven and non-event-driven systems is the most widely used division and is the best for defining the state of the art of these systems.

The former involves the detection of an event (understood as an appliance turning on, turning off, or switching to a different consumption state) and then classifying it based on the features associated with the appliance that caused it. This type of approach can therefore be divided into three basic steps: event detection, feature extraction, and load identification. In particular, the last step is performed in most cases using ML algorithms that work well as classification systems. Numerous supervised ML algorithms have been proposed in the literature, including K Nearest Neighbor (KNN) [17], naïve Bayes [18], Decision Tree (DT) [19], Support Vector Machine (SVM) [20], Principal Component Analysis (PCA) [21], and Artificial Neural Network (ANN) [22,23]. Finally, unsupervised [24] and semi-supervised learning algorithms [25], as well as those related to graph signal processing [26], have also been proposed.

On the other hand, non-event-based systems are NILM systems that do not have an event-detection phase. In these cases, the concept of the "signature"/"features" of an appliance is also lost, as the only feature used by the models is the aggregate power profile. They use a window of samples of the aggregate signal (therefore time series data) as input; the samples are processed continuously without waiting for the occurrence of events. For this reason, this type of system is particularly suitable for low-frequency signals. Indeed, it was developed precisely to allow the processing of signals acquired with reduced frequencies, for which the detection of events is more difficult. In some cases, the disaggregation problem is formulated as a blind source separation (BSS) problem—that is, the problem of recovering a signal from a set of mixed signals. Numerous approaches have been proposed in the last decade, the most significant being those based on Combinatorial Optimization [27], Discriminative Sparse Coding [28], Hidden Markov Model Approaches [9,29–35], and Deep Learning (DL) [36–43].

NILM systems are used in a wide range of applications. Among these, very promising are the applications in Ambient Assisted Living, i.e., systems that make it possible to meet the needs of elderly or disabled users, allowing them to live independently [44]. In fact, knowing the changes in the status of the various appliances in a relatively short time, it is possible to infer the Activities of Daily Living (ADL) of the occupants.

This paper presents a measurement system for nonintrusive monitoring (it does not require modifications to the electrical system) based on the injection of a variable frequency sinusoidal signal and the characterization of the system based on the response to it. This technique is called Sweep Frequency Response Analysis (SFRA) and is widely used in diagnostics and fault-finding in transformers and electric motors.

The proposed solution is very different from the other solutions proposed in the literature, which provide for the analysis of time-varying electrical signals through different approaches. Following these approaches, the focus is generally on transients of the absorbed current, which indicate a change in the connection state. The measurement of the current in static conditions does not allow the identification of active devices, except in very special simple cases.

The approach proposed in this paper makes it possible to identify which appliances are inserted through a measurement performed in static conditions (not in the connection/disconnection transient). It allows for the detection of a sort of signature that is unique and independent of the absorbed current. This approach, as illustrated below, allows us to overcome the typical problems of NILM systems in identifying multi-state or continuous variable load household appliances (or, in general, electrical loads).

All the SFRA apparatuses available on the market can only work on single devices that are switched off and disconnected from the grid. The SFRA system proposed in this article can operate online [45], thus allowing it to extend its operating range to systems for continuous diagnostics on devices while supplied by the mains; no functioning interruptions or disconnection operations are needed for the standard SFRA apparatuses.

The proposed system is based on a machine-learning algorithm, the Support Vector Machine (SVM), which is capable of determining the status of individual household appliances starting from the measurement obtained by the SFRA system. It was installed on a home test system and acquired and processed the data locally.

Extensive measurements were made in order to verify the operational characteristics. The results obtained from field applications are also included and discussed.

2. Frequency Response Analysis of Household Appliances

SFRA has been successfully used to perform diagnostics on the windings of electric machines during the production process [46,47]. An electric machine can be considered a complex electrical network of capacitances, inductances, and resistors. As shown in Figure 1, the SFRA instrument injects a sinusoidal excitation voltage (the typical amplitude is 10 Vpp) with a continuously increasing frequency into one end of the transformer winding and measures the signal returning from the other end. This test is conducted with the machine disconnected from the power line. More details are reported in another study [46].

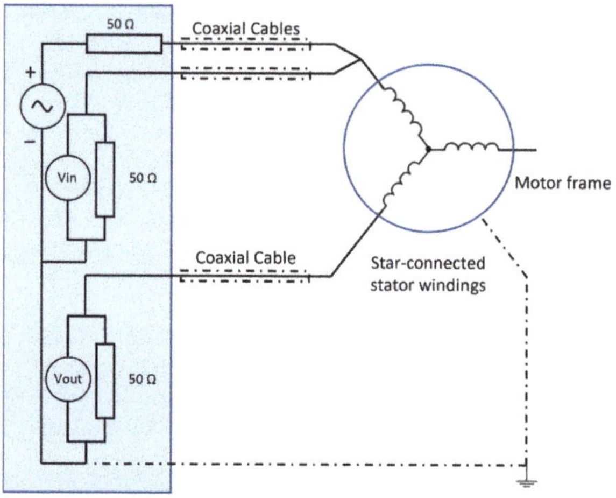

Figure 1. SFRA applied to a star-connected electric machine.

The comparison of input and output signals generates a frequency response, which can be compared with reference data. Degradation of the insulating materials or a change in the shape of the windings will result in a change in the RLC components of the network and, consequently, in the frequency response curve. Faults can therefore be detected by processing correlation indices between different curves.

In the proposed application, shown in Figure 2, the SFRA technique is applied to the electrical system supplied by the mains in order to obtain a signature that allows for discriminating different power supply conditions of a domestic system. The applied signal and the output signal, between the terminal of the neutral conductor and the ground, are acquired and processed by the system. The proposed measurement system can therefore be conveniently installed on a standard domestic socket.

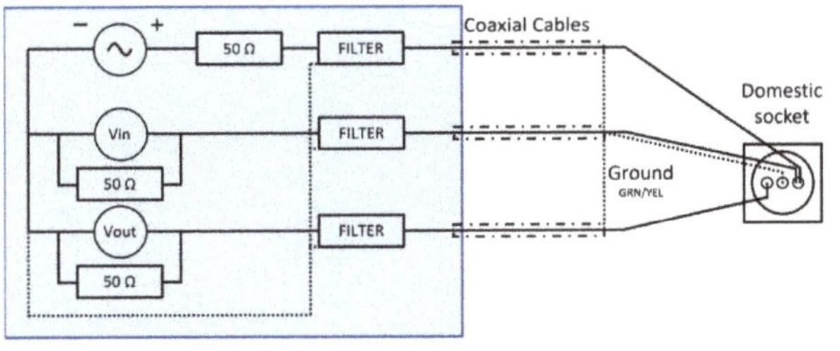

Figure 2. SFRA system.

A low-voltage (±5 Vpp) sinusoidal signal with variable frequency (from 2 kHz to 1.5 MHz) is superimposed on the supply voltage (240 Vrms and 50 Hz) and applied between the power phase conductor terminal and ground.

The signal generator is coupled to the network by means of a band-pass filter that allows only the passage of the test signal. The two input channels of the measurement circuit are also decoupled from the power supply by two other band-pass filters. The filters block both the fundamental frequency (50 Hz) and the harmonic components (up to 2 kHz) [48].

As the first part of the work, the system's response was evaluated over a fairly wide frequency range and by acquiring a sufficiently high number of points.

The frequency response was obtained by injecting a signal generated at 100 MS/s. In order to optimize the memory, the sampling frequencies to acquire both applied and output signals were adapted according to the frequency to be analyzed. In detail, the sampling frequency was chosen as being equal to 25 times the analyzed frequency. To obtain a better resolution, the FFT was performed by fixing a frequency bin at the frequency of the generated sinusoid. The FFT was also performed on the output signal and the sample at the same bin was considered.

A Hanning window with a width equal to the acquisition time (corresponding to 64 cycles of the generated frequency) was used to process the FFT. Downstream of the FFT processing, the system calculated the V_{out}/V_{in} ratio. For example, the 1 kHz response is achieved by injecting a 1 kHz sinusoidal signal generated at a frequency of 100 MS/s. The applied signal and the output signal were sampled at a sampling rate of 25 × 1000 = 25,000 Hz. A time window of (1/1000) × 64 = 0.064 s was considered for the processing of the FFTs, corresponding to 1600 samples. This process was repeated for all the frequencies of interest. The block diagram of the LabVIEW code is shown in Figure S1.

In order to evaluate the validity of the signature for different frequency ranges, four sub-bands were defined:

(1) 2–10 kHz;
(2) 10–100 kHz;
(3) 100 kHz–1 MHz;
(4) 1–1.5 MHz.

For each sub-band, 200 points were initially acquired. These sub-divisions were obtained considering the possible response to this type of excitation signal. Figure 3 schematically shows the installation of the SFRA system in the test system. From the knowledge in the literature about SFRA [48], the low-frequency response (2–10 kHz) is characterized by an ohmic-inductive behavior in which the characteristics of the grid upstream of the system are predominant; therefore, the contribution of the loads is usually not significant. The medium-frequency response (10 kHz–1 MHz) is characterized by resonance phenomena. As this band is generally the most interesting in terms of the effect of loads on the response, it has been split into two sub-bands to increase resolution. The high-frequency response (1–1.5 MHz) is characterized by capacitive effects due both to the network and the user loads and the connection of the measuring instrument itself, which generally determine a poor reproducibility of the measurement.

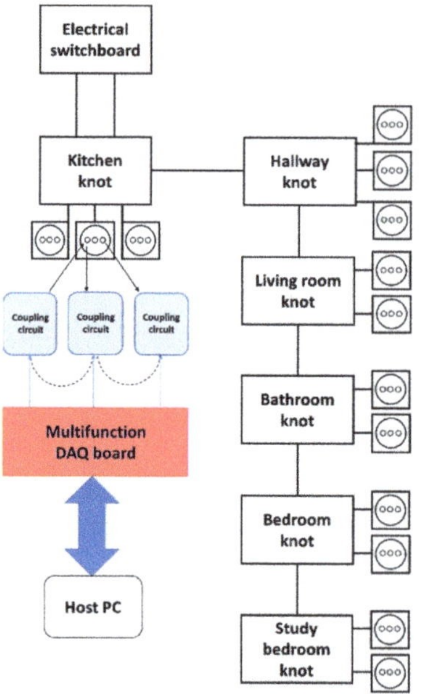

Figure 3. Installation of the SFRA in the test system.

The sinusoidal test signal introduces no problems to the system. This is essentially due to the reduced amplitude of the test signal with respect to the line voltage (1.54%), which is fully within the limits imposed by the standard [49].

During the tests, it was verified that the signal does not create problems in intelligent automation systems operating with conveyed waves [50]. This is also because these systems adopt sophisticated signal-modulation algorithms that encode the data transmitted with different sub-carriers or that widen the transmission band (Spread Spectrum), obtaining a better resistance to interference and noise. Other systems adopt Orthogonal Frequency Division Multiplexing (OFDM) modulation techniques, which are even more effective.

Several tests were performed at a residential test facility. A wide variety of loads were taken into consideration, powering them individually or simultaneously and under different working conditions:

(1) Hairdryer;
(2) Microwave oven;
(3) Lamp;
(4) Laptop;
(5) Induction hob;

(6) Heater;
(7) Drill;
(8) TV.

Figure 4 shows the frequency response of these appliances when powered individually. The measurements were conducted in 24 different power supply scenarios, as summarized in Table 1. It is important to note that Scenario 1 represents the case in which none of the appliances was powered (condition indicated with "Open Circuit" in Figure 1). Scenarios 2 to 9 represent the single power supply conditions of household appliances. Scenarios 10 to 24 represent the simultaneous power conditions.

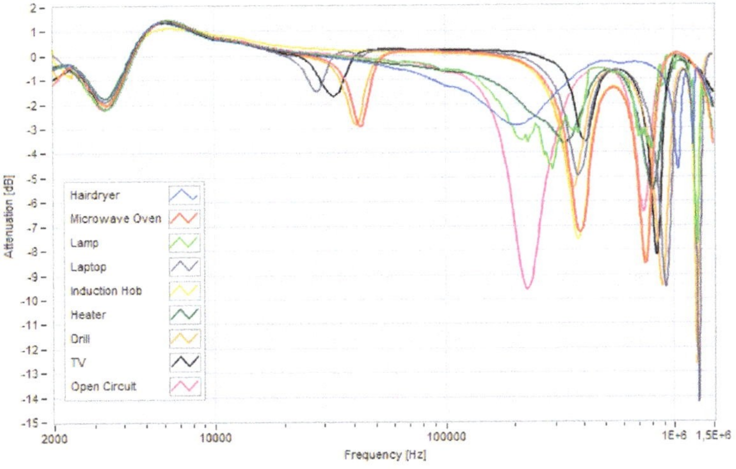

Figure 4. Frequency response of individually powered household appliances.

Table 1. Power supply scenarios.

	Hairdryer	Microwave Oven	Lamp	Laptop	Induction hob	Heater	Drill	TV
1								
2	x							
3		x						
4			x					
5				x				
6					x			
7						x		
8							x	
9								x
10			x	x				
11	x					x		
12		x			x			
13			x	x				x
14	x					x		x
15		x			x			x
16			x	x			x	
17	x					x	x	
18		x			x		x	
19	x		x	x		x		
20		x	x	x	x			
21				x			x	x
22			x	x			x	x
23	x					x	x	x
24		x			x		x	x

To support an objective evaluation, Figure 5 shows the lower and upper envelopes of the traces obtained in the presence and absence of each of the eight considered appliances, obtained following the measurements performed for the different scenarios. Measurements were performed for each of the 24 scenarios reported in Table 1, thus obtaining 24 SFRA traces. For each envelope (related to each appliance), the traces were divided into two groups according to the presence or absence of the appliance in the power supply scenario. The envelopes were then obtained by considering the maximum and minimum values of each of the two groups for each frequency bin. From these envelopes, it is immediately evident that the contribution of the low-frequency measurement (2–10 kHz) is not influenced by the different load configurations; therefore, in the rest of the work, we will only refer to the other three sub-bands.

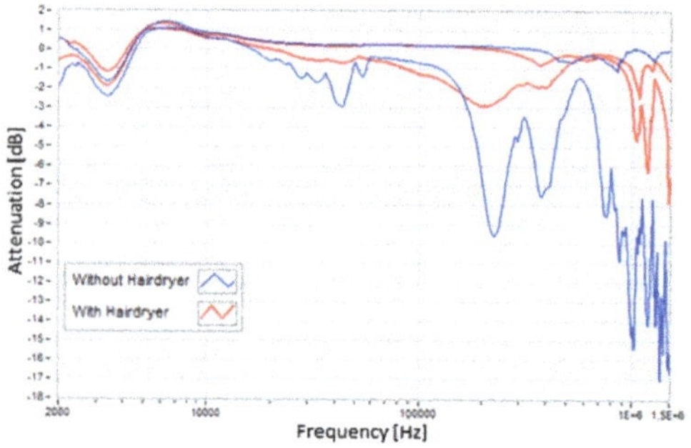

(a)

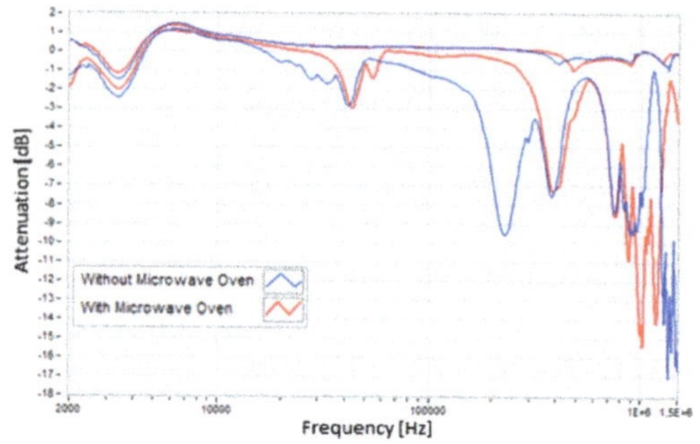

(b)

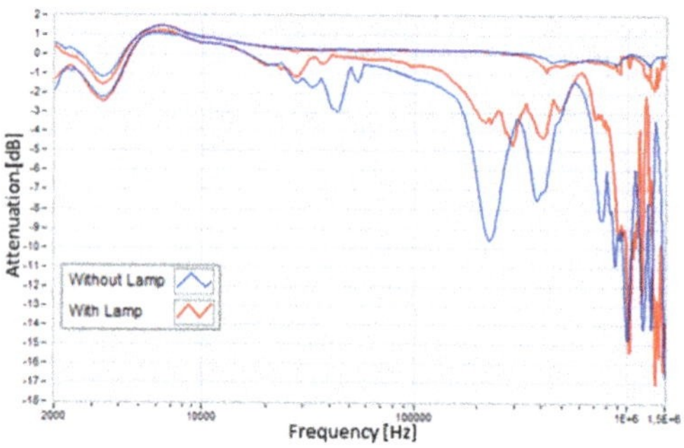

(c)

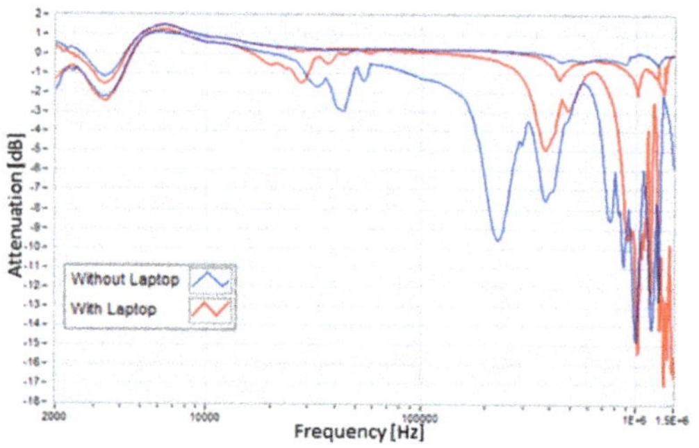

(d)

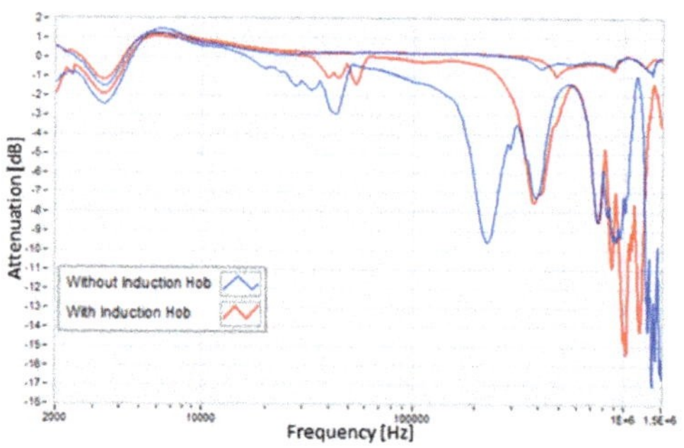

(e)

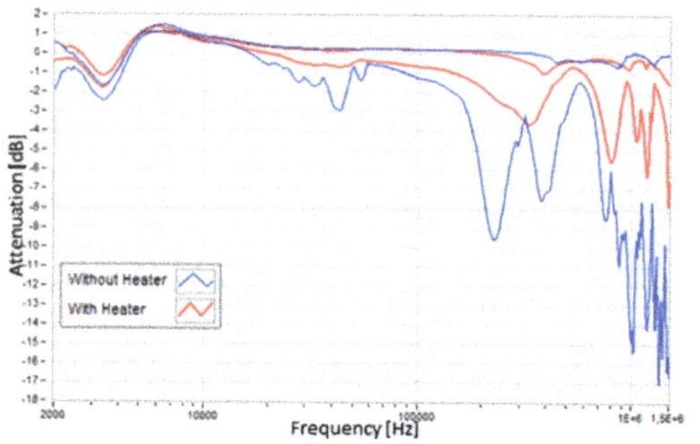

(f)

Figure 5. *Cont.*

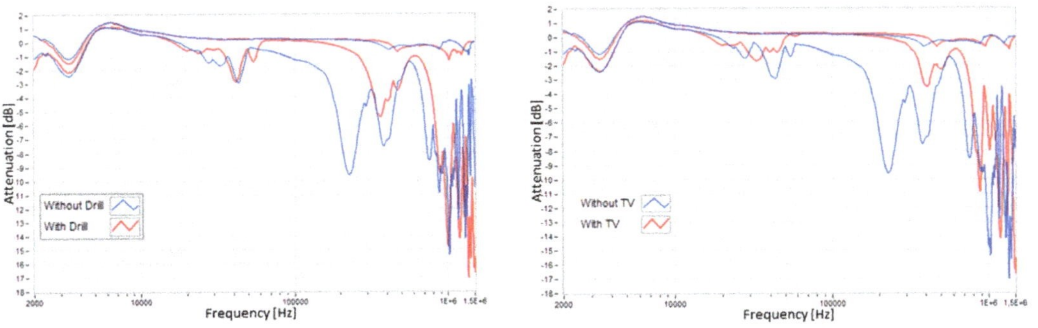

(g) **(h)**

Figure 5. Envelopes of the traces obtained in the presence and absence of the: (**a**) hairdryer, (**b**) microwave oven, (**c**) lamp, (**d**) laptop, (**e**) induction hob, (**f**) heater, (**g**) drill, and (**h**) TV.

These traces were used as inputs to a machine-learning-based classification algorithm, the Support Vector Machine (SVM), to determine the correct combination of powered appliances. A NILM system based on this type of input is easy to install, as it can be connected to a standard domestic socket, such as any household appliance. Traditional NILM systems, on the other hand, measure the aggregate power upstream of the plant and therefore require a more difficult installation.

The measurement obtained represents the transfer function of the equivalent RLC circuit [23]. Therefore, the result is mainly influenced by the physical characteristics of the appliances rather than by their power absorption. This represents a great advantage for the discrimination of multi-state or continuously variable load appliances (such as drills) whose identification is often critical for systems based on the analysis of power consumption.

The transfer function is minimally influenced by the choice of the socket in which to install the measuring system. Tests were carried out in all the sockets shown in Figure 3; all of the possible positions of the instrument on the various sockets allow the maximum reproducibility of the measurement. Regardless, the instrument is meant to be used on a single socket. The proposed algorithm is described in Section 3.

3. Machine-Learning Systems

Machine learning is the field of study that allows computers to learn without being explicitly programmed [51]. Unlike traditional programming, which provides a list of more or less complex rules defined by the programmer to obtain certain outputs, machine learning automatically learns patterns and correlations to solve extremely complex problems. In problems where existing solutions require a lot of manual adjustments or long lists of rules, a machine-learning algorithm can often simplify the code and achieve better performance. Sometimes they allow us to find solutions to problems that otherwise would not be solved through traditional approaches. These algorithms are used to process large amounts of data in order to discover patterns that are not immediately apparent. They are also used in situations where the algorithm needs to dynamically adapt to new patterns in the data or when the data itself is generated as a function of time, such as stock price prediction; in this case, we speak of online learning.

Machine-learning algorithms can be classified into supervised learning, unsupervised learning, semi-supervised learning, and reinforcement learning. This classification is made in relation to the quantity of data available during the training phase and the type of supervision during the training.

Specifically, in supervised learning, the training data provided to the algorithm include desired solutions called labels. Supervised learning solves two types of problems: classification and regression.

Classification is the problem of cataloging data into two or more classes; so, by providing input to the machine-learning system, it must return its class of belonging.

On the other hand, regression interpolates data to associate two or more features with each other. By providing the algorithm with an input feature, the regressor returns the other feature. A system of estimating the price of houses starting from features, such as size, number of rooms, and area, is a regression system.

The most popular supervised-learning algorithms are k-Nearest Neighbors, linear regression, logistic regression, Support Vector Machine (SVM), Decision Trees, Random Forests, and Neural Networks.

The NILM problem can be set up either as a regression problem—for example, when the algorithm is called to estimate the power absorbed by the single appliance starting from the aggregate power measurement [52]—or as a classification problem [53], as in the case in which starting from the aggregate power measurement is necessary to determine which appliances are powered and which are not.

The system proposed in this manuscript solves a multi-label classification problem since, starting from an SFRA trace, it is possible to identify several powered appliances simultaneously. The algorithm used is the SVM; the system configuration and its operation are illustrated in the following paragraphs.

3.1. Support Vector Machine

A SVM is one of the most popular models in machine learning, as it is very powerful and versatile [51]. SVMs are best suited for classifying complex but small- to medium-sized datasets. While classic classification algorithms discriminate based on characteristics common to each class, the SVM algorithms build the model based on the most difficult samples to discriminate, i.e., the most similar samples belonging to different classes. In this sense, the only samples used in the construction of the model are called support vectors. The other samples are therefore useless.

Based on the support vectors, the algorithm finds the optimal hyperplane that separates them, which can then be used to discriminate new samples. In other words, adding more formation samples far from the hyperplane (therefore not particularly complex to classify) will not affect the decision boundary, which will be completely determined by the samples located at the edge of the hyperplane.

Consider a case in which the samples to be classified are defined by only two features.

This case can be represented on a two-dimensional plane, as shown in Figure 6. A SVM algorithm looks for the line capable of maximizing the margin between the most similar samples belonging to different classes, i.e., the support vectors.

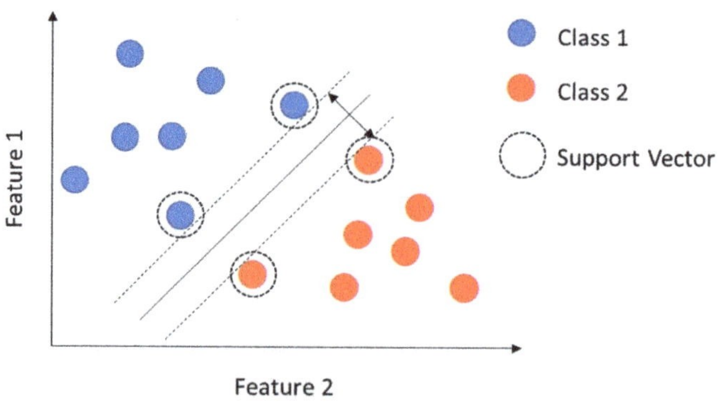

Figure 6. Representation of a linear classification problem in which the samples are defined by only two features.

Consider a linear classification problem in which n-dimensional inputs $\boldsymbol{X}$ are divided into two classes $y \in \{-1, 1\}$. The classifier can be formulated as follows:

$$f_{(x)} = \boldsymbol{w}^T \phi_{(x)} + b, \tag{1}$$

where $\boldsymbol{w}$ is the vector of weights, b is the bias, and $\phi_{(x)}$ is the feature space of the inputs. The sign of f(x) will be the output y_i of the classification.

Since the inputs are linearly separable, it will be possible to choose several linear decision boundaries, each of which will not produce classification errors in the training data.

Training a SVM model positions the boundary to maximize the margin—that is, the distance from the hyperplane to the nearest data point in either class. More specifically, we want to optimize the following objective function:

$$\max_{\boldsymbol{w},b} \min_{i} dist(\boldsymbol{x}_i, \boldsymbol{w}, b) \quad \Big| \quad \forall i \; y_i\left(\boldsymbol{w}^T \phi_{(x_i)} + b\right) \geq 0, \tag{2}$$

where $dist(\boldsymbol{x}, \boldsymbol{w}, b)$ is the Euclidean distance from the feature point $\phi_{(x)}$ to the hyperplane defined by $\boldsymbol{w}$ and b. With this objective function, the distance from the decision boundary $\boldsymbol{w}^T \phi_{(x)} + b = 0$ to the nearest point i is maximized. The constraints force finding a decision boundary that correctly classifies all the training data. In other words, for the classifier, a correct training point y_i and $\boldsymbol{w}^T \phi_{(x_i)} + b$ must have the same sign, in which case their product must be positive.

It is known from Euclidean geometry that the distance between the point $\phi_{(x_i)}$ and the hyperplane $\boldsymbol{w}^T \phi_{(x)} + b = 0$ can be defined as $\frac{|w^T \phi_{(x_i)} + b|}{||w||}$. Since y_i is the sign of $f_{(x_i)}$, it can be written as follows:

$$\max_{\boldsymbol{w},b} \min_{i} \frac{y_i(\boldsymbol{w}^T \phi_{(x_i)} + b)}{||\boldsymbol{w}||} \quad | \quad \forall i \; y_i(\boldsymbol{w}^T \phi_{(x_i)} + b) \geq 0, \tag{3}$$

We can observe that, due to the normalization of $||\boldsymbol{w}||$ in (3), the scale of $\boldsymbol{w}$ is arbitrary in this objective function. That is, if $\boldsymbol{w}$ and b are multiplied by a real scalar α, the factors of α in the numerator and denominator will cancel each other out. Now, suppose we choose the scale so that the point closest to the hyperplane, x_i satisfies $y_i(\boldsymbol{w}^T \phi_{(x_i)} + b) = 1$. With this assumption, the $\min_i$ in Equation (3) becomes redundant and can be removed. The objective function and constraint can be rewritten as:

$$\max_{\boldsymbol{w},b} \frac{1}{||\boldsymbol{w}||} \quad | \quad \forall i \; y_i\left(\boldsymbol{w}^T \phi_{(x_i)} + b\right) \geq 0, \tag{4}$$

Finally, we convert the problem into a quadratic program (QP). In this way, the objective function is quadratic in the unknowns and all constraints are linear in the unknowns. A QP has a single global minimum, which can be found efficiently with current optimization packages [54].

$$\max_{\boldsymbol{w},b} \frac{1}{2} ||\boldsymbol{w}||^2 \quad | \quad \forall i \; y_i\left(\boldsymbol{w}^T \phi_{(x_i)} + b\right) \geq 0, \tag{5}$$

However, not all classification problems are linear; in fact, in some cases, it is not possible to separate the classes with a straight line; therefore, we speak of non-linear classification. The kernel trick [55] solves non-linear classification problems with SVM algorithms.

In more detail, a polynomial kernel was used to determine the presence, or absence, of an appliance starting from the SFRA traces. Using a polynomial kernel means determining similarity, not only by processing the features of the input samples but also by their combinations, as shown in Figure 7.

Moreover, in real scenarios, data belonging to different classes overlap. As a result, it will not be possible to satisfy all the constraints in (5). One way to deal with this problem and still train useful classifiers is to relax some constraints by introducing so-called slack variables [56]. Normally, a Lagrangian transformation addresses the optimization problem, which allows the constrained optimization problem expressed in (5) to be reformulated into a non-constrained optimization problem.

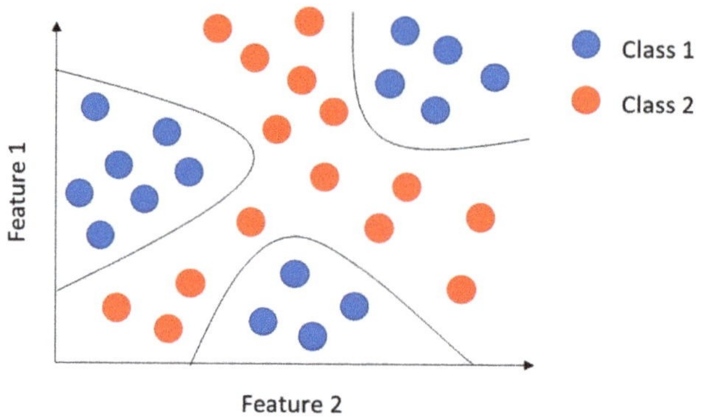

Figure 7. Representation of a non-linear classification problem in which the examples are defined by only two features.

The Lagrangian for the SVM objective function in (5), with Lagrange multipliers $a_i \geq 0$, is:

$$L_{(a_{1:N})} = \sum a_i - \frac{1}{2}\sum_i \sum_j a_i a_j y_i y_j k_{(x_i, x_j)}, \quad (6)$$

where $k_{(x_i, x_j)}$ is called a kernel function. For example, if we used the basic linear features, i.e., $\phi_{(x)} = x$, then $k_{(x_i, x_j)} = x_i{}^T x_j$. Instead, because a polynomial kernel has been chosen in the implemented SVM classifier, it will be defined as:

$$k_{(x_i, x_j)} = \left(\mathrm{a} + x_i{}^T x_j\right)^b, \quad (7)$$

3.2. The Proposed Structure

In the proposed system, the input is the trace obtained from the SFRA system; thus, each point of the trace represents a feature of the SVM. The algorithm must have a number of input functions equal to the number of bins of the measured frequency response.

The problem is also attributable to a multi-label classification problem, where a single sample can belong to multiple defined classes, unlike in multi-class classification, where each sample can uniquely belong to only one class.

In fact, the purpose of the system is to determine the status (ON or OFF) of the appliances. This means that the number of classes is equal to that of the appliances and the belonging of an SFRA trace to a certain class will indicate the ON state of that appliance. A single SFRA trace must therefore be able to be associated with multiple classes (or labels), as the system must be able to recognize the loads even under simultaneous power supply conditions. SVMs are not natively capable of performing multi-class or multi-label classifications since, as explained above, a SVM defines a hyperplane that separates classes equidistantly in order to guarantee the maximum margin. When the number of classes rises to three or more, thus passing from a binary classification to multi-class, it is possible to guarantee equidistance only between two of the classes, discarding this property with all the other classes.

To solve this classification problem, which involves assigning multiple labels to an instance, we converted it to multiple binary classification problems. A SVM was therefore associated with each household appliance, performing a binary classification in order to determine its ON or OFF status, starting from the SFRA trace. The proposed structure is shown in Figure 8.

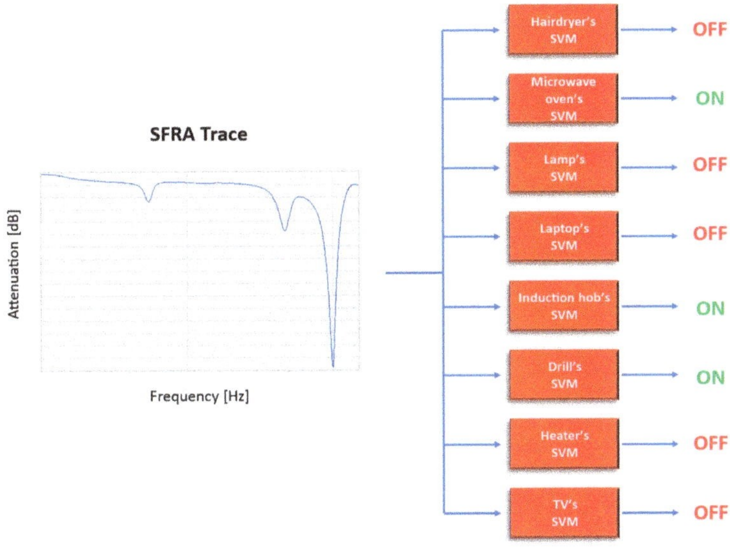

Figure 8. The proposed structure.

4. Experimental Results

As part of the development phase, the proposed algorithm was implemented and tested to evaluate its performance with real data.

4.1. The Proposed System Setup

As explained in Section 2, the SFRA technique was performed by plugging the instrument into a standard household socket. As previously discussed, the input signal is a variable frequency sinusoidal signal applied between the phase conductor terminal and ground, while the output signal is the measured signal between the neutral conductor terminal and ground. Both signals are acquired and processed. Figure 9 shows the measurement system used.

The measurement system must be connected to the test system by means of cables with suitable bandwidth and the same characteristic impedance of the generator to avoid reflection and signal mismatch and to improve the sensitivity, repeatability, and reliability of the measurement.

The input signal and related acquisition for the SFRA were performed using the Digilent Analog Discovery 2 NI Edition card with a BNC adapter.

The control system was developed using LabVIEW and run on a PC; this software automatically programs the Discovery FPGA at startup, with a configuration file designed to implement the measurement application. Once programmed, the integrated FPGA communicates with the PC via a USB 2.0 connection. The PC enables the creation of the user interface to access the data and process them in the experimental phase. A final NILM system can bypass the PC by integrating post-processing directly into the system.

The Discovery FPGA has a $\pm$25 V input range, a 14-bit resolution, a 100 MS/s sampling frequency, and a 30 MHz bandwidth. It is equipped with an arbitrary function generator with an output range of $\pm$5 V, a bandwidth of 20 MHz, and a sampling rate of 100 MS/s.

For appropriate interfacing with the network, the instrument is equipped with a coupling circuit for each of the three channels (one for generation and two for acquisition), as shown in Figure 9. The coupling circuit includes a third-order Butterworth filter with a flat passband and high attenuation outside the desired frequency range. The generation section and acquisition section coupling circuits both involve a 50 Ω resistor in series and parallel, respectively, to allow impedance adaptation. In addition, all coupling circuits are provided with a high-voltage ac blocking capacitor, connected in series with a 1:1 pulse transformer. The features of the filters developed for the SFRA apparatus are shown in Figures 10 and 11.

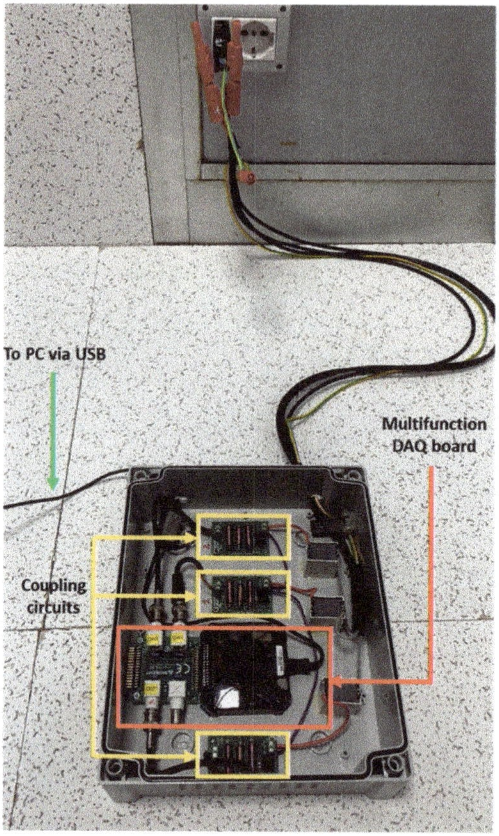

Figure 9. The SFRA measurement system.

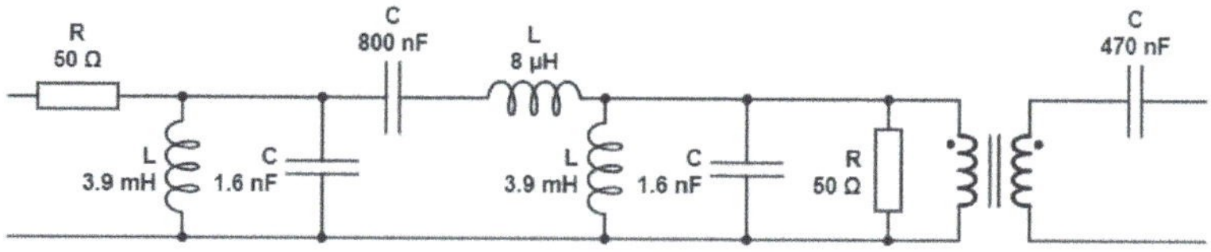

Figure 10. Coupling circuit for the signal generation section.

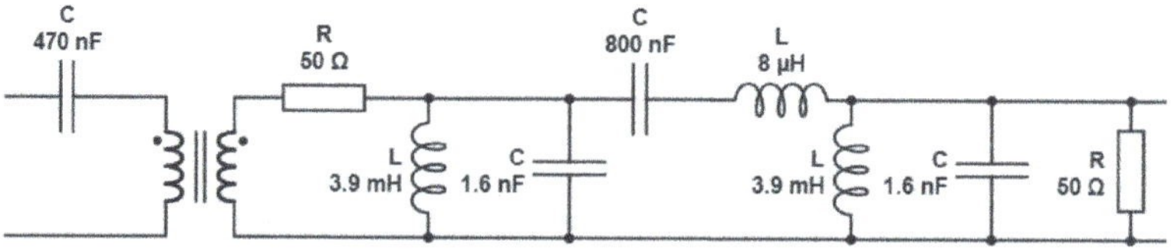

Figure 11. Coupling circuit for the signal acquisition section.

In order to avoid unwanted over-voltages due to resonance phenomena at high frequencies, the amplitude of the applied signal must not exceed a few volts (5 Vpp in the present case). The accuracy of the adopted measurement system, as discussed in a previous paper [57], has been evaluated using a reference parallel LCR circuit. This circuit consists of a 50 Ω resistive adapter, a fixed inductance, and a variable capacitance. The referenced values of the circuit impedance were measured with a Keysight E4980AL precision LCR meter. The estimated accuracy of the Vout/Vin ratio was better than ±0.2 dB in the interval from +5 to −25 dB and in the frequency range of 5 kHz to 1.5 MHz.

The SVM was implemented on a desktop computer (based on the Windows 10 × 64-bit operating system) using the open-source Python 3.7 from Anaconda [58]; the machine-learning algorithm was developed using the Scikit-learn library. Python is the programming language mostly used in artificial intelligence (AI) applications due to the availability of numerous libraries for continuous data acquisition and processing.

4.2. The Achieved Results

The proposed measurement technique is innovative and does not appear to have been tested by other authors. Due to the specificity of the acquired data (frequency response), there are no public datasets used by other authors against which to compare the performance of the proposed algorithm [59].

The measurement system was installed on a test facility, which was designed to generate electrical loads created by domestic users as part of the "non-intrusive infrastructure for monitoring loads in residential users" research project. The facility, located in the Electrical Engineering Laboratory of the University of L'Aquila (I), allows for the generation of electrical loads in a single or simultaneous way.

During the test phase, various parameters were evaluated in order to define the most significant sub-bands, the number of measurement points to be acquired, and the number of training examples needed to obtain a satisfactory performance. To this end, the precision, recall, and F1-Score during classification were evaluated [60]. These parameters were obtained using the numbers of true positive (TP), false positive (FP), true negative (TN), and false negative (FN) as follows:

$$Precision = \frac{TP}{TP + FP}, \tag{8}$$

$$Recall = \frac{TP}{TP + FN}, \tag{9}$$

$$F1 - score = 2 \times \frac{precision \times recall}{precision + recall}, \tag{10}$$

The concept of positive has been attributed to the ON state of household appliances and that of negative to the OFF state. Precision indicates all of the times the system has provided an indication of the ON state of an appliance and how many times the prediction has been correct. Precision does not take FNs into account. On the other hand, Recall indicates how many times the system has provided a correct indication about the ON state of the appliance compared to all of the samples in which the appliance was actually in the ON state. Recall does not take FPs into account. To have a metric capable of taking into account both FPs and FNs, the F1-Score is used, which is a harmonic mean of Precision and Recall.

Since, as already explained above, each appliance is associated with a SVM algorithm that reveals its presence, or not, the performance of each SVM was evaluated individually.

We started by acquiring 20 samples for each of the 24 scenarios, for a total of 480 training samples. Each sample consisted of an SFRA trace in which 200 points were acquired for each of the 3 sub-bands. Performance was evaluated on a test set consisting of 50 samples for each scenario, for a total of 1200 test samples. The obtained results, shown in Table 2, are already excellent, as 480 training samples is a relatively low number considering that acquiring a single sample takes about 40 s. The system does not make mistakes for five of the eight appliances analyzed and also shows high performance regarding the other three appliances. To define which of the three sub-bands made the most significant contribution to the identification of household appliances, the system's performance was evaluated by providing the three sub-bands separately as input to the machine-learning system. The results are reported in Table 3 and a graphical comparison is provided in Figure 12.

Table 2. The results obtained with 480 training samples and 200 points for each sub-band.

	Total Errors	FP	FN	Precision	Recall	F1-Score
Hairdryer	0	0	0	1.00	1.00	1.00
Microwave Oven	0	0	0	1.00	1.00	1.00
Lamp	27	0	27	1.00	0.92	0.96
Laptop	0	0	0	1.00	1.00	1.00
Induction Hob	0	0	0	1.00	1.00	1.00
Heater	5	5	0	0.98	1.00	0.99
Drill	29	29	0	0.93	1.00	0.97
TV	0	0	0	1.00	1.00	1.00

Table 3. Performance evaluation for each sub-band.

		10–100 kHz				
	Total Errors	**FP**	**FN**	**Precision**	**Recall**	**F1-Score**
Hairdryer	98	98	0	0.75	1.00	0.86
Microwave Oven	50	0	50	1.00	0.83	0.91
Lamp	110	96	14	0.78	0.96	0.86
Laptop	51	51	0	0.89	1.00	0.94
Induction Hob	0	0	0	1.00	1.00	1.00
Heater	61	61	0	0.83	1.00	0.91
Drill	48	48	0	0.89	1.00	0.94
TV	0	0	0	1.00	1.00	1.00
		100 kHz–1 MHz				
	Total Errors	**FP**	**FN**	**Precision**	**Recall**	**F1-Score**
Hairdryer	0	0	0	1.00	1.00	1.00
Microwave Oven	0	0	0	1.00	1.00	1.00
Lamp	141	30	111	0.89	0.68	0.77
Laptop	9	0	9	1.00	0.98	0.99
Induction Hob	59	9	50	0.97	0.83	0.89
Heater	5	5	0	0.98	1.00	0.99
Drill	116	106	10	0.79	0.98	0.87
TV	0	0	0	1.00	1.00	1.00
		1–1.5 MHz				
	Total Errors	**FP**	**FN**	**Precision**	**Recall**	**F1-Score**
Hairdryer	2	2	0	0.99	1.00	0.99
Microwave Oven	93	85	8	0.77	0.97	0.86
Lamp	115	0	115	1.00	0.67	0.80
Laptop	71	6	65	0.98	0.84	0.90
Induction Hob	79	74	5	0.80	0.98	0.88
Heater	29	29	0	0.91	1.00	0.95
Drill	90	76	14	0.84	0.97	0.90
TV	48	39	9	0.91	0.98	0.94

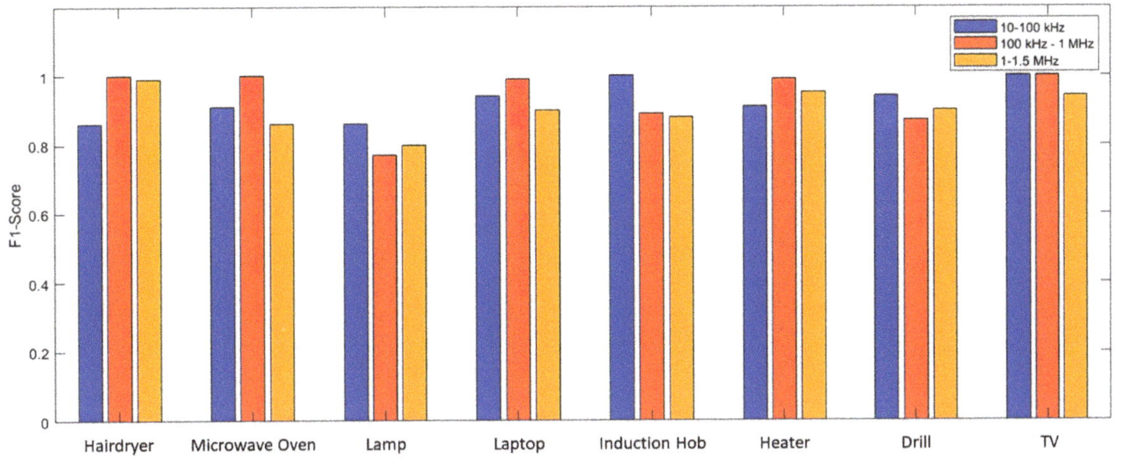

Figure 12. F1-Scores obtained for each considered sub-band.

In light of these results, it was decided that we would consider only the sub-bands of 10–100 kHz and 100 kHz–1 MHz in order to reduce the time required for the measurement. In fact, it is evident from Figure 12 that the 1–1.5 MHz band never allows for appliance discrimination that outperforms the previous bands. This reduces the time it takes to acquire a single trace to 22.56 s. Table 4 reports the performance evaluation using only the first two sub-bands as input.

Table 4. The results obtained with 480 training samples and 200 points for each sub-band, using only the first two sub-bands.

	Total Errors	FP	FN	Precision	Recall	F1-Score
Hairdryer	0	0	0	1.00	1.00	1.00
Microwave Oven	0	0	0	1.00	1.00	1.00
Lamp	29	0	29	1.00	0.92	0.96
Laptop	4	4	0	0.99	1.00	0.99
Induction Hob	0	0	0	1.00	1.00	1.00
Heater	3	3	0	0.99	1.00	0.99
Drill	7	7	0	0.98	1.00	0.99
TV	0	0	0	1.00	1.00	1.00

Comparing the results with those of Table 2, it can be seen that the system's performance has remained roughly unchanged. However, there is a significant improvement in the detection of the drill, highlighting that the 1–1.5 MHz sub-band introduced useless randomness for identification purposes. In this way, 400 points are acquired in the 10 kHz–1 MHz frequency band.

The possibility of decreasing the number of acquired points has been evaluated. Therefore, in Table 5, the performances obtained for 200, 134, and 100 points are reported. Furthermore, Figure 13 shows a graphical comparison of the impact of the number of acquired points on the F1-Score.

Table 5. Performance evaluation as the points acquired decrease.

	Total Errors	FP	FN	Precision	Recall	F1-Score
10 kHz–1 MHz (200 Points)						
Hairdryer	0	0	0	1.00	1.00	1.00
Microwave Oven	0	0	0	1.00	1.00	1.00
Lamp	39	0	39	1.00	0.89	0.94
Laptop	0	0	0	1.00	1.00	1.00
Induction Hob	0	0	0	1.00	1.00	1.00
Heater	2	2	0	0.99	1.00	1.00
Drill	5	5	0	0.99	1.00	0.99
TV	0	0	0	1.00	1.00	1.00
10 kHz–1 MHz (134 Points)						
	Total Errors	**FP**	**FN**	**Precision**	**Recall**	**F1-Score**
Hairdryer	0	0	0	1.00	1.00	1.00
Microwave Oven	0	0	0	1.00	1.00	1.00
Lamp	26	0	26	1.00	0.93	0.96
Laptop	0	0	0	1.00	1.00	1.00
Induction Hob	0	0	0	1.00	1.00	1.00
Heater	5	5	0	0.98	1.00	0.99
Drill	5	5	0	0.99	1.00	0.99
TV	0	0	0	1.00	1.00	1.00
10 kHz–1 MHz (100 Points)						
	Total Errors	**FP**	**FN**	**Precision**	**Recall**	**F1-Score**
Hairdryer	0	0	0	1.00	1.00	1.00
Microwave Oven	0	0	0	1.00	1.00	1.00
Lamp	42	0	42	1.00	0.88	0.94
Laptop	0	0	0	1.00	1.00	1.00
Induction Hob	0	0	0	1.00	1.00	1.00
Heater	6	6	0	0.98	1.00	0.99
Drill	2	2	0	0.99	1.00	0.99
TV	0	0	0	1.00	1.00	1.00

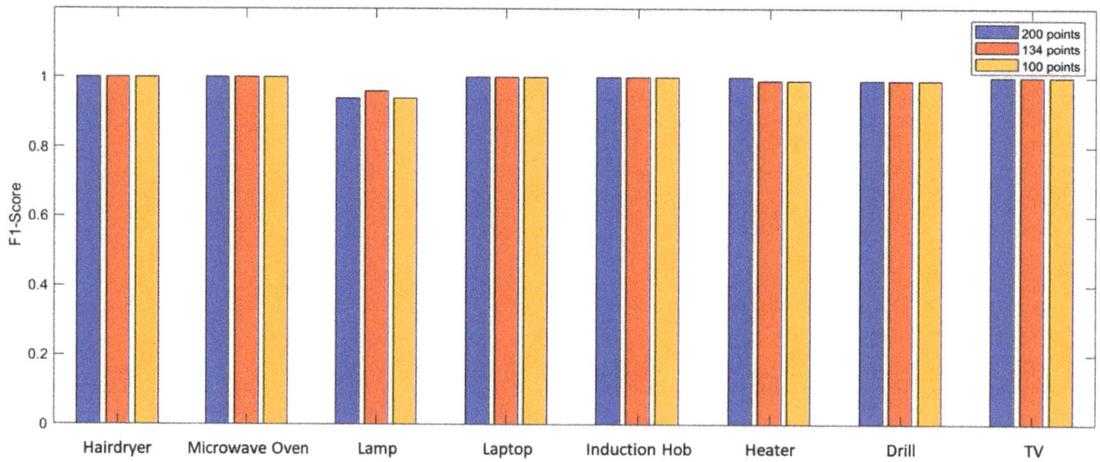

Figure 13. Graphical comparison of the impact of the number of acquired points on the F1-Score.

The performance proved to be very good, even when only using 100 measurement points as a system input. In these conditions, in fact, the system made errors only for three of the eight appliances analyzed while maintaining a minimum F1-Score of 0.94. This reduction allowed a decrease in the execution time of the measurement system from 22.56 s to 6.09 s. The performances shown so far always foresaw 480 training samples (20 for each of the 24 scenarios). As a final analysis, the impact of the number of training samples on performance was evaluated as shown in Figure 14. Table 6 reports the results obtained using an SFRA trace consisting of 100 points acquired in the 10 kHz–1 MHz frequency band, reducing the number of samples used in the training phase.

Table 6. Performance evaluation as training samples decrease.

	10 kHz–1 MHz (100 Points. 15 Samples for Each Scenario)					
	Total Errors	**FP**	**FN**	**Precision**	**Recall**	**F1-Score**
Hairdryer	0	0	0	1.00	1.00	1.00
Microwave Oven	0	0	0	1.00	1.00	1.00
Lamp	42	0	42	1.00	0.88	0.94
Laptop	0	0	0	1.00	1.00	1.00
Induction Hob	0	0	0	1.00	1.00	1.00
Heater	6	6	0	0.98	1.00	0.99
Drill	3	3	0	0.99	1.00	0.99
TV	0	0	0	1.00	1.00	1.00
	10 kHz–1 MHz (100 Points. 10 Samples for Each Scenario)					
	Total Errors	**FP**	**FN**	**Precision**	**Recall**	**F1-Score**
Hairdryer	0	0	0	1.00	1.00	1.00
Microwave Oven	0	0	0	1.00	1.00	1.00
Lamp	59	0	59	1.00	0.83	0.91
Laptop	2	0	2	1.00	0.99	0.99
Induction Hob	0	0	0	1.00	1.00	1.00
Heater	5	5	0	0.98	1.00	0.99
Drill	23	21	2	0.95	1.00	0.97
TV	0	0	0	1.00	1.00	1.00
	10 kHz–1 MHz (100 Points. 5 Samples for Each Scenario)					
	Total Errors	**FP**	**FN**	**Precision**	**Recall**	**F1-Score**
Hairdryer	5	5	0	0.98	1.00	0.99
Microwave Oven	0	0	0	1.00	1.00	1.00
Lamp	71	0	71	1.00	0.80	0.89
Laptop	26	0	26	1.00	0.94	0.97
Induction Hob	0	0	0	1.00	1.00	1.00
Heater	16	16	0	0.95	1.00	0.97
Drill	31	29	2	0.93	0.99	0.96
TV	0	0	0	1.00	1.00	1.00
	10 kHz–1 MHz (100 Points. 1 Sample for Each Scenario)					
	Total Errors	**FP**	**FN**	**Precision**	**Recall**	**F1-Score**
Hairdryer	5	5	0	0.98	1.00	0.99
Microwave Oven	1	1	0	0.99	1.00	0.99
Lamp	125	0	125	1.00	0.64	0.78
Laptop	95	2	93	0.99	0.77	0.87
Induction Hob	0	0	0	1.00	1.00	1.00
Heater	17	17	0	0.95	1.00	0.97
Drill	53	53	0	0.88	1.00	0.94
TV	8	8	0	0.98	1.00	0.94

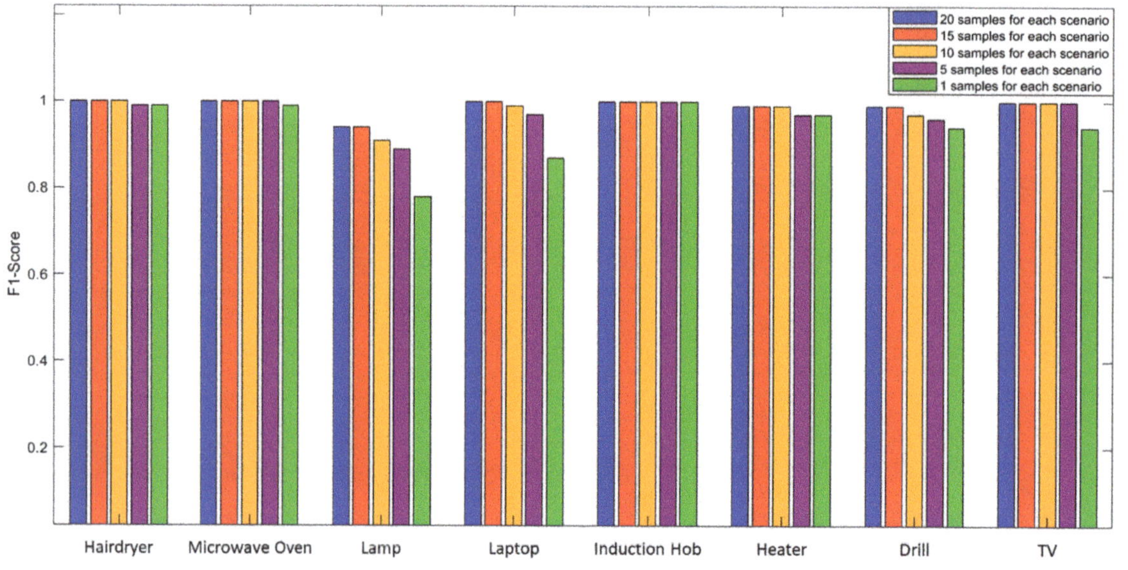

Figure 14. Graphical comparison of the impact of the number of training samples on the F1-Score.

The system maintains interesting performances even when trained with only one training sample for each scenario (therefore with 24 total training samples). This is mainly because the SVM natively suffers more from the quality of the training samples rather than the quantity, which is precisely because it builds a model based only on the most difficult samples to discriminate.

Lower performance was found in the detection of the Lamp, Laptop, and Drill. In the case of the Lamp, this is due to the insignificance of its related load compared to the overall network, while in the case of the Laptop and Drill, it is due to the extreme variability of their working conditions. However, F1-Score values of 0.78, 0.87, and 0.94, respectively, can be considered largely satisfactory for a trained system with such a small number of samples.

In order to provide an overall assessment of the system's performance, metrics widely used for multi-label classification systems were used, including micro-average and macro-average. As reported in (11)–(13), in the micro-average, all TPs, TNs, FPs, and FNs are summed for all of the labels and subsequently averaged:

$$Precision_{micro-averaging} = \frac{\sum_{n=1}^{N} TP_n}{\sum_{n=1}^{N} TP_n + FP_n}, \quad (11)$$

$$Recall_{micro-averaging} = \frac{\sum_{n=1}^{N} TP_n}{\sum_{n=1}^{N} TP_n + FN_n}, \quad (12)$$

$$F1 - score_{micro-averaging} = \frac{2 \times Precision_{micro-averaging} \times Recall_{micro-averaging}}{Precision_{micro-averaging} + Recall_{micro-averaging}}, \quad (13)$$

On the other hand, the macro-average, as reported in (14)–(16), is simply the average of the Precision and Recall for each label:

$$Precision_{macro-averaging} = \frac{\sum_{n=1}^{N} Precision_n}{N}, \quad (14)$$

$$Recall_{macro-averaging} = \frac{\sum_{n=1}^{N} Recall_n}{N}, \quad (15)$$

$$F1 - score_{macro-averaging} = \frac{2 \times Precision_{macro-averaging} \times Recall_{macro-averaging}}{Precision_{macro-averaging} + Recall_{macro-averaging}}, \quad (16)$$

The difference between the two lies is the fact that the micro-average reflects any imbalances in the dataset. Unbalance means there are test samples in a greater number of one or more classes than the others. In other words, having more samples for a given scenario, the macro-average, by creating a simple average of Precision, Recall, and F1-Score, does not consider this imbalance. On the contrary, the micro-average takes these situations into account.

In the case in question, the dataset is balanced; therefore, both averages are functional and adequate for verifying the performance of this system. Table 7 reports the micro-averages and macro-averages calculated based on the values reported in Table 6.

Table 7. Impact of the size of the training set on multi-label classification.

	Micro-Average			Macro-Average		
Training Samples for Each Scenario	**Precision**	**Recall**	**F1-Score**	**Precision**	**Recall**	**F1-Score**
20	0.99	0.98	0.99	0.99	0.98	0.99
15	0.99	0.98	0.99	0.99	0.98	0.99
10	0.99	0.97	0.98	0.99	0.97	0.98
5	0.98	0.96	0.97	0.98	0.96	0.97
1	0.96	0.92	0.94	0.97	0.92	0.94

An additional consideration needs to be made to integrate the proposed system into an electrical system. As explained above, there is no interference with the normal operation of the devices during system operation. Furthermore, the system poses no problems to the EMI filters, which are the input stage of the monitored devices, as the powers involved—which can be associated with the test signal—are extremely low.

To analyze the operating conditions of the measurement system in detail, it was simulated in a SPICE environment.

Specifically, the simulation was oriented to analyze the effects produced by the test signal on commercial EMI filters that could be connected (to other devices) in proximity to the system being tested. The analysis was extended to the entire range of frequencies involved; as a reference, a commercial EMI filter family was considered [61] for standard use in commercial and residential apparatuses for AC currents up to 16 A_{rms} in single-phase systems.

The analysis was extended to the entire range of frequencies involved. Figure 13 summarizes the scheme considered for the simulation. The resistance R_{Load} equal to 50 Ω was chosen in order to simulate the load of a generic household appliance (230 V_{rms}/50 Ω = 4.6 A_{rms}).

The system's response was evaluated by varying the frequency in the range in which the proposed system operates in the final configuration (10 kHz–1 MHz). The frequency response of the current entering the EMI filter was evaluated. Several simulations were carried out by varying the RLC parameters of the EMI filter. The current was found to be harmless across the entire spectrum. As an example, Figure 15 shows the input current response obtained with the RLC parameters reported in Figure 16. The spectrum shows two resonance peaks and a maximum current draw of 4.64 mA.

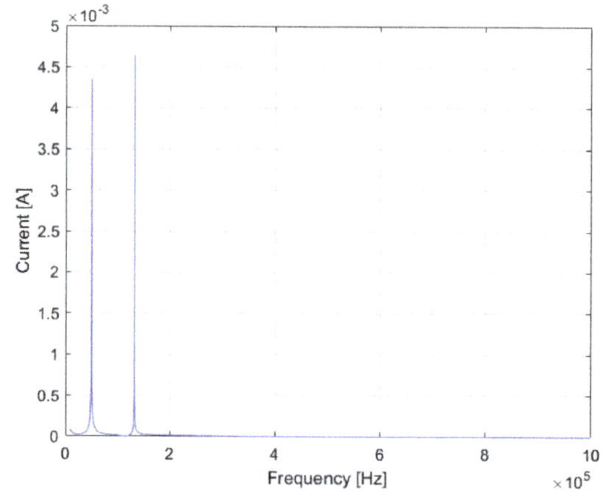

Figure 15. The frequency response of the input current to the EMI filter.

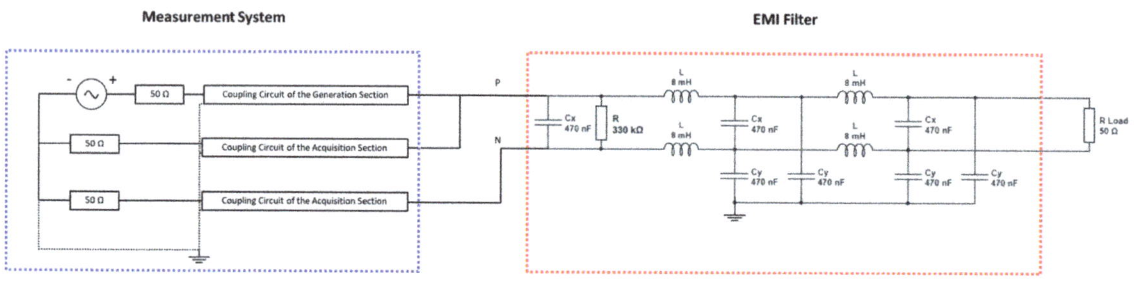

Figure 16. The scheme used for SPICE simulation.

The reduced value of this peak current does not lead to overheating of the filter components since the associated dissipated power is reduced. Furthermore, such verification is pejorative for the following reasons:

(1) The proposed system adopts a Digilent Analog Discovery 2 board, which has a limitation on the maximum output current that can be supplied by the DAC channels at 4 mA.

(2) In our simulation, the measurement system is only connected to the device being tested. In the real case, the generator is connected to a generic socket of the electrical system; therefore, the current that can be supplied (4 mA) is distributed in the various parallel branches of the other connected devices, greatly reducing the intensity of the portion that could affect the EMI filters.

5. Conclusions and Final Remarks

Modern home, energy, and assisted environment management systems require non-intrusive monitoring of the power supply status of the various loads, regardless of information related to their consumption. This parameter is not easy to obtain from NILM systems. The SFRA technique, already widely used in the diagnostics of transformers and asynchronous motors, has been applied here to characterize household appliances from the point of view of their influence in modifying the frequency response of the electrical system. The obtained signature, influenced by the physical characteristics of the loads, has been used as input for a machine-learning algorithm, the SVM. The proposed algorithm has been implemented in Python's open-source development environment, thus reducing the cost of the system.

A large campaign of measurements was carried out on a test facility, during which eight different electrical loads were powered individually and simultaneously. In particular, variable consumption loads, such as a drill and a laptop, were considered, which are generally among the most difficult for NILM systems to discriminate. The proposed system demonstrated excellent performance, even when trained with a minimum number of samples. In order to provide a comparison against other pre-published literature in the field, works that used similar metrics [62–64] were considered. The performances achieved by the cited works, by evaluating the F1-Score, were 91.5%, 93.2%, and 98.0%, respectively. The proposed system outperforms all three systems, as when all training data were provided (20 training samples for each scenario), the F1-Score achieved was 99.0%. It is important to note that the systems proposed in previous studies [62,63] were outperformed, even when the system was trained with the minimum number of samples when the system performance was 94.0%. The system is designed for local operation and is thus oriented toward edge implementation. The final system can be conveniently installed at any household outlet by detecting the presence of appliances connected to the system autonomously and providing data externally, for example, through wireless communication or the ability to download data histories via an SD card. The latter part will therefore be the subject of future research developments. Furthermore, the proposed system allows us to obtain information on which loads are powered in extremely short times (6.09 s in the final configuration of the system). These times were evaluated by considering both the time required to perform the measurement through the SFRA instrument and the time required to perform the prediction via the SVM classifier. Therefore, to ensure real-time operation, the edge system must incorporate multitasking capabilities. Two main tasks can be identified: in the first task, the system acquires and process the data to obtain the SFRA signature; in the second task, the system executes the SVM classifier and become ready to transfer the data over the WiFi network. The task of acquiring data and obtaining the signature, or SFRA trace, takes approximately 6 s, while the time required for processing the signature using the SVM classifier and transferring the data over WiFi (e.g., via an ESP32 module) is negligible and estimated to be around 10 ms based on experimental evaluations. This second task can be performed during the acquisition time of the first task. In fact, considering the first two signature-defining frequencies in the final configuration of the proposed system, namely 10,000 Hz and 11,350 Hz, the time needed for acquiring these initial points of the signature, as described in Section 2, amounts to 12 ms. Thus, under these conditions, the system can maintain real-time operation while meeting the requirements for post-processing and data transmission.

Supplementary Materials: The following supporting information can be downloaded at: https://www.mdpi.com/article/10.3390/s23115226/s1, Figure S1: The block diagram of the LabVIEW code.

Author Contributions: Conceptualization, S.M.; software, S.M.; validation, S.M.; investigation, S.M.; writing—original draft preparation, S.M.; writing—review and editing, G.B.; visualization, A.F.; supervision, F.C. and E.F.; project administration, G.B. All authors have read and agreed to the published version of the manuscript.

Funding: This research received no external funding.

Conflicts of Interest: The authors declare no conflict of interest.

References

1. Hill, J. The Smart Home: A Glossary Guide for the Perplexed. *T3. Retrieved at 27 March 2017*, 12 September 2015.
2. Hart, G.W. Non-intrusive appliance load monitoring. *Proc. IEEE* **1992**, *80*, 1870–1891. [CrossRef]
3. Bucci, G.; Ciancetta, F.; Fiorucci, E.; Mari, S. Load identification system for residential applications based on the NILM technique. In Proceedings of the IEEE Instrumentation and Measurement Technology Conference I2MTC 2020, Dubrovnik, Croatia, 20–25 May 2020.
4. Dong, R.; Ratliff, L.J.; Ohlsson, H.; Sastry, S.S. Energy disaggregation via adaptive filtering. In Proceedings of the 2013 51st Annual Allerton Conference on Communication, Control, and Computing (Allerton), Monticello, IL, USA, 2–4 October 2013; pp. 173–180. [CrossRef]

5. Egarter, D.; Bhuvana, V.P.; Elmenreich, W. PALDi: Online load disaggregation via particle filtering. *IEEE Trans. Instrum. Meas.* **2015**, *64*, 467–477. [CrossRef]
6. Barker, S.; Kalra, S.; Irwin, D.; Shenoy, P. Powerplay: Creating virtual power meters through online load tracking. In Proceedings of the 1st ACM Conference on Embedded Systems for Energy-Efficient Buildings, Memphis, TN, USA, 4–6 November 2014; pp. 60–69.
7. Tang, G.; Wu, K.; Lei, J.; Tang, J. A simple model-driven approach to energy disaggregation. In Proceedings of the 2014 IEEE International Conference on Smart Grid Communications (SmartGridComm), Venice, Italy, 3–6 November 2014; pp. 566–571. [CrossRef]
8. Jia, R.; Gao, Y.; Spanos, C.J. A fully unsupervised non-intrusive load monitoring framework. In Proceedings of the 2015 IEEE International Conference on Smart Grid Communications (SmartGridComm), Miami, FL, USA, 2–5 November 2015; pp. 872–878. [CrossRef]
9. Kolter, J.Z.; Jaakkola, T. Approximate inference in additive factorial hmms with application to energy disaggregation. In Proceedings of the 15th International Conference on Artificial Intelligence and Statistics (AISTATS) 2012, La Palma, Canary Islands, Spain, 21–23 April 2012; pp. 1472–1482.
10. Bucci, G.; Ciancetta, F.; Fiorucci, E.; Mari, S.; Fioravanti, A. Measurements for non-intrusive load monitoring through machine learning approaches. *Acta IMEKO* **2021**, *10*, 90–96. [CrossRef]
11. Bucci, G.; Ciancetta, F.; Fiorucci, E.; Mari, S.; Fioravanti, A. Deep Learning applied to SFRA Results: A Preliminary Study. In Proceedings of the 7th International Conference on Computing and Artificial Intelligence ICCAI 2021, Tianjin, China, 23–26 April 2021.
12. Figueiredo, M.; Ribeiro, B.; de Almeida, A. Electrical signal source separation via non-negative tensor factorization using on site measurements in a smart home. *IEEE Trans. Instrum. Meas.* **2014**, *63*, 364–373. [CrossRef]
13. Ciancetta, F.; Bucci, G.; Fiorucci, E.; Mari, S.; Fioravanti, A. A New Convolutional Neural Network-Based System for NILM Applications. *IEEE Trans. Instrum. Meas.* **2020**, *70*, 9246573. [CrossRef]
14. Bucci, G.; Ciancetta, F.; Fiorucci, E.; Mari, S.; Fioravanti, A. Multi-State Appliances Identification through a NILM System Based on Convolutional Neural Network. In Proceedings of the IEEE Instrumentation and Measurement Technology Conference I2MTC 2021, Glasgow, UK, 17–21 May 2021.
15. Cannas, B.; Carcangiu, S.; Carta, D.; Fanni, A.; Muscas, C. Selection of Features Based on Electric Power Quantities for Non-Intrusive Load Monitoring. *Appl. Sci.* **2021**, *11*, 533. [CrossRef]
16. Ruzzelli, A.G.; Nicolas, C.; Schoofs, A.; O'Hare, G.M.P. Real-time recognition and profiling of appliances through a single electricity sensor. In Proceedings of the 2010 7th Annual IEEE Communications Society Conference on Sensor, Mesh and Ad Hoc Communications and Networks (SECON), Boston, MA, USA, 21–25 June 2010; pp. 1–9.
17. Athanasiadis, C.L.; Papadopoulos, T.A.; Doukas, D.I. Real-time non-intrusive load monitoring: A light-weight and scalable approach. *Energy Build.* **2021**, *253*, 111523. [CrossRef]
18. Yang, C.; Soh, C.; Yap, V. A systematic approach in appliance disaggregation using k-nearest neighbours and naive Bayes classifiers for energy efficiency. *Energy Effic.* **2018**, *11*, 239–259. [CrossRef]
19. Guedes, J.; Ferreira, D.; Barbosa, B. A non-intrusive approach to classify electrical appliances based on higher-order statistics and genetic algorithm: A smart grid perspective. *Electr. Power Syst. Res.* **2016**, *140*, 65–69. [CrossRef]
20. Wang, A.L.; Chen, B.X.; Wang, C.G.; Hua, D. Non-intrusive load monitoring algorithm based on features of V–I trajectory. *Electr. Power Syst. Res.* **2018**, *157*, 134–144. [CrossRef]
21. Hassan, T.; Javed, F.; Arshad, N. An empirical investigation of VI trajectory based load signatures for non-intrusive load monitoring. *IEEE Trans. Smart Grid* **2013**, *5*, 870–878. [CrossRef]
22. Chang, H.H.; Chen, K.L.; Tsai, Y.P.; Lee, W.J. A new measurement method for power signatures of nonintrusive demand monitoring and load identification. *IEEE Trans. Ind. Appl.* **2012**, *48*, 764–771. [CrossRef]
23. Chang, H.H.; Lian, K.L.; Su, Y.C.; Lee, W.J. Power-spectrum-based wavelet transform for nonintrusive demand monitoring and load identification. *IEEE Trans. Ind. Appl.* **2014**, *50*, 2081–2089. [CrossRef]
24. Ducange, P.; Marcelloni, F.; Antonelli, M. A novel approach based on finite-state machines with fuzzy transitions for nonintrusive home appliance monitoring. *IEEE Trans. Ind. Inform.* **2014**, *10*, 1185–1197. [CrossRef]
25. Gillis, J.M.; Morsi, W.G. Non-intrusive load monitoring using semi-supervised machine learning and wavelet design. *IEEE Trans. Smart Grid* **2016**, *8*, 2648–2655. [CrossRef]
26. He, K.; Stankovic, L.; Liao, J.; Stankovic, V. Non-intrusive load disaggregation using graph signal processing. *IEEE Trans. Smart Grid* **2016**, *9*, 1739–1747. [CrossRef]
27. Batra, N.; Dutta, H.; Singh, A. INDiC: Improved Non-intrusive Load Monitoring Using Load Division and Calibration. In Proceedings of the 2013 12th International Conference on Machine Learning and Applications, Miami, FL, USA, 4–7 December 2013; pp. 79–84. [CrossRef]
28. Kolter, J.Z.; Batra, S.; Ng, A.Y. Energy Disaggregation via Discriminative Sparse Coding. In Proceedings of the Advances in Neural Information Processing Systems, Vancouver, BC, Canada, 6–9 December 2010.
29. Parson, O.; Ghosh, S.; Weal, M.; Rogers, A. Non-intrusive load monitoring using prior models of general appliance types. In Proceedings of the National Conference on Artificial Intelligence, Toronto, ON, Canada, 22–26 July 2012; pp. 1–8. Available online: http://eprints.soton.ac.uk/id/eprint/336812 (accessed on 1 May 2022).

30. Kim, H.; Marwah, M.; Arlitt, M.; Lyon, G.; Han, J. Unsupervised disaggregation of low frequency power measurements. In Proceedings of the 2011 SIAM International Conference on Data Mining, SIAM, Mesa, AZ, USA, 28–30 April 2011; pp. 747–758.
31. Paradiso, F.; Paganelli, F.; Giuli, D.; Capobianco, S. Context-based energy disaggregation in smart homes. *Future Internet* **2016**, *8*, 4. [CrossRef]
32. Bonfigli, R.; Principi, E.; Fagiani, M.; Severini, M.; Squartini, S.; Piazza, F. Non-intrusive load monitoring by using active and reactive power in additive Factorial Hidden Markov Models. *Appl. Energy* **2017**, *208*, 1590–1607. [CrossRef]
33. Makonin, S.; Popowich, F.; Bajić, I.; Gill, B.; Bartram, L. Exploiting HMM sparsity to perform online real-time nonintrusive load monitoring. *IEEE Trans. Smart Grid* **2015**, *7*, 2575–2585. [CrossRef]
34. Aiad, M.; Lee, P.H. Unsupervised approach for load disaggregation with devices interactions. *Energy Build.* **2016**, *116*, 96–103. [CrossRef]
35. Li, Y.; Peng, Z.; Huang, J.; Zhang, Z.; Son, J.H. Energy disaggregation via hierarchical factorial hmm. In Proceedings of the 2nd International Workshop on Non-Intrusive Load Monitoring, Austin, TX, USA, 3 June 2014; Volume 3, pp. 1–4.
36. Xia, M.; Liu, W.; Wang, K.; Zhang, X.; Xu, Y. Non-Intrusive Load Disaggregation Based on Deep Dilated Residual Network. *Electr. Power Syst. Res.* **2019**, *170*, 277–285. [CrossRef]
37. Kaselimi, M.; Protopapadakis, E.; Voulodimos, A.; Doulamis, N.; Doulamis, A. Multi-Channel Recurrent Convolutional Neural Networks for Energy Disaggregation. *IEEE Access* **2019**, *7*, 81047–81056. [CrossRef]
38. Zhang, C.; Zhong, M.; Wang, Z.; Goddard, N.; Sutton, C. Sequence-to-point learning with neural networks for non-intrusive load monitoring. In Proceedings of the AAAI Conference on Artificial Intelligence, New Orleans, LA, USA, 2–7 February 2018; pp. 2604–2611.
39. Ahmed, S.; Bons, M. Edge Computed NILM: A Phone-Based Implementation Using MobileNet Compressed by Tensorflow Lite. In Proceedings of the 5th International Workshop on Non-Intrusive Load Monitoring (NILM'20), Association for Computing Machinery, New York, NY, USA, 18 November 2020; pp. 44–48.
40. Kukunuri, R.; Aglawe, A.; Chauhan, J.; Bhagtani, K.; Patil, R.; Walia, S.; Batra, N. EdgeNILM: Towards NILM on Edge Devices. In Proceedings of the 7th ACM International Conference on Systems for Energy-Efficient Buildings, Cities, and Transportation (BuildSys '20), Association for Computing Machinery, New York, NY, USA, 18–20 November 2020; pp. 90–99.
41. Bonfigli, R.; Felicetti, A.; Principi, E.; Fagiani, M.; Squartini, S.; Piazza, F. Denoising Autoencoders for Non-Intrusive Load Monitoring: Improvements and Comparative Evaluation. *Energy Build.* **2018**, *158*, 1461–1474. [CrossRef]
42. Kong, W.; Dong, Z.; Wang, B.; Zhao, J.; Huang, J. A Practical Solution for Non-Intrusive Type II Load Monitoring Based on Deep Learning and Post-Processing. *IEEE Trans. Smart Grid* **2020**, *11*, 148–160. [CrossRef]
43. Murray, D.; Stankovic, L.; Stankovic, V.; Lulic, S.; Sladojevic, S. Transferability of Neural Network Approaches for Low-Rate Energy Disaggregation. In Proceedings of the ICASSP 2019 IEEE International Conference on Acoustics, Speech and Signal Processing (ICASSP), Brighton, UK, 12–17 May 2019; pp. 8330–8334.
44. Noury, N.; Berenguer, M.; Teyssier, H.; Bouzid, M.; Giordani, M. Building an index of activity of inhabitants from their activity on the residential electrical power line. *IEEE Trans. Inf. Technol. Biomed.* **2011**, *15*, 758–766. [CrossRef] [PubMed]
45. Bucci, G.; Ciancetta, F.; Fiorucci, E. Apparatus for Online Continuous Diagnosis of Induction Motors Based on the SFRA Technique. *IEEE Trans. Instrum. Meas.* **2020**, *69*, 4134–4144. [CrossRef]
46. *IEC 60076-18:2012*; Power Transformers—Part 18: Measurement of Frequency Response. IEC: Geneva, Switzerland, 2012.
47. *IEEE Std C57.149-2012*; IEEE Guide for the Application and Interpretation of Frequency Response Analysis for Oil-Immersed Transformers. IEEE: New York, NY, USA, 2013; pp. 1–72. [CrossRef]
48. Fioravanti, A.; Prudenzi, A.; Bucci, G.; Fiorucci, E.; Ciancetta, F.; Mari, S. Non intrusive electrical load identification through an online SFRA based approach. In Proceedings of the 2020 International Symposium on Power Electronics, Electrical Drives, Automation and Motion (SPEEDAM), Sorrento, Italy, 24–26 June 2020.
49. *CEI EN 50160*; Power Quality Standard. CEI: Blue Ash, OH, USA, 2020.
50. D'Innocenzo, F.; Bucci, G.; Dolce, S.; Fiorucci, E.; Ciancetta, F. Power line communication, overview of standards and applications. In Proceedings of the XXI IMEKO World Congress "Measurement in Research and Industry", Prague, Czech Republic, 30 August–4 September 2015.
51. Géron, A. *Hands-on Machine Learning with Scikit-Learn, Keras and TensorFlow: Concepts, Tools, and Techniques to Build Intelligent Systems*, 2nd ed.; O'Reilly Media, Inc.: Sebastopol, CA, USA, 2019.
52. Kelly, J.; Knottenbelt, W. Neural NILM: Deep Neural Networks Applied to Energy Disaggregation. In Proceedings of the 2nd ACM International Conference on Embedded Systems for Energy-Efficient Built Environments (BuildSys '15). Association for Computing Machinery, New York, NY, USA, 4–5 November 2015; pp. 55–64. [CrossRef]
53. Bucci, G.; Ciancetta, F.; Fiorucci, E.; Mari, S.; Fioravanti, A. A Non-Intrusive Load Identification System Based on Frequency Response Analysis. In Proceedings of the 2021 IEEE International Workshop on Metrology for Industry 4.0 & IoT (MetroInd4.0 & IoT), Rome, Italy, 7–9 June 2021; pp. 254–258. [CrossRef]
54. Bishop, C.M. *Pattern Recognition and Machine Learning*; Springer: New York, NY, USA, 2006.
55. Hofmann, T.; Schölkopf, B.; Smola, A.J. Kernel methods in machine learning. *Ann. Statist.* **2008**, *36*, 1171–1220. [CrossRef]
56. Xu, X.; Tsang, I.W.; Xu, D. Soft Margin Multiple Kernel Learning. *IEEE Trans. Neural Netw. Learn. Syst.* **2013**, *24*, 749–761. [CrossRef]

57. Dolce, S.; Fiorucci, E.; Bucci, G.; D'Innocenzo, F.; Ciancetta, F.; Di Pasquale, A. Test instrument for the automatic compliance check of cast resin insulated windings for power transformers. *Measurement* **2017**, *100*, 50–61. [CrossRef]
58. Anaconda Inc. Anaconda Software Distribution. 2020. Available online: https://www.anaconda.com/distribution/ (accessed on 13 February 2022).
59. Mari, S.; Bucci, G.; Ciancetta, F.; Fiorucci, E.; Fioravanti, A. Advanced Architecture for Training and Testing NILM Systems. In Proceedings of the IEEE Instrumentation and Measurement Technology Conference I2MTC 2022, Ottawa, ON, Canada, 16–19 May 2022.
60. Makonin, S.; Popowich, F. Nonintrusive load monitoring (NILM) performance evaluation. *Energy Effic.* **2014**, *8*, 809–814. [CrossRef]
61. Schaffner Group. *Very High Performance Single-Phase Filters*; FN 2090 datasheet; Schaffner Group: Luterbach, Switzerland, 2017.
62. Aiad, M.; Lee, P.H. Energy disaggregation of overlapping home appliances consumptions using a cluster splitting approach. *Sustain. Cities Soc.* **2018**, *43*, 487–494. [CrossRef]
63. Jain, A.K.; Ahmed, S.S.; Sundaramoorthy, P.; Thiruvengadam, R.; Vijayaraghavan, V. Current peak based device classification in NILM on a low-cost embedded platform using extra-trees. In Proceedings of the 2017 IEEE MIT Undergraduate Research Technology Conference (URTC), Cambridge, MA, USA, 3–5 November 2017; pp. 1–4.
64. Cannas, B.; Carcangiu, S.; Carta, D.; Fanni, A.; Muscas, C.; Sias, G.; Canetto, B.; Fresi, L.; Porcu, P. Real-Time Monitoring System of the Electricity Consumption in a Household Using NILM Techniques. In Proceedings of the 24th IMEKO TC4 International Symposium and 22nd International Workshop on ADC and DAC Modelling and Testing, Palermo, Italy, 14–16 September 2020; pp. 90–95.

Disclaimer/Publisher's Note: The statements, opinions and data contained in all publications are solely those of the individual author(s) and contributor(s) and not of MDPI and/or the editor(s). MDPI and/or the editor(s) disclaim responsibility for any injury to people or property resulting from any ideas, methods, instructions or products referred to in the content.

sensors

MDPI

Article

Human Behavior Recognition via Hierarchical Patches Descriptor and Approximate Locality-Constrained Linear Coding

Lina Liu [1,2], Kevin I-Kai Wang [2], Biao Tian [3], Waleed H. Abdulla [2], Mingliang Gao [1] and Gwanggil Jeon [1,4,*]

1 College of Electrical and Electronic Engineering, Shandong University of Technology, Zibo 255000, China
2 Department of Electrical, Computer, and Software Engineering, Faculty of Engineering, The University of Auckland, 20 Symonds St, Auckland 1010, New Zealand
3 Science and Technology Cooperation and Exchange Center of Zouping, Zouping 256200, China
4 Department of Embedded Systems Engineering, Incheon National University, Incheon 22012, Republic of Korea
* Correspondence: gjeon@inu.ac.kr

Abstract: Human behavior recognition technology is widely adopted in intelligent surveillance, human–machine interaction, video retrieval, and ambient intelligence applications. To achieve efficient and accurate human behavior recognition, a unique approach based on the hierarchical patches descriptor (HPD) and approximate locality-constrained linear coding (ALLC) algorithm is proposed. The HPD is a detailed local feature description, and ALLC is a fast coding method, which makes it more computationally efficient than some competitive feature-coding methods. Firstly, energy image species were calculated to describe human behavior in a global manner. Secondly, an HPD was constructed to describe human behaviors in detail through the spatial pyramid matching method. Finally, ALLC was employed to encode the patches of each level, and a feature coding with good structural characteristics and local sparsity smoothness was obtained for recognition. The recognition experimental results on both Weizmann and DHA datasets demonstrated that the accuracy of five energy image species combined with HPD and ALLC was relatively high, scoring 100% in motion history image (MHI), 98.77% in motion energy image (MEI), 93.28% in average motion energy image (AMEI), 94.68% in enhanced motion energy image (EMEI), and 95.62% in motion entropy image (MEnI).

Keywords: human behavior recognition; energy image species; hierarchical patches descriptor; approximate locality-constrained linear coding algorithm

Citation: Liu, L.; Wang, K.I.-K.; Tian, B.; Abdulla, W.H.; Gao, M.; Jeon, G. Human Behavior Recognition via Hierarchical Patches Descriptor and Approximate Locality-Constrained Linear Coding. *Sensors* **2023**, *23*, 5179. https://doi.org/10.3390/s23115179

Academic Editor: Daniele Cenni

Received: 28 March 2023
Revised: 23 May 2023
Accepted: 24 May 2023
Published: 29 May 2023

Copyright: © 2023 by the authors. Licensee MDPI, Basel, Switzerland. This article is an open access article distributed under the terms and conditions of the Creative Commons Attribution (CC BY) license (https://creativecommons.org/licenses/by/4.0/).

1. Introduction

The proliferation of interconnected devices has led to the scenario of the Internet of Everything (IoE), which enables many intelligent and context-aware applications. Human behavior recognition, which is broadly applied to intelligent surveillance, human–machine interaction, video retrieval, etc. [1–3], has recently attracted more attention in computer vision. At present, most of the research on behavior recognition is based on video sequence analysis. Despite the significant progress made in this area, this remains a complex and challenging task. There are significant variations caused by subject behavior, viewpoint variations, occlusions, camera motion, cluttered background, the similarity between different behaviors, and even movement variability of the same behavior. Due to the aforementioned factors, researchers have put forward many different countermeasures.

Human behavior recognition contains two main tasks, namely behavior feature extraction and behavior pattern recognition. Feature extraction is the dominant step. With a given human behavior recognition framework, the performance of human behavior recognition depends on the quality of the feature extraction [4,5]. Human behavior recognition approaches based on vision can be separated into two categories: the traditional artificial-feature-based approach and the learning-feature-based approach [4,6]. The artificial features are dependent on the predesigned feature detectors and descriptors, which are relatively simple and easy to implement. However, they are difficult to interpret intuitively and have the problem of low recognition accuracy. The learning-feature-based approach is divided into two major categories, some of these approaches being sparse-representation and dictionary-learning-based methods, and others being deep-learning-based models. Dictionary learning employs the sparse representation of the input data, which is applicable to image or video-based classification tasks. Although the sparse-representation and dictionary-learning-based approaches have obtained good performance on several public datasets, rapidly constructing an effective dictionary-learning model for behavior recognition remains challenging. As it needs to solve norm optimization problems repeatedly during the model optimization process, this process has high computation cost and execution time.

Meanwhile, with the boom in artificial intelligence, deep learning has made remarkable achievements in the computer vision area. In many real-world applications, there may not exist enough large-scale datasets for training a deep-learning model. Therefore, especially for small-scale datasets, it is still a challenge to improve the recognition accuracy and robustness. Some of the problems include:

(1) Traditional hand-crafted representation-based features are difficult to interpret intuitively and have the problem of low recognition accuracy;
(2) Learning an effective dictionary-learning model is computationally expensive and time-consuming;
(3) For small-scale datasets, it is still a challenge to improve the recognition accuracy and robustness.

To address the aforementioned constraints, in this paper, a unique human behavior recognition approach is proposed based on a hierarchical patches descriptor (HPD) and ALLC algorithm. The main contributions of this article are as follows:

(1) Five energy image species are utilized to describe human behavior in a global manner. These are statistical features based on motion information. Moreover, an HPD is constructed to obtain detailed local feature descriptions for recognition. Combining local features with global features can better describe behavioral features, which can improve recognition accuracy.
(2) The proposed method is based on the ALLC algorithm for fast coding, which is computationally efficient because it has a closed-form analytical solution and it does not need to solve the norm optimization repeatedly.
(3) We demonstrate the superior performance of the proposed method in comparison with state-of-the-art alternatives by conducting experiments on both Weizmann and DHA datasets.

The remainder of this paper is organized as follows: Related work is presented in Section 2. The framework of the proposed approach, human behavior feature extraction, and a human behavior recognition scheme is presented in Section 3. Section 4 analyses experimental results and Section 5 presents the discussion. The paper is concluded in Section 6.

2. Related work

2.1. Traditional Artificial-Feature-Based Approach

The traditional artificial-feature-based approach is dependent on the predesigned feature detectors and descriptors, such as the bag-of-words (BoW) model [7], scale-invariant feature transform (SIFT) [8,9] and weighted hierarchical features [10], histogram of oriented gradients (HoG) [11,12] and pyramid histogram of oriented gradients (PHOG) [13], and local binary pattern (LBP) [14]. These features are relatively simple and easy to implement, but are difficult to interpret intuitively and have the problem of low recognition accuracy.

2.2. Learning-Feature-Based Learning Approach

Unlike the handcrafted-feature-based approaches, with the help of the concepts of a trainable feature extractor and classifier, feature-learning-based approaches can automatically learn features from the input data. Some of these approaches are based on sparse-representation and dictionary learning, and others are based on deep-learning models. Dictionary learning employs the sparse representation of the input data, which is applicable to image- or video-based classification tasks. Dictionary learning has been widely employed in computer vision areas, such as image classification [15–19] and action recognition [13,20–22]. Wright et al. [15] were one of the pioneers that used sparse representation for face recognition and achieved good results. The sparse coding [16] and locality-constrained linear coding (LLC) algorithms were widely used to deal with image classification [18], multiview facial expression recognition [19], and view-invariant action recognition [23]. Wang et al. [20] proposed to divide the 3D skeleton sequence into multiple non-interrelated sub-sequences, and used the coordinated representation of the motion density trajectories of the sub-sequences for behavior recognition.

Aiming to deepen the image sequence, Gao et al. [22] proposed a multi-feature mapping and dictionary-learning model (MMDLM) to obtain the correlation of different features, where MMDLM is a typical multi-modality dictionary-learning algorithm for feature fusion. The multi-modality joint representation and recognition (MMJRR) [12] is also a typical multi-modality algorithm for action recognition. Moreover, an RGBD action recognition approach based on a collaborative sparse representation (CSR) learning model was proposed in [22], where BoW features were extracted for RGB and depth modality, respectively. Then, they were weighted together by the CSR learning algorithm, and the collaborative reconstruction error was applied for classification.

Meanwhile, with the boom in artificial intelligence, deep learning has made remarkable achievements in the computer vision area. In particular, convolutional neural network (CNN)- and recurrent neural network (RNN)-based approaches have been widely used in human behavior feature extraction [24–27]. Wang et al. [24] proposed a three-stream CNN to learn behavior descriptors by feeding weighted layer depth motion maps to the network. Sharif et al. [25] proposed a hand-crafted and deep CNN feature fusion and selection strategy, and HOG features as the input of the CNN model for recognition. Bhatt et al. [26] summarized CNN variants for computer vision from five aspects: history, architecture, application, challenges, and future scope. Patel et al. [27] proposed a dimension-based generic convolution block for object recognition. Due to overfitting caused by the lack of training data, learning an effective deep neural network for action recognition remains a challenge. Therefore, data augmentation [24] and synthetic depth images [25] were used to reduce the possibility of overfitting. The introduction of some large-scale RGBD-based datasets [28–30] made it possible to develop more effective action recognition approaches based on deep learning.

Inspired by the above research, this work combines an artificial-feature-based approach and a feature-learning-based approach to describe the human behavior feature in a more detailed manner. Furthermore, a fast-coding method is utilized to improve the efficiency of recognition.

3. The Proposed Methods

3.1. Framework of the Proposed Human Behavior Recognition Approach

Aimed at improving the accuracy and robustness of human behavior recognition, a unique human behavior recognition approach based on HPD and ALLC is proposed. In the proposed technique, five energy image species for each human behavior video sequence are first calculated to describe human behavior in a global manner. The energy image species include motion energy image (MEI) and motion history image (MHI) [31], average motion energy image (AMEI), enhanced motion energy image (EMEI), and motion entropy image (MEnI) [32]. However, these energy image species cannot describe the local human behavior in detail, and HPD is proposed to analyze the energy image species at different scales for describing the local details of human behavior. Thus, we encode the HPD by using an ALLC algorithm for fast coding to acquire effective coding for human behavior recognition.

The framework of the proposed human behavior recognition approach is illustrated in Figure 1. The overall process consists of three major steps: human body segmentation, human behavior feature extraction, and behavior pattern recognition.

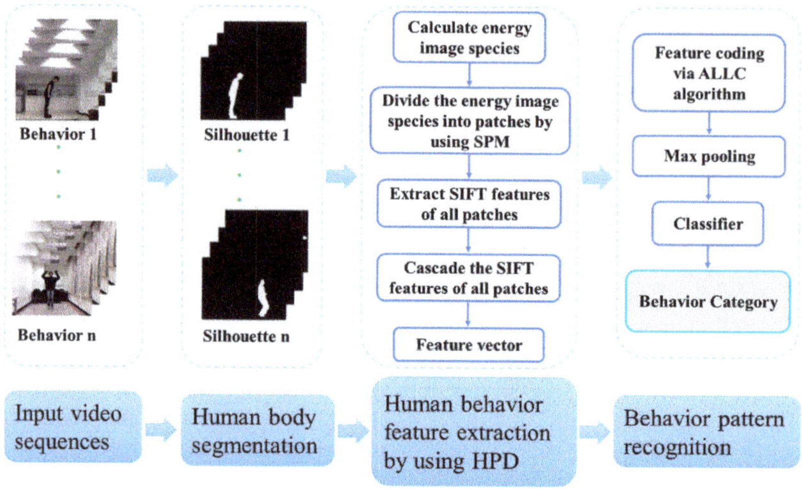

Figure 1. Framework of the proposed human behavior recognition approach.

The details of each individual step are as follows.

(1) Human body segmentation. In the input video sequences, there often exists a large amount of background information, which significantly reduces the computation efficiency and affects the human motion feature extraction. Thus, segmentation is an essential step to ensure that critical behavior information can be retained while unnecessary background information can be removed. In this paper, human behavior recognition is targeted at the whole body behavior, instead of the actions of specific human body parts. Therefore, the human body silhouette is segmented from the background as the input data for the feature extraction step.

(2) Human behavior feature extraction. To describe the human behavior information in detail, a combined strategy of global and local feature extraction is utilized in the paper. For each video sequence, several energy image species of the human body silhouette images are calculated as global feature descriptors of the human behavior. The advantage of this method is that it can describe the global human behavior information well in a statistical manner by using one image per video, which can greatly reduce the computational load of local feature extraction in the following processes. However, it cannot express the local human behavior information well. Therefore, after calculating each energy image species, an HPD is constructed to describe the local feature information of the targeted human behavior, which contains three steps.

Firstly, the energy image species is divided into patches on different resolutions by adopting the spatial pyramid matching (SPM) algorithm. Secondly, the BoW model with spatial–temporal features is employed to analyze the energy image species at different scales for local descriptions of human behavior. In view of obtaining local features that are scale-invariant, the SIFT features of all patches are extracted, which will generate numerous features that can densely cover the image in the whole scale and location range, which is beneficial to describe the local human behavior information. Finally, the SIFT features of all patches are cascaded together to form a vector for recognition.

(3) Behavior pattern recognition. After extraction of human behavior features from the video sequences, different human behaviors are learned individually from the training video sequences of each class by using the ALLC algorithm and max-pooling. Each testing video sequence is then attributed to a predefined class according to its corresponding feature. At this stage, the HPD feature vectors are encoded together by the ALLC algorithm, which is a simple, yet effective, fast coding algorithm.

Since the ALLC algorithm has better constructability and local smooth sparsity, the correlations between similar descriptors can be obtained easily by ensuring similar patches have similar codes, which is beneficial for human behavior recognition. In addition, it has an analytical solution and does not need to solve the norm optimization repeatedly, as in a sparse coding algorithm. Therefore, it has higher computational efficiency and needs less storage space in the process of objective function optimization, making it an effective and simple fast coding algorithm.

The coding results of all HPD feature vectors are in matrix form, which makes it difficult to construct eigenvectors for recognition. Therefore, it is necessary to pool all codes together and cascade them together to form a final feature vector for recognition. Considering that max-pooling almost always performs better than average pooling, especially with a linear SVM [33,34], max-pooling is used in the proposed approach.

3.2. Human Behavior Feature Extraction

3.2.1. Environmental Modelling and Human Body Segmentation

Human body segmentation is the basis of behavior recognition, and it aims to extract the body silhouette from an image sequence. In this paper, the background difference method [35] is employed to extract the human body silhouette. This method assumes that the background changes slowly or tends to be stationary, but in reality, there often exist factors such as light changes, background disturbances, and camera jitter. Therefore, it is necessary to model the background. However, if the initial frame used for modeling contains a moving target, the previous foreground target will be taken as background in the foreground determination step, which will lead to the so-called ghost area appearing in the pedestrian detection results of the current frame, as shown in Figure 2.

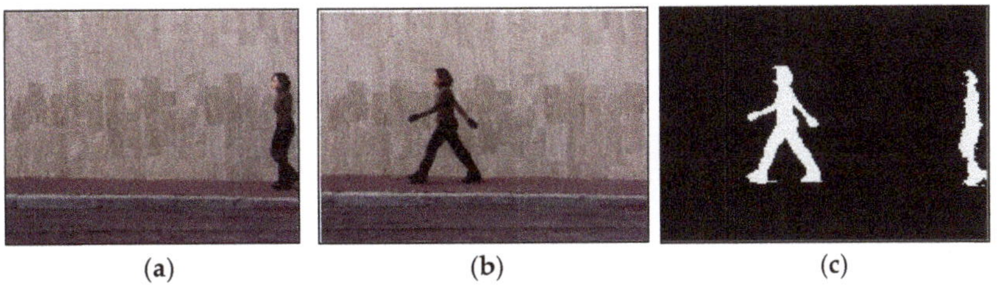

(a) (b) (c)

Figure 2. Pedestrian detection results in a single-frame image. (**a**) The initial frame, (**b**) the current frame, and (**c**) the detection result of ghost area.

To remove the ghost area, the VIBE background modelling method is adopted to extract the moving human target contour [36]. VIBE has the characteristics of less computational cost, fast speed, and less memory. By randomly selecting images, the temporal correlation can be improved, and the actual scene can be better coped with. By randomly selecting neighborhood locations, the spatial correlation can be improved, and the camera jitter can be dealt with, thus, the ghost area can be eliminated as soon as possible.

Figure 3 shows the pedestrian detection results in the 0–20th frame (every 5 frames) of a walking video by background-updating strategy. In Figure 3, we can observe that those ghost areas remain in the contours of target detection in subsequent frames since the initial frame contains a moving target. However, the ghost area residues gradually disappear with the updating strategy. By the 20th frame, the outline of the human body has become very clear. Therefore, VIBE can utilize the spatial propagation advantages of the pixels to gradually diffuse the background model outward and quickly eliminate the ghost areas.

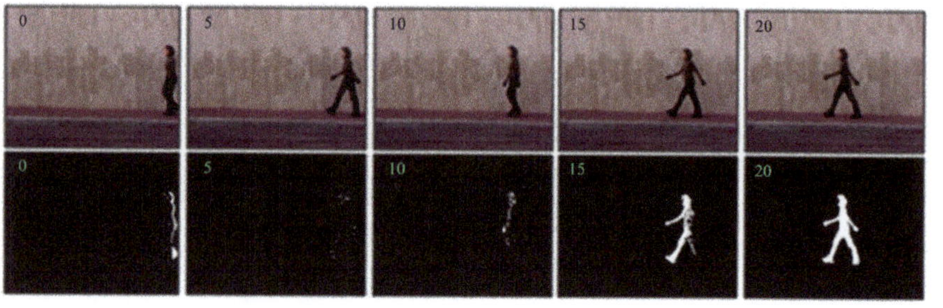

Figure 3. Pedestrian detection results of ghost area elimination by background updating strategy, where the numbers represent the number of frames.

3.2.2. Calculation of the Energy Image Species

Energy image species is one type of global feature, which is commonly used to statistically represent the spatial–temporal information of behavior. It mainly targets object contour images and has the advantages of simple calculation and not being sensitive to the background and movement time [31]. In this paper, five energy image species are utilized to represent human behavior, namely, MEI, MHI [31], AMEI, EMEI, and MEnI [32]. Although the five energy image species are all global descriptions, they are still slight differences because they are focused on different contents. The MEI and MHI focus on the change of human motion with time and the motion that happened at an earlier time, respectively. AMEI focuses on the overall movement by using binary contours, while EMEI is extracted to highlight the dynamic parts, and MEnI is defined by computing the Shannon entropy of the average motion energy image, trying to reflect the dynamic process from a microscopic perspective.

Let $\mathbf{I}_{seq}(\mathbf{x},\mathbf{y},t)$ denote an image sequence and $\mathbf{D}_{dif}(\mathbf{x},\mathbf{y},t)$ represent a binary image sequence, which indicates the motion regions of $\mathbf{I}_{seq}(\mathbf{x},\mathbf{y},t)$, and can be calculated by image differentiating, i.e., $\mathbf{D}_{dif}(\mathbf{x},\mathbf{y},t) = \mathbf{I}_{seq}(\mathbf{x},\mathbf{y},t+1) - \mathbf{I}_{seq}(\mathbf{x},\mathbf{y},t)$, where t, $1 \leq t \leq N$ represents the t-th frame, and N is the duration of the considered image sequence. Specific calculations of the five energy image species are as follows:

(1) MEI and MHI: The binary MEI $\mathbf{E}_{MEI}(\mathbf{x},\mathbf{y},t)$ and MHI $\mathbf{E}_{MHI}(\mathbf{x},\mathbf{y},t)$ can be calculated by Equations (1) and (2), respectively.

$$\mathbf{E}_{MEI}(\mathbf{x},\mathbf{y},t) = \bigcup_{i=0}^{\tau-1} \mathbf{D}_{dif}(\mathbf{x},\mathbf{y},t-i). \tag{1}$$

$$\mathbf{E}_{MHI}(\mathbf{x},\mathbf{y},t) = \begin{cases} \tau, & \text{if } \mathbf{D}_{dif}(\mathbf{x},\mathbf{y},t) = 1 \\ \max(0, \mathbf{E}_{MHI}(\mathbf{x},\mathbf{y},t-1) - 1), & \textit{otherwise} \end{cases}. \tag{2}$$

where τ is the motion duration, which is crucial in defining the temporal range of behavior.

(2) AMEI, EMEI, and MEnI: For the whole motion sequence of N frames, the average value of the binary contour is calculated as AMEI, which is shown in Equation (3).

$$\mathbf{E}_{AMEI}(\mathbf{x},\mathbf{y}) = \frac{1}{N} \sum_{t=1}^{N} \mathbf{I}_{seq}(\mathbf{x},\mathbf{y},t) \tag{3}$$

EMEI is calculated by:

$$\mathbf{E}_{EMEI}(\mathbf{x},\mathbf{y}) = \frac{1}{N}\sum_{t=1}^{N}\left\|\mathbf{I}_{seq}(\mathbf{x},\mathbf{y},t) - \mathbf{E}_{AMEI}(\mathbf{x},\mathbf{y})\right\|. \tag{4}$$

MEnI can be computed by:

$$\begin{aligned}\mathbf{E}_{MEnI}(\mathbf{x},\mathbf{y}) = & -\tfrac{1}{N}\sum_{t=1}^{N}\mathbf{I}_{seq}(\mathbf{x},\mathbf{y},t) \times \log_2(\tfrac{1}{N}\sum_{t=1}^{N}\mathbf{I}_{seq}(\mathbf{x},\mathbf{y},t) + \lambda) \\ & -(1 - \tfrac{1}{N}\sum_{t=1}^{N}\mathbf{I}_{seq}(\mathbf{x},\mathbf{y},t)) \times \log_2(1 - \tfrac{1}{N}\sum_{t=1}^{N}\mathbf{I}_{seq}(\mathbf{x},\mathbf{y},t) + \lambda)\end{aligned} \tag{5}$$

where λ is a small positive parameter, which is introduced to avoid the zero value for a logarithmic function.

As can be seen from Figure 3, the object contour images have an obvious black background. Therefore, the energy species of such an image will also have a black background, which does not express any behavior information and varies in size depending on the silhouettes of different performers. When we extract features from such energy species, it will not only increase the computation load but also affect the recognition results. Therefore, to remove the black background area, we extract the minimum bounding rectangle of the target contour region, i.e., the region of interest (ROI). Several samples of the energy image species on the Weizmann and DHA datasets are shown in Figure 4.

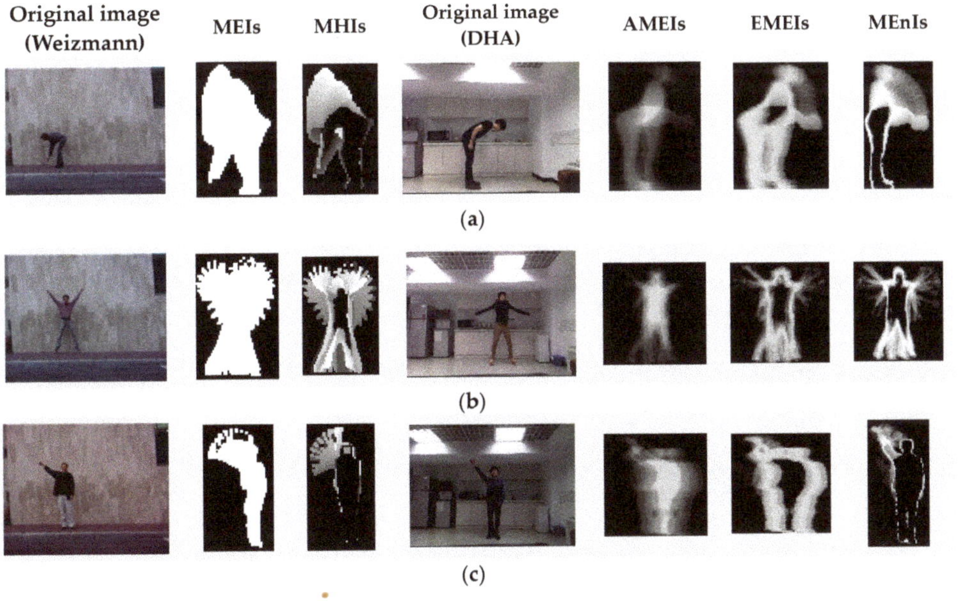

Figure 4. Some samples of the energy image species on the Weizmann and DHA datasets. (**a**) Bend behavior, (**b**) jack behavior, and (**c**) one-hand wave (wave1) behavior.

3.2.3. Construction of the Hierarchical Patches Descriptor (HPD)

By comparing with the original motion images in the leftmost column of Figure 4, we can see that the energy image species can represent the motion information in a global manner for most behaviors, such as the motion of body parts, the action area of the trunk, and the motion range of limbs. However, it cannot describe the details of local motion information very well. Taking one-hand wave behavior as an example, we find that the static trunk is clearly presented using AMEI and EMEI. In contrast, the waving hand and arm parts are shown as a vague shape area, which is very likely to lead to confusion with other similar behavior, such as a typical motion in tai chi. Therefore, it is necessary to extract local detailed features for more accurate recognition.

Recently, BoW has been one of the most successful methods used to describe the detailed features of images. The investigation of many extension methods of BoW shows that SPM [37] reports the most successful results. Therefore, in this paper, an HPD is constructed by using the SPM-based BoW model, and the algorithm flow is shown in Algorithm 1.

Algorithm 1 Construction Process of HPD

Input Energy image species $\mathbf{E}_{MEI}(\mathbf{x},\mathbf{y},t)$, $\mathbf{E}_{MHI}(\mathbf{x},\mathbf{y},t)$, $\mathbf{E}_{AMEI}(\mathbf{x},\mathbf{y})$, $\mathbf{E}_{EMEI}(\mathbf{x},\mathbf{y})$, **and** $\mathbf{E}_{MEnI}(\mathbf{x},\mathbf{y})$;
Output HPD feature vector **X**:
Step 1: Obtain SIFT descriptors. For each energy species, the SIFT descriptors of 31×31 patches calculated over a grid with a spacing of 16 pixels are extracted from each key point or patch as local features. This is realized by using a difference-of-Gaussian function:
$\mathbf{D}_{sift}(\mathbf{x},\mathbf{y},\sigma) = (\mathbf{G}(\mathbf{x},\mathbf{y},k\sigma) - \mathbf{G}(\mathbf{x},\mathbf{y},\sigma))^{*}\mathbf{E}(\mathbf{x},\mathbf{y}) = \mathbf{L}(\mathbf{x},\mathbf{y},k\sigma) - \mathbf{L}(\mathbf{x},\mathbf{y},\sigma)$.
where $\mathbf{G}(\mathbf{x},\mathbf{y},\sigma) = \frac{1}{2\pi\sigma^2}\exp^{-\frac{(x^2+y^2)}{2\sigma^2}}$.
Step 2: Generate a codebook with M channels by sparse coding [8]. To improve the computational efficiency, the K-means clustering method can be used to compute the cluster centers.
Step 3: Encode the descriptors. Each SIFT descriptor is encoded into a code vector with codewords in the codebook and each descriptor is transferred to an $\mathbf{R}^M$ code.
Step 4: Spatial feature pooling.

(a) Segment the image into finer spatial subregions by using SPM method;
(b) Construct a histogram by pooling multiple codes of each subregion together after averaging and normalizing operations;
(c) Cascade the histograms of all patches in different spatial pyramid segmentation levels to form the HPD feature vector **X**.

Get the HPD feature vector.

Figure 5 shows a simple schematic of structuring a three-level spatial pyramid. We assume that the energy image species have three feature types, expressed in circles, rhombuses, and stars. First, the image is divided into three different levels of scale. Second, the features that fall in each spatial bin are counted for each level of the scale channel. Last, on the basis of a spatial-pyramid match kernel function, each spatial histogram is weighted together; that is

$$\mathbf{K}^L(\mathbf{X},\mathbf{Y})^{\mathrm{T}} = \frac{1}{2^{L-l}}\mathbf{I}^0 + \sum_{l=0}^{L-1}\frac{1}{2^{L-l+1}}\mathbf{I}^l. \tag{6}$$

The spatial-pyramid match kernel is a Mercer kernel, which allows processing of Gaussian variables.

From Figure 5, we can see that the image is segmented into finer spatial subregions, and then the histograms of each subregion are computed as the local features. Generally, $2^l \times 2^l$ (where $l = 0,1,2$) sub-regions are typically used. In this case, for L segmentation levels and M channels, the dimensionality of the final feature vector for human behavior recognition is

$$Dim_{final} = M \times \sum_{l=0}^{L} 4^L = M \times \frac{1}{3} \times (4^{L+1} - 1). \tag{7}$$

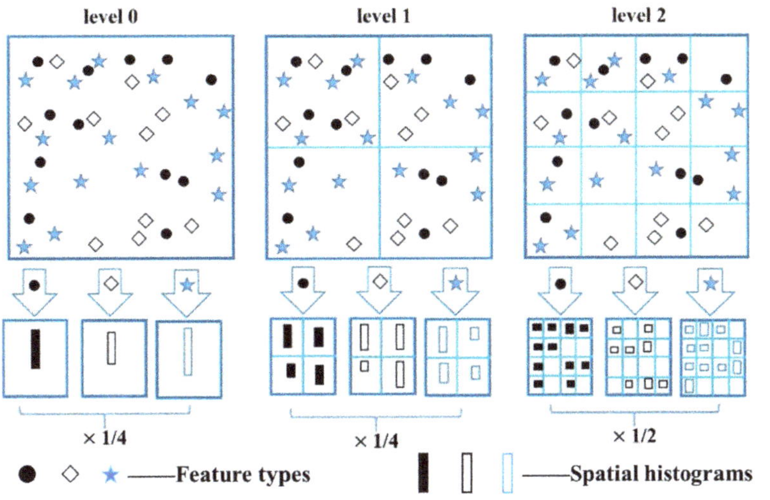

Figure 5. A simple schematic of structuring a three-level spatial pyramid.

3.3. *Human Behavior Recognition Scheme Based on LLC Algorithm*

The SPM method utilizes the vector-quantization (VQ) coding strategy for coding, whose code has only non-zero coefficients following the non-zero constraint condition. To improve its scalability, Yang et al. [16] proposed the sparse-coding-based SPM (ScSPM) approach, where a sparse-coding algorithm was used to a encode nonlinear code. Yu et al. [38] proposed a local coordinate coding algorithm and verified that locality is more critical than sparsity under certain assumptions. Although both coding algorithms have achieved superior performance on several benchmarks, they all need to solve the ℓ_1 norm optimization, which leads to a higher computational expense. Based on this knowledge, in this paper, we employ the ALLC algorithm, which has an analytical solution and its computational cost efficiency is lower than the sparse coding and local coordinate coding. In this section, the recognition scheme based on the ALLC algorithm will be introduced in detail.

3.3.1. Problem Formulation

Let $\mathbf{X} = [\mathbf{x}_1, \mathbf{x}_2, \cdots, \mathbf{x}_N] \in \mathbf{R}^{D\times N}$ represent a set of local features with D dimensionality, which is extracted from energy image species; $\mathbf{B} = [\mathbf{b}_1, \mathbf{b}_2, \cdots, \mathbf{b}_M] \in \mathbf{R}^{D\times M}$, $\mathbf{b}_j \in \mathbf{R}^{D\times 1}$ denote a codebook with M codewords; and $\mathbf{C} = [\mathbf{c}_1, \mathbf{c}_2, \cdots, \mathbf{c}_N] \in \mathbf{R}^{M\times N}$, $\mathbf{c}_i \in \mathbf{R}^{M\times 1}$ represent the coding vector for feature $\mathbf{X}$ based on codebook $\mathbf{B}$. The purpose of feature coding is to obtain the coding vector $\mathbf{C}$ by using different coding algorithms.

For most coding algorithms, only a part of codewords will be chosen for feature representation, and its coefficients are non-zero. However, most codewords are not chosen, and their corresponding coefficients are equal to zero. Therefore, the coding vector $\mathbf{C}$ is usually sparse.

3.3.2. The LLC Algorithm

The traditional SPM algorithm uses the VQ coding method, and the coding vector $\mathbf{C}$ is obtained by finding the constrained least squares fitting solution. The objective function is:

$$<\mathbf{C}>= \arg\min_{\mathbf{C}} \sum_{i=1}^{N} \|\mathbf{x}_i - \mathbf{B}\mathbf{c}_i\|^2, \; s.t. \; \|\mathbf{c}_i\|_{\ell_0} = 1, \|\mathbf{c}_i\|_{\ell_1} = 1, \mathbf{c}_i \geq 0, \forall i. \tag{8}$$

where the cardinality constrained condition $\|\mathbf{c}_i\|_{\ell_0} = 1$ expresses that each coding vector $\mathbf{c}_i$ contains only one non-zero element, corresponding to the quantitative ID of $\mathbf{x}_i$. By searching for the nearest neighbor of its neighborhood, the single non-zero element can be obtained. The non-negative constrained term $\|\mathbf{c}_i\|_{\ell_1} = 1, \mathbf{c}_i \geq 0$ denotes that the coding weight of $\mathbf{x}_i$ is 1.

To reduce the vector loss of the VQ algorithm, the cardinality constraint condition $\|\mathbf{c}_i\|_{\ell_0} = 1$ can be relaxed by utilizing the sparse regularization term, and its objective function is rewritten as

$$< \mathbf{C} >= \arg\min_{\mathbf{C}} \sum_{i=1}^{N} \|\mathbf{x}_i - \mathbf{B}\mathbf{c}_i\|^2 + \lambda\|\mathbf{c}_i\|_{\ell_1}, \; s.t. \; \|\mathbf{b}_m\| \leq 1, \; \forall m = 1, 2, \cdots, M. \tag{9}$$

where the sparse constrained term has three functions: (1) due to the codebook being over-complete, i.e., $M > D$, it is necessary to add an ℓ_1 regularization term to make sure of the uniqueness of solution for the under-determined system; (2) it allows the obtained representation to acquire a salient pattern of local descriptors; and (3) compared with VQ algorithm, the quantization error is reduced.

According to the suggestion of the local coordinate coding algorithm, the locality is more significant than sparsity. Therefore, the LLC algorithm utilizes the locality-constrained term to replace the sparsity constrained term in Equation (9), and its objective function can be written as:

$$< \mathbf{C} >= \arg\min_{\mathbf{C}} \sum_{i=1}^{N} (\|\mathbf{x}_i - \mathbf{B}\mathbf{c}_i\|^2 + \lambda\|\mathbf{d}_i \odot \mathbf{c}_i\|^2), \; s.t. \; \mathbf{1}^T\mathbf{c}_i = 1, \; \forall i. \tag{10}$$

where $\mathbf{1} \in \mathbf{R}^{M\times 1}$ is a column vector with all elements as ones, $\odot$ expresses an element-wise multiplication operator, and $\mathbf{d}_i \in \mathbf{R}^M$ denotes a locality adaptor, and is calculated by Equation (11),

$$\mathbf{d}_i = \exp(\frac{\mathbf{D}_{dis}(\mathbf{x}_i, \mathbf{B})}{\sigma}). \tag{11}$$

where $\mathbf{D}_{dis}(\mathbf{x}_i, \mathbf{B}) = [\mathbf{D}_{dis}(\mathbf{x}_i - \mathbf{b}_1), \cdots, \mathbf{D}_{dis}(\mathbf{x}_i - \mathbf{b}_M)]^T$ and $\mathbf{D}_{dis}(\mathbf{x}_i - \mathbf{b}_j)$ express the Euclidean distance between $\mathbf{x}_i$ and each codeword and σ is a tune parameter to adjust the speed of weight decay. Moreover, compared with sparse coding and local coordinate coding, the constraint condition $\mathbf{1}^T\mathbf{c}_i = 1$ of LLC is more crucial than sparsity, which follows the shift-invariant requirements.

The LLC algorithm has a closed-form analytical solution

$$\tilde{\mathbf{c}}_i = (\mathbf{C}_i + \lambda diag(\mathbf{d}_i)) \backslash \mathbf{1}. \tag{12}$$

$$\mathbf{c}_i = \mathbf{c}_i / \mathbf{1}^T\mathbf{c}_i. \tag{13}$$

where $c_i = (\mathbf{B} - \mathbf{1}\mathbf{x}_i^T)(\mathbf{B} - \mathbf{1}\mathbf{x}_i^T)^T$ is a covariance matrix.

3.3.3. ALLC Algorithm for Fast Coding

In the process of solving object function (10), a local coordinate system is constructed on the local basis of each descriptor. Moreover, without solving the objective function (10) directly, the K-nearest neighbors (where $K < D < M$) of $\mathbf{x}_i$ in the codebook can be simply used as the local bases $\mathbf{B}_i$, then the coding vector $\mathbf{C}$ is computed by solving a much smaller linear system, and its objective function is

$$< \mathbf{C} >= \arg\min_{\mathbf{C}} \sum_{i=1}^{N} \|\mathbf{x}_i - \mathbf{c}_i\tilde{\mathbf{B}}_i\|^2, \; s.t. \; \mathbf{1}^T\mathbf{c}_i = 1, \forall i. \tag{14}$$

Because $\mathbf{B}_i$ is the K-nearest neighbor code-word for $\mathbf{x}_i$, and $K \ll M$, the approximate algorithm can reflect the locality and sparsity simultaneously. In addition, the computational complexity declined from $O(M^2)$ to $O(M + K^2)$, which greatly reduces the computation cost. Its coding process is illustrated in Figure 6.

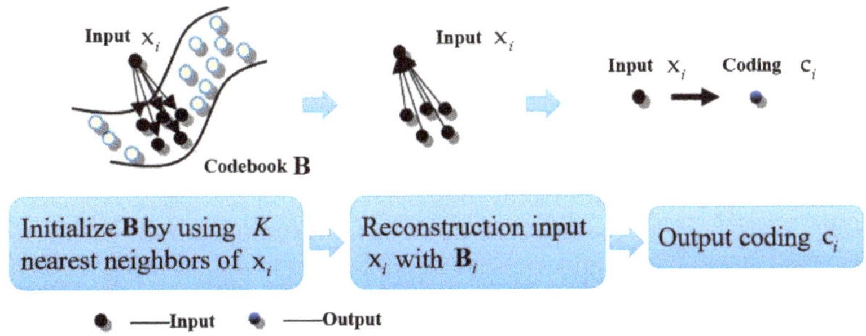

Figure 6. The coding process of ALLC algorithm.

3.3.4. Max-Pooling

The ALLC algorithm is used to encode all patches on each level of the SPM in matrix form, which makes it difficult to construct eigenvectors. Therefore, it is necessary to pool all codes and normalize them to form a final feature vector for behavior recognition. In this paper, the max-pooling method is used, which is as follows:

$$\mathbf{c}_{out} = \max(\mathbf{c}_{in1}, \cdots, \mathbf{c}_{in2}). \quad (15)$$

where the max function is pooled in rows and the dimension of the returned vector is the same as $\mathbf{c}_{in1}$. Moreover, the pooling feature is normalized by ℓ_2 norm:

$$\mathbf{c}_{out} = \mathbf{c}_{in} / \|\mathbf{c}_{in}\|_2. \quad (16)$$

4. Experimental Results

4.1. Experimental Settings and Descriptions

The experiments reported in this section were conducted on two public human behaviour datasets, namely, the classical Weizmann dataset [39], and DHA dataset [40]. Different from the Weizmann dataset, the DHA dataset is more challenging. It contains RGB and depth data, with more variations in background, illumination fluctuations, and behavior complexity, and it is a multi-modality dataset. The details are as follows:

(1) Weizmann dataset: The Weizmann dataset consists of 10 human behavior categories, every behavior was completed by nine performers in a similar environment. Each video sequence has a different length. Following the database instructions of literature [7,41], nine behaviors were selected for MEI and MHI, which were bend, jump, jack, side, run, walk, skip, wave1 (one-hand wave), and wave2 (two-hand wave).

(2) DHA dataset: The DHA dataset contains 23 categories of human behavior (e.g., bend, jump, pitch, and arm-swing), where every behavior contains 21 performers (12 males and 9 females). The duration of the video sequences also varies. Following [10] and the database instructions, 14 behaviors were selected for MEnI, including bend, jump, jack, run, skip, walk, side, wave1, wave2, side-box, arm-swing, tai chi, and leg-kick, and 17 behaviors were selected for AMEI and EMEI, including bend, jump, jack, pjump, run, walk, skip, side, wave1, wave2, arm-swing, leg-lick, front-lap, side-box, side-box, rod-swing, and tai chi.

For convenience of comparison, the leave-one-video-out evaluation strategy was adopted to assess the approach performance. The proposed approach was compared with some existing techniques mainly on three aspects: different combined features, feature-coding algorithms, and different data modality-based approaches. For each comparison, the parameter setting was provided with the reported results on the two public datasets. The confusion matrix analysis was also conducted for the proposed approach.

All the experiments were performed on a computer with an 11th Gen Intel(R) Core(TM) i7-1165G7 @ 2.80GHz CPU and Windows 11 Professional edition operating system using Matlab 2018b software.

4.2. Parameter Selection

Following the parameter selection scheme of FDDL and LCKSVD, we evaluated all parameters by using the five-fold cross-validation. There were four parameters in the proposed approach that needed to be adjusted, namely, the size of codebook M, parameter K for the K-means clustering algorithm, regularization parameter c for linear SVM, and the segmentation of subregions for SPM. A codebook with 1024 bases was pre-trained for the two datasets and three-level 4 × 4, 2 × 2, and 1 × 1 subregions were used for SPM. Therefore, the dimensions of the final feature vectors were 21,504 according to Equation (7). According to [35,36], two trade-off parameters λ_1 and λ_2 of FDDL and four parameters (dictionary size, sparsity, and two trade-off parameters α and β) of LCKSVD were set. The parameter selections of the benchmark approaches are summarized in Table 1.

Table 1. Parameters selection of different approaches.

Approach	Ours				FDDL		LCKSVD			
Parameters	M	K	c	L	λ_1	λ_2	Dictionary Size	Sparsity	α	β
MEI	1024	5	7	2	0.05	0.5	60	8	0.05	0.001
MHI	1024	3	13	2	0.05	0.5	60	8	0.01	0.001
MEnI	1024	3	10	2	0.05	0.5	150	10	0.01	0.001
AMEI	1024	3	13	2	0.005	0.05	-	-	-	-
EMEI	1024	3	13	2	0.005	0.05	-	-	-	-

4.3. Experimental Results and Comparative Analysis on Weizmann Dataset

4.3.1. Comparison of Different Feature Combinations

We evaluated the proposed feature extraction strategy with several existing feature combinations, which contained MHI+BoW [7], MEI+PHOG [41], MHI+PHOG [41], MEI+R [41], and MHI+R [41]. An SVM classifier with a linear kernel function was employed for the aforementioned feature combinations, except for MHI+BoW [7], which used a KNN classifier. The results of different combined features on the Weizmann dataset are shown in Table 2.

Table 2. Testing results of different combined feature comparisons on the Weizmann dataset.

Features	Accuracy Rate (%)
MHI+BoW [7]	90
MEI+PHOG [41]	82.7
MHI+PHOG [41]	92.6
MEI+R [41]	86.4
MHI+R [41]	81.5
Our MEI+HPD	100
Our MHI+HPD	98.77

The proposed feature combinations (i.e., with HPD) obtained better recognition results, and the accuracy was substantially higher than the other methods. These results demonstrate that the combination of energy image species and HPD is an effective strategy, as it combines the global features and local features to better describe human behavior for recognition.

4.3.2. Comparison of Feature-Coding Algorithms

The ALLC algorithm was evaluated by comparing it against two other state-of-the-art feature-coding algorithms: Fisher discrimination dictionary learning (FDDL) [42] algorithm and label consistent K-SVD (LCKSVD) [43] algorithm. To get reliable results under different features, we need to indicate that if the subregions segmentation parameter for SPM is set to be 0, i.e., $l = 0$, means that the original energy image species would not be segmented, and the MEI+HPD and MHI+HPD features will be reduced to MEI and MHI features, respectively. Therefore, we compared the ALLC algorithm with the aforementioned two feature-coding methods under the same features. The testing results of the feature-coding algorithm comparison on the Weizmann dataset are shown in Table 3.

Table 3. Testing results of feature-coding algorithm comparison on the Weizmann dataset.

Features	Feature-Coding Algorithms	Accuracy Rate (%)
MEI	LCKSVD1	92.6
MEI	LCKSVD2	95.07
MHI	LCKSVD1	93.83
MHI	LCKSVD2	96.3
MEI	FDDL	96.3
MHI	FDDL	95.06
MEI	Our ALLC	95.06
MHI	Our ALLC	93.83

Obviously, the performance of the ALLC algorithms was comparable to the other two feature-coding algorithms: FDDL and LCKSVD. For example, considering the MEI feature, the accuracy of the ALLC algorithm was 95.06%, while the accuracy of FDDL, LCKSVD1, and LCKSVD2 were 96.3%, 92.6%, and 95.07%, respectively. There were similar results for the energy image species MHI. These results prove that the ALLC feature-coding method can achieve a comparable result while being more computationally efficient.

4.3.3. Comparison with Other Behavior Recognition Approaches

The evaluation results of the proposed approach with other existing approaches, including 3D-SIFT [10], HOGS [11], and HOG+CNN [24] are summarized in Table 4.

Table 4. Testing results of some competitive approaches on the Weizmann dataset.

Features	Classifiers	Accuracy Rate (%)
3D-SIFT [10]	KNN	97.84
HOGS [11]	KNN	99.65
HOG+CNN [24]	SVM	99.4
Our MEI+HPD+ALLC	SVM	100
Our MHI+HPD+ALLC	SVM	98.77

Referring to Table 4, we can see that the accuracy of MHI+HPD+ALLC was 98.77%, which is a little lower than HOGS [11] and HOG+CNN [24] with an accuracy of 99.65% and 99.4%, respectively. However, the proposed approach of MEI+HPD+ALLC achieved the highest accuracy of 100%. Thus, the proposed approach is comparable with the existing state-of-the-art approaches, especially with small-scale datasets, such as the Weizmann dataset. Here, an exciting result is that the proposed approach reached a comparable accuracy to the HOG+CNN approach [24]. This indicates that the proposed approach can offer comparable results to CNN-based approaches in targeted behavior recognition scenarios. Literature [24] also indicated that the training/testing ratio gives scope for a significant role in achieving greater accuracy; it reported that a 70:30 (70: training, 30: testing) ratio is considered optimal, however, with 80:20 and 50:50, the results tend to reduce. Therefore, CNN-based approaches are sensitive to the training/testing ratio. In comparison, our approach does not need to consider the training/testing ratio more.

4.4. Experimental Results and Comparative Analysis on DHA Dataset

4.4.1. Comparison of Different Feature Combinations

The proposed feature extraction strategy was compared with different combined features, which contain HOGS [11], depth multi-perspective projections and PHOG features (DMPP+PHOG) [13], depth-limited RGB multi-perspective projection and PHOG features (DLRMPP+PHOG) [13], fusion of the RGB and depth features of DMPP and DLRMPP (DMPP+DLRMPP+PHOG) [13], GIST feature combined with space–time interest points from depth videos (GIST+DSTIPs) [21], and human pose representation model and temporal modeling representation (HPM+TM) [22]. The comparison results of different feature combinations on the DHA dataset are shown in Table 5.

Table 5. Testing results of different feature combination comparisons on the DHA dataset.

Features	Accuracy Rate (%)
HOGS [11]	99.39
DMPP+PHOG [13]	95
DLRMPP+PHOG [13]	95.6
DMPP+DLRMPP+PHOG [13]	98.2
GIST+DSTIPs [21]	93
HPM+TM [22]	90.8
Our AMEI+HPD	95.52
Our EMEI+HPD	96.08
Our MEnI+HPD	97.61

From Table 5, one can see that, for RGB data modality, the HOGS [11] feature has achieved the highest recognition rate 99.39%. The proposed approach with 3 different energy image species (AMEI+HPD, EMEI+HPD, and MEnI+HPD) achieves a comparable recognition rate between 95% and 97%. The results further prove that the proposed strategy of combing energy image species and HPD can represent the human behavior well for recognition.

4.4.2. Comparison of Feature-Coding Algorithms

The ALLC algorithm was evaluated and compared with four other existing feature-coding algorithms: SRC, CSR, FDDL, and LCKSVD. We also need to indicate that the subregions segmentation parameter for SPM was also set to be 0, i.e., $l = 0$, and the three different combined features (AMEI+HPD, EMHI+HPD, and MEnI+PHD) will reduce to the original energy image species (AMEI, EMHI, and MEnI). The results of the feature-coding algorithm comparison on the DHA dataset are detailed in Table 6.

Table 6. Testing results of feature coding algorithm comparison on the DHA dataset.

Features	Feature-Coding Algorithms	Accuracy Rate (%)
GIST+DSTIPs [17]	SRC	93
HPM+TM [18]	SRC	93
HPM+TM [18]	CSR	98.6
AMEI	FDDL	89.09
EMEI	FDDL	91.32
MEnI	LCKSVD1	92.88
MEnI	LCKSVD2	94.58
Our AMEI+HPD	Our ALLC	93.28
Our EMEI+HPD	Our ALLC	94.68
Our MEnI+HPD	Our ALLC	95.92

Taking MEnI+HPD features, the proposed approach can achieve an improvement of 1% to 4% compared with most of the benchmark methods and also achieves a comparable result with the HPM+TM approach.

4.4.3. Comparison of Different Multi-Modality Fusion Methods

RGB is an essential channel of RGB-D data, which includes rich information features, e.g., color, shape, and texture. While depth images could provide information about the distance from the surface of the scene object of the viewpoint. Aiming to get higher accuracy and robustness in human behavior recognition, many researchers focus on depth-modality data-based approaches and multi-modality data-based approaches. Even the proposed approach mainly targets the RGB data, and was evaluated against some competitive single modality (RGB or depth)-based approaches and multimodality-based approaches. The testing results of different modality data-based approaches on the DHA dataset are summarized in Table 7.

Table 7. Testing results of different modality data-based approaches on the DHA dataset.

Data Modality	Features	Accuracy Rate (%)
RGB	HOGS [11]	99.39
RGB	DLRMPP+PHOG [13]	95.6
RGB	HPM+TM [22]	91.9
RGB	Our AMEI+HPD	95.52
RGB	Our EMEI+HPD	96.08
RGB	Our MEnI+HPD	97.61
Depth	DMPP+PHOG [13]	95
Depth	GIST+DSTIPs [21]	94
Depth	HPM+TM [22]	90.8
RGB+Depth	DMPP+DLRMPP+PHOG [13]	98.2
RGB+Depth	MMDJM+GIST+DSTIP [21]	97
RGB+Depth	HPM+TM+CSR [22]	98.6
RGB+Depth	HPM+TM+SRC [22]	94.4

The proposed approach achieved better performance compared to the depth modality data-based approach, with about 2–7% improvement in recognition accuracy. Compared with RGB modality data-based approaches, it was better than HPM+TM [22] and DLRMPP+PHOG [13], but had a little lower accuracy than HOGS [11]. In comparison with the multi-modality fusion approaches, DMPP+DLRMPP+PHOG [13] and HPM+TM+CSR [18] obtained marginally higher accuracy of 98.2% and 98.6%, respectively, while MMDJM_GIST_DSTIP [21] and HPM+TM+SRC [22] achieved a slightly lower recognition accuracy than the proposed approaches.

It is worth mentioning that in Tables 3 and 5, Tables 6 and 7 the results consist of three parts. For FDDL [42] and LC-KSVD [43], we implemented the publicly available code provided by the authors on the datasets. For HOGS [11], GIST+DSTIPs [20], and HPM+TM [21], the results are cited directly from their original work. The rest are the results of the proposed methods.

4.4.4. Confusion Matrix Analysis

To make further analyses of the recognition performance, a correlation analysis was carried out by using the confusion matrix. In this section, the confusion matrices of two energy image species (AMEI and EMEI) are presented in Figure 7a,b, respectively. According to the confusion matrices and analysis results, the following conclusions can be drawn:

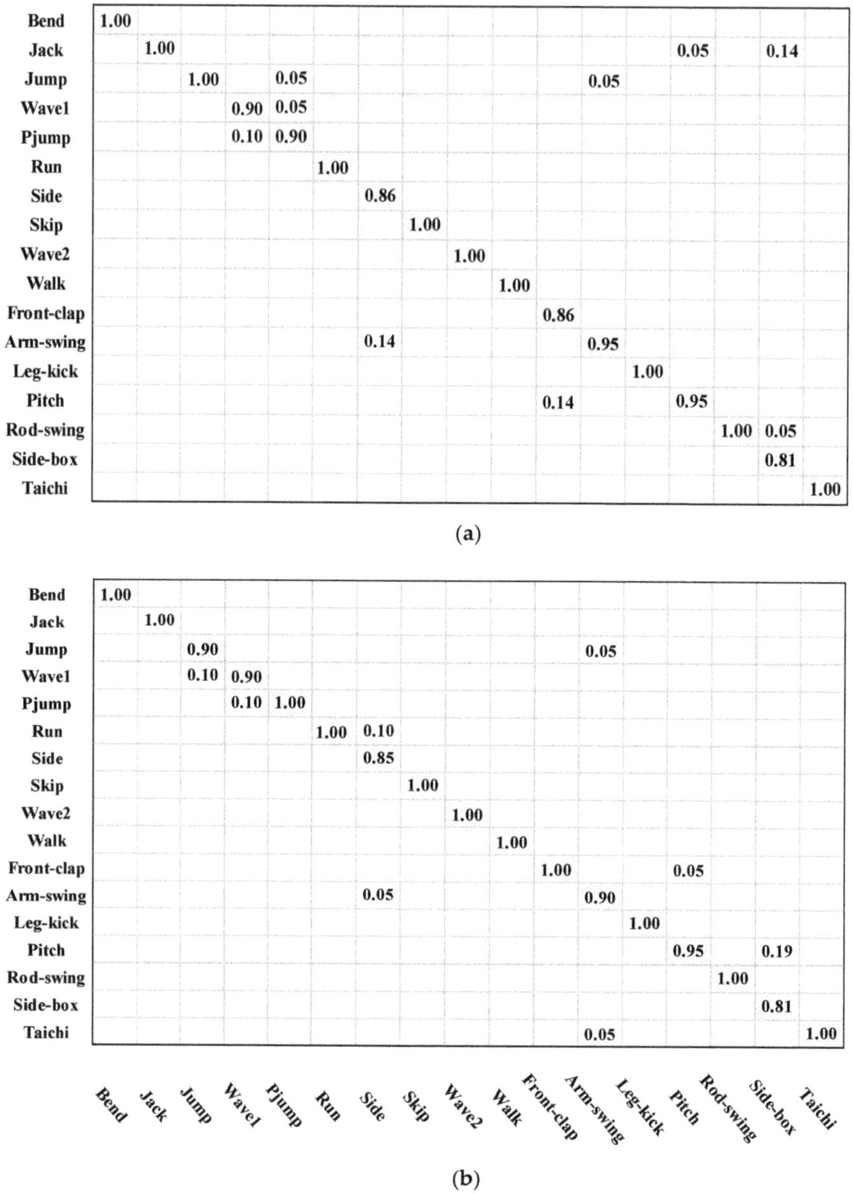

Figure 7. The confusion matrices for different energy image species descriptions. (**a**) Confusion matrix of DHA dataset based on AMEI and (**b**) confusion matrix of DHA dataset based on EMEI.

(1) The lowest correct recognition rate was 81% for both AMEI and EMEI on the DHA dataset; 10 and 11 out of 17 types of behaviors achieved 100% accuracy in recognition, respectively.

(2) Through analysing the confusion matrix, we can observe that certain behaviors were similar and may have caused confusion with each other; for example, wave1 and pjump; skip and jump; walk, skip and run; wave2 and leg-kick; pjump and jump; arm-swing and tai chi; side-box, jack, and pitch. Especially for side-box behavior,

owing to the different motion ranges, angles, and boxing directions of the different performers, the accuracy was only 81%.

(3) For behaviors with high similarity and involving position change, such as run, pjump, front-clap, side, the recognition results were worse than the other behaviors. One possible reason is that those behaviors all contain leg and arm movements, however, their motion directions and positions may vary between image frames. Although HPD was constructed based on different energy image species for obtaining detailed motion features, they could not describe the depth information well. Therefore, it was difficult to identify these types of behaviors correctly.

5. Discussion

From Sections 4.3 and 4.4, the experimental results prove that the proposed energy image species combined with the HPD feature extraction approach can better describe human behavior information than classical methods, because it describe the local and global features together. Meanwhile, the ALLC algorithm is a fast coding method, superior than the multi-modality algorithms which are computationally more expensive. One possible reason is that it has an analytical solution, thus it is more computationally efficient than some competitive feature-coding algorithms and multi-modality fusion approaches. Meanwhile, through sharing local bases, the ALLC algorithm could obtain the correlations between descriptors with similarity and make sure that patches with higher similarity have similar codes, which is very beneficial for feature recognition.

The research conducted in this work benefits other researchers that require automatic and robust extraction of self-learning features for human behavior recognition from video sequences in different ambient intelligence applications. Thus, it leads to us assume that this field may also quickly and effectively achieve good results in the case of insufficient data. However, there are still certain behaviors that usually contain depth information with a high degree of similarity, and the HPD could not describe the depth of information well.

6. Conclusions

Overall, many studies have been done on dictionary-learning-based approaches to human behavior recognition, and the present work adds other unique architectures involving energy images, the hierarchical patches descriptor (HPD), and the approximate locality-constrained linear coding (ALLC) algorithm. Experimental results and comparative analyses using the Weizmann and DHA datasets were demonstrated to be superior to some state-of-the-art approaches. In future work, to improve the robustness, we will consider constructing a human behavior model by fusing the RGB and depth information. In addition, in the case of large-scale data, deep-learning-based approaches need to be considered, such as multi-modality-based improved CNN and RNN.

Author Contributions: Conceptualization, L.L. and K.I.-K.W.; methodology, W.H.A.; software, B.T.; validation, L.L. and K.I.-K.W.; formal analysis, M.G.; investigation, G.J.; resources, L.L.; data curation, K.I.-K.W. and W.H.A.; writing—original draft preparation, L.L.; writing—review and editing, K.I.-K.W. and M.G.; visualization, G.J.; supervision, M.G.; project administration, B.T.; funding acquisition, L.L. and M.G. All authors have read and agreed to the published version of the manuscript.

Funding: This research was funded by the Natural Science Foundation of Shandong Province of China (grant number ZR2016FP04).

Institutional Review Board Statement: Not applicable.

Informed Consent Statement: Not applicable.

Data Availability Statement: Data will be made available on demand.

Acknowledgments: This work was supported by the Shandong Provincial Government Public Study Abroad Project and the Shandong University of Technology Doctoral Research Start-up Fund Project.

Conflicts of Interest: The authors declare no conflict of interest.

References

1. Khaire, P.; Kumar, P. Deep learning and RGB-D based human action, human-human and human-object interaction recognition: A survey. *JVCIR* **2022**, *86*, 103531. [CrossRef]
2. Wang, Z.H.; Zheng, Y.F.; Liu, Z.; Li, Y.J. A survey of video human behaviour recognition methodologies in the perspective of spatial-temporal. In Proceedings of the 2022 2nd International Conference on Intelligent Technology and Embedded Systems, Chengdou, China, 23–26 September 2022; pp. 138–147.
3. Chen, A.T.; Morteza, B.A.; Wang, K.I. Investigating fast re-identification for multi-camera indoor person tracking. *Comput. Electr. Eng.* **2019**, *77*, 273–288. [CrossRef]
4. Yue, R.J.; Tian, Z.Q.; Du, S.Y. Action recognition based on RGB and skeleton data sets: A survey. *Neurocomputing* **2022**, *512*, 287–306. [CrossRef]
5. Yao, G.L.; Tao, L.; Zhong, J.D. A review of convolutional neural network based action recognition. *Pattern Recogn. Lett.* **2019**, *118*, 14–22.
6. Kumar, D.; Kukreja, V. Early recognition of wheat powdery mildew disease based on mask RCNN. In Proceedings of the 2022 International Conference on Data Analytics for Business and Industry (ICDABI), Sakheer, Bahrain, 25–26 October 2022; pp. 542–546.
7. Plizzari, C.; Cannici, M.; Matteucci, M. Skeleton-based action recognition via spatial and temporal transformer networks. *Comput. Vis. Image Und.* **2021**, *208*, 103219. [CrossRef]
8. Lowe, D.G. Distinctive image features from scale-invariant key points. *Int. J. Comput. Vis.* **2004**, *60*, 91–110. [CrossRef]
9. Kumar, D.; Kukreja, V. MRISVM: A object detection and feature vector machine based network for brown mite variation in wheat plant. In Proceedings of the 2022 International Conference on Data Analytics for Business and Industry (ICDABI), Sakheer, Bahrain, 25–26 October 2022; pp. 707–711.
10. Zeng, M.Y.; Wu, Z.M.; Chang, T.; Fu, Y.; Jie, F.R. Fusing appearance statistical features for person re-identification. *J. Electron. Inf. Technol.* **2014**, *36*, 1844–1851.
11. Obaidi, S.A.; Abhayaratne, C. Temporal salience based human action recognition. Proceeedings of the 2019 International Conference on Acoustics, Speech and Signal Processing (ICASSP), Bradu, UK, 12–17 May 2019; pp. 2017–2021.
12. Patel, C.I.; Labana, D.; Pandya, S. Histogram of oriented gradient-based fusion of features for human action recognition in action video sequences. *Sensors* **2020**, *20*, 7299. [CrossRef]
13. Gao, Z.; Zhang, H.; Xu, G.P.; Xue, Y.B. Multi-perspective and multi-modality joint representation and recognition model for 3D action recognition. *Neurocomputing* **2015**, *151*, 554–564. [CrossRef]
14. Chen, C.; Liu, M.Y.; Zhang, B.C. 3D action recognition using multi-temporal depth motion maps and fisher vector. In Proceedings of the 2016 International Conference on Artificial Intelligence, New York, NY, USA, 15 July 2016; pp. 3331–3337.
15. Wright, J.; Yang, A.Y.; Ganesh, A.; Sastry, S.S.; Ma, Y. Robust face recognition via sparse representation. *IEEE T-PAMI* **2009**, *31*, 210–227. [CrossRef]
16. Yang, J.C.; Yu, K.; Gong, Y.H.; Huang, T.S. Linear spatial pyramid matching using sparse coding for image classification. In Proceedings of the 2009 IEEE Conference on Computer Vision and Pattern Recognition (CVPR), Miami, FL, USA, 22–25 June 2009; pp. 1794–1800.
17. Kumar, D.; Kukreja, V. Application of PSPNET and fuzzy Logic for wheat leaf rust disease and its severity. In Proceedings of the 2022 International Conference on Data Analytics for Business and Industry (ICDABI), Sakheer, Bahrain, 25–26 October 2022; pp. 547–551.
18. Wang, J.J.; Yang, J.C.; Yu, K.; Lv, F.; Gong, Y. Locality-constrained linear coding for image classification. In Proceedings of the 2010 IEEE Conference on Computer Vision and Pattern Recognition (CVPR), San Francisco, CA, USA, 13–18 June 2010; pp. 3267–3360.
19. Wu, J.L.; Lin, Z.C.; Zheng, W.M.; Zha, H. Locality-constrained linear coding based bi-layer model for multi-view facial expression recognition. *Neurocomputing* **2017**, *239*, 143–152. [CrossRef]
20. Wang, L.; Zhao, X.; Liu, Y.C. Skeleton feature based on multi-stream for action recognition. *IEEE Access* **2018**, *6*, 20788–20800. [CrossRef]
21. Gao, Z.; Zhang, H.; Liu, A.A.; Xu, G.; Xue, Y. Human action recognition on depth dataset. *Neural Comput. Appl.* **2016**, *27*, 2047–2054. [CrossRef]
22. Gao, Z.; Li, S.H.; Zhu, Y.J.; Wang, C.; Zhang, H. Collaborative sparse representation leaning model for RGB-D action recognition. *J. Vis. Commun. Image R* **2017**, *48*, 442–452. [CrossRef]
23. Yan, Y.; Ricci, E.; Subramanian, R.; Liu, G.W.; Sebe, N. Multitask linear discriminant analysis for view invariant action recognition. *IEEE Trans. Image Process.* **2014**, *23*, 5599–5611. [CrossRef]
24. Wang, P.C.; Li, W.Q.; Gao, Z.M. Action recognition from depth maps using deep convolutional neural networks. *IEEE T. Hum. Mach. Syst.* **2016**, *46*, 498–509. [CrossRef]
25. Sharif, M.; Akram, T.; Raza, M. Hand-crafted and deep convolutional neural network features fusion and selection strategy: An application to intelligent human action recognition. *Appl. Soft Comput.* **2020**, *87*, 105986.
26. Bhatt, D.; Patel, C.I.; Talsania, H. CNN variants for computer vision: History, architecture, application, challenges and future scope. *Electronics* **2021**, *10*, 2470. [CrossRef]
27. Patel, C.I.; Bhatt, D.; Sharma, U. DBGC: Dimension-based generic convolution block for object recognition. *Sensors* **2022**, *22*, 1780. [CrossRef]

28. Xue, F.; Ji, H.B.; Zhang, W.B.; Cao, Y. Attention based spatial temporal hierarchical ConvLSTM network for action recognition in videos. *IET Comput. Vis.* **2019**, *13*, 708–718. [CrossRef]
29. Rocha, A.; Lopes, S.I.; Abreu, C. A Cost-effective infrared thermographic system for diabetic foot screening. In Proceedings of the 10th International Workshop on E-Health Pervasive Wireless Applications and Services, Thessaloniki, Greece, 10–12 October 2022; pp. 106–111.
30. Kumar, D.; Kukreja, V. A symbiosis with panicle-SEG based CNN for count the number of wheat ears. In Proceedings of the 10th International Conference on Reliability, Infocom Technologies and Optimization (Trends and Future Directions) (ICRITO) Amity University, Noida, India, 13–14 October 2022; pp. 1–5.
31. Bobick, A.F.; Davis, J.W. The recognition of human movement using temporal templates. *IEEE TPAMI* **2001**, *23*, 257–267. [CrossRef]
32. Bashir, K.; Tao, X.; Gong, S. Gait recognition using gait entropy image. In Proceedings of the 2010 International Conference on Crime Detection and Prevention, London, UK, 3 December 2009; pp. 1–5.
33. Patel, C.I.; Garg, S.; Zaveri, T.; Banerjee, A.; Patel, R. Human action recognition using fusion of features for unconstrained video sequences. *Comput. Electr. Eng.* **2018**, *70*, 284–301. [CrossRef]
34. Du, T.; Wang, H.; Torresani, L.; Ray, J.; Paluri, M. A closer look at spatiotemporal convolutions for action recognition. In Proceedings of the 2018 IEEE Conference on Computer Vision and Pattern Recognition (CVPR), Salt Lake City, UT, USA, 18–21 June 2018; pp. 6450–6459.
35. Zhang, K.; Yang, K.; Li, S.Y.; Chen, H.B. A difference-based local contrast method for infrared small target detection under complex background. *IEEE Access* **2019**, *7*, 105503–105513. [CrossRef]
36. Barnich, O.; Droogenbroeck, M.V. ViBe: A universal background subtraction algorithm for video sequences. *IEEE T. Image Process* **2011**, *20*, 1709–1724. [CrossRef] [PubMed]
37. Lazebnik, S.; Schmid, C.; Ponce, J. Beyond bags of features: Spatial pyramid matching for recognizing natural scene categories. In Proceedings of the 2006 IEEE Conference on Computer Vision and Pattern Recognition (CVPR), New York, NY, USA, 17–22 June 2006; pp. 2169–2178.
38. Yu, K.; Zhang, T.; Gong, Y. Nonlinear learning using local coordinate coding. In Proceedings of the 2009 International Conference on Neural Information Processing Systems, Vancouver, BC, Canada, 20–22 May 2009; pp. 2223–2231.
39. Blank, M.; Gorelick, L.; Shechtman, E.; Irani, M.; Basri, R. Actions as space-time shapes. In Proceedings of the 2005 IEEE International Conference on Computer Vision (ICCV), Beijing, China, 17–21 October 2005; pp. 1395–1402.
40. Lin, Y.C.; Hu, M.C.; Cheng, W.H.; Hsieh, Y.H.; Chen, H.M. Human action recognition and retrieval using sole depth information. In Proceedings of the 2012 ACM MM, Nara, Japan, 5–8 September 2012; pp. 1053–1056.
41. Liu, L.N.; Ma, S.W.; Fu, Q. Human action recognition based on locality constrained linear coding and two-dimensional spatial-temporal templates. In Proceedings of the 2017 China Automation Conference (CAC), Jinan, China, 20–22 October 2017; pp. 1879–1883.
42. Yang, M.; Zhang, L.; Feng, X.C.; Zhang, D. Fisher discrimination dictionary learning for sparse representation. In Proceedings of the 2011 IEEE International Conference on Computer Vision (ICCV), Barcelona, Spain, 6–13 November 2011; pp. 543–550.
43. Jiang, Z.L.; Lin, Z.; Davis, L.S. Label consistent K-SVD: Learning a discriminative dictionary for recognition. *IEEE Trans. Pattern Anal. Mach. Intell.* **2013**, *35*, 2651–2664. [CrossRef] [PubMed]

Disclaimer/Publisher's Note: The statements, opinions and data contained in all publications are solely those of the individual author(s) and contributor(s) and not of MDPI and/or the editor(s). MDPI and/or the editor(s) disclaim responsibility for any injury to people or property resulting from any ideas, methods, instructions or products referred to in the content.

sensors

MDPI

Article

Towards a Smart Environment: Optimization of WLAN Technologies to Enable Concurrent Smart Services

Ali Mohd Ali [1], Mohammad R. Hassan [1], Ahmad al-Qerem [2], Ala Hamarsheh [3], Khalid Al-Qawasmi [2], Mohammad Aljaidi [2], Ahmed Abu-Khadrah [4], Omprakash Kaiwartya [5,*] and Jaime Lloret [6]

1 Communications and Computer Engineering Department, Faculty of Engineering, Al-Ahliyya Amman University, Amman 19328, Jordan
2 Computer Science Department, Faculty of Information Technology, Zarqa University, Zarqa 13110, Jordan
3 Computer Systems Engineering, Faculty of Engineering, Arab American University, Jenin P.O Box 240, Palestine
4 College of Computing & Informatics, Saudi Electronic University, Riyadh 11673, Saudi Arabia
5 Department of Computer Science, Nottingham Trent University, Nottingham NG11 8NS, UK
6 Instituto de Investigación para la Gestión Integrada de Zonas Costeras, Universitat Politècnica de València, Camino Vera s/n, 46022 Valencia, Spain
* Correspondence: omprakash.kaiwartya@ntu.ac.uk

Abstract: In this research paper, the spatial distributions of five different services—Voice over Internet Protocol (VoIP), Video Conferencing (VC), Hypertext Transfer Protocol (HTTP), and Electronic Mail—are investigated using three different approaches: circular, random, and uniform approaches. The amount of each service varies from one to another. In certain distinct settings, which are collectively referred to as mixed applications, a variety of services are activated and configured at predetermined percentages. These services run simultaneously. Furthermore, this paper has established a new algorithm to assess both the real-time and best-effort services of the various IEEE 802.11 technologies, describing the best networking architecture as either a Basic Service Set (BSS), an Extended Service Set (ESS), or an Independent Basic Service Set (IBSS). Due to this fact, the purpose of our research is to provide the user or client with an analysis that suggests a suitable technology and network configuration without wasting resources on unnecessary technologies or requiring a complete re-setup. In this context, this paper presents a network prioritization framework for enabling smart environments to determine an appropriate WLAN standard or a combination of standards that best supports a specific set of smart network applications in a specified environment. A network QoS modeling technique for smart services has been derived for assessing best-effort HTTP and FTP, and the real-time performance of VoIP and VC services enabled via IEEE 802.11 protocols in order to discover more optimal network architecture. A number of IEEE 802.11 technologies have been ranked by using the proposed network optimization technique with separate case studies for the circular, random, and uniform geographical distributions of smart services. The performance of the proposed framework is validated using a realistic smart environment simulation setting, considering both real-time and best-effort services as case studies with a range of metrics related to smart environments.

Keywords: smart environment; real-time applications; QoS performance analysis; IEEE technologies

Citation: Ali, A.M.; Hassan, M.R.; al-Qerem, A.; Hamarsheh, A.; Al-Qawasmi, K.; Aljaidi, M.; Abu-Khadrah, A.; Kaiwartya, O.; Lloret, J. Towards a Smart Environment: Optimization of WLAN Technologies to Enable Concurrent Smart Services. *Sensors* **2023**, *23*, 2432. https://doi.org/10.3390/s23052432

Academic Editor: Daniele Cenni

Received: 2 January 2023
Revised: 14 February 2023
Accepted: 15 February 2023
Published: 22 February 2023

Copyright: © 2023 by the authors. Licensee MDPI, Basel, Switzerland. This article is an open access article distributed under the terms and conditions of the Creative Commons Attribution (CC BY) license (https://creativecommons.org/licenses/by/4.0/).

1. Introduction

There has been a continuous increase in the use of networking technologies in newer application domains, such as smart homes, intelligent transport systems, online commerce, and medical domains, as a result of advances in communication and Internet technology. Wireless Local Area Networks (WLANs) and mobile networks have been the primary technologies for wireless communication. WLANs and mobile networks are becoming more common technologies for smart environments as they have more use cases and are easy to install while being relatively inexpensive. Wi-Fi networks employ a standard known as IEEE 802.11 as their Media Access Control (MAC) protocol. It also makes it possible for people who are thousands of kilometers apart to share information with one another in the form of papers, photographs, and movies across the globe. All of these services and applications can be carried out using a WLAN as a transmission channel. There is a large number of physical layer communication technologies to choose from, making it difficult to determine which one would provide the best performance for a given use case in a smart environment [1]. IEEE technology in smart industrial communication networks at its peak performance, in contrast to previous technologies, is not always guaranteed and should not be considered a default answer without confirmation from various types of studies that provide an in-depth investigation of these technologies [2,3]. In other words, determining the best way to employ IEEE technology in industrial communication networks is quite similar to determining the best way to use an older piece of technology. It is important to note that each IEEE 802.11 standard has a unique set of advantages and disadvantages. For example, 802.11a is less likely than 802.11b or 802.11g to cause radio frequency interference (RF). In densely populated locations, the preferable technique is 802.11b owing to its ability to provide interactive audio, video, and picture services. Although this is true, the range is inferior to that of 802.11b, and thus the two cannot be used in conjunction with one another. In light of this, the goal of this research is to determine which technologies and networks are most beneficial to end users and customers.

To enable smart environments, there are further considerations that need to be taken into account before picking a technology and network design that will be most effective when put into practice, including the number of access points, the number of nodes, and the type of data being communicated. QoS measures must therefore be used to ensure consumer satisfaction during selection. IEEE 802.11b/g/n and IEEE 802.11a are both wireless networking technologies; however, they operate in different frequency ranges. It is possible to use IEEE 802.11n to operate in the 5 GHz frequency spectrum if so desired. IEEE 802.11ac, on the other hand, can only operate at a frequency of 5 GHz because of technical constraints [4]. In order to use older hardware after a new software update, backward compatibility between IEEE 802.11 technology generations is needed. For the first time, new technologies can be implemented on a large scale because IEEE 802.11ac and 802.11n nodes are backward compatible. To put it another way, wireless AC (the router implementing the wireless networking protocol 802.11ac) can only be used to its full capacity when communicating from an IEEE 802.11ac device to another IEEE 802.11ac device, though nodes supporting all three standards can coexist in an 802.11 wireless LAN. Specifically, this is due to the router's 802.11ac wireless networking technology. There will be limitations in terms of overall performance because of the previous standard. As a result, IEEE 802.11ac technology must be used by both the router and the devices. It was also found in [5] that the data rate performance of nodes in both the 802.11ac and IEEE 802.11a/n standards significantly decreased compared to a single network. Nodes from both sets of nodes were combined and simulated. Both sorts of nodes were shown to work together in this situation. As a result, it is critical to look for ways to improve the multistranded efficiency of IEEE 802.11 WLANs. Because we are looking for the finest technology and the best network design for a variety of technologies, we are interested in investigating mixed network topologies. Research on Internet apps as stand-alone services, in which every design and configuration is tailored to a single application, has now been expanded for mixed applications [6–9]. A wide range of services, as well as more nodes and IEEE technologies, are performed at specified percentages in these scenarios. In our previous work [6–9], we concentrated on the installation of Internet applications as an independent service. Configuration is used throughout this project. For each IEEE 802.11 standard, we studied the effects of node distribution (circular, random, or uniform) on network performance. Circular, random, and uniform node distribution all affected network performance; therefore, this study examined them all.

The management of this multiservice on wireless networks while maintaining QoS is already a significant difficulty; therefore, traffic measurements such as latency and jitter must be considered, acknowledged, and applied. Implementing QoS characteristics such as delay, jitter, and packet loss across real-time networks is likewise seen as a significant challenge. Choosing which technology to utilize and execute in a WLAN business from several physical layer technologies necessitates concurrent scientific analysis. On the other hand, it is now more challenging to decide which network configuration is ideal for allocating wireless network resources to deliver high quality due to the existence of three different network architectures, which include three tiers of service that have been referred to as the Basic Service Set (BSS), Independent Basic Service Set (IBSS), and Extended Service Set (ESS). In recent years, there has been an increase in high-quality digital content, as well as a change in end-user usage patterns. This fact, along with the adaptability, affordability, and digital media capabilities of the IEEE 802.11 standard, has caused Wi-Fi technology to dominate the market and created barriers to network efficiency and usability. The development of digital media distribution and streaming apps has been aided by the emergence of media platforms such as YouTube, Netflix, and others. If not appropriately handled, each of these services has a substantial influence on the level of consumer experience relating to data transfer rate, delay, and jitter [10]. The major contributions in this paper are listed below:

1. A network prioritization algorithm for enabling smart environments has been developed to determine an appropriate WLAN standard (or a combination of standards) that best supports a specific set of smart network applications in a specified environment.
2. A network QoS modeling technique for smart services has been derived for the assessment of best-effort HTTP and FTP and the real-time performance of VoIP and VC services enabled via IEEE 802.11 protocols in order to discover optimal network design structures.
3. A number of IEEE 802.11 technologies have been ranked through the use of the proposed network optimization technique with separate case studies for the circular, random, and uniform geographical distributions of smart services.
4. The performance of the proposed framework was validated using a realistic smart environment simulation setting that considered both real-time services and bad best-effort services as case studies, which included a range of metrics related to the smart environment.

In this paper, a novel algorithm was developed to compare the performance of the BSS, ESS, and IBSS nodes while providing best-effort services, such as HTTP and FTP, and real-time services, such as VoIP and VC, across various IEEE 802.11 technologies. The proposed algorithm will provide a ranking of the various IEEE 802.11 technologies. Further, this work provides its own case study of the analysis of these services for three spatial distributions (circular, random, uniform). In addition, we discuss how different geographical configurations influence the performance of each WLAN technology. This study considers various factors and provides the client with a menu of options. A compromise will have to be made between speed and cost. In many cases, the maximum data rate is unreasonably expensive for customers, so they should not be assumed to always be the best option. Clients are interested in seeing cost–performance data so they can choose a service with rates they are comfortable with at a price they are willing to pay.

This study's novel contributions consist of (a) a framework/algorithm for analyzing network performance and (b) a method for implementing that analysis to determine the most effective network configuration, given the state of the art; additionally, the study aims to identify which IEEE technologies and network architectures can be used for future web-based programs and services. The performance of five distinct services (applications) was measured and analyzed in light of contextual variables such as geographic dispersion, network topology, and node density. Several quality of service (QoS) metrics were used and analyzed as part of the creation of a novel algorithm for comparing the performance of best-effort services, such as HTTP, FTP, and E-mail, to real-time services, such as VoIP and VC, across a number of different IEEE 802.11 technologies. These include latency, jitter, throughput, packet loss, download time, and page load time. The study's overarching goal is to devise a weighted coefficient for each application's metric parameters that can be used to rank the current IEEE 802.11 standards, using both stand-alone and mixed-use cases.

The purpose of this study is to identify which WLAN standard (or combination of standards) will provide the best overall performance for a certain smart environment scenario and for a set of applications. This research provides the consumer with a list of options after accepting a number of parameters. You might have to choose between speed and cost. This is not always the case because clients' budgets might not allow for the fastest data rate. Customers want to know how much a service will cost them relative to how well it performs so that they are able to select a plan that offers the speeds they need at a cost they can afford. Because it infrequently corresponds to the actual delivered rates, the maximum data rate has little value to a prospective customer. Ordinary 802.11e is useless because no one ever achieves the theoretical maximum data rate (54 Mbps). It is not clear which network design is best for optimizing wireless network allocation and efficiency; furthermore, IBSS, BSS, and ESS have added to the uncertainty. This study examines the implications of varying the node count and the deployment of IEEE physical layer technologies across various spatial distributions. We want to highlight that four major types of topologies, including mesh, ring, uniform, and random topologies, were used to test network architecture and communication protocol performance for best-effort applications and real-time applications.

The rest of this article is divided into the following sections. Section 2 critically reviews the literature related to smart network environment prioritization. The proposed network selection algorithm for a mixture of services and related QoS derivations is presented in Section 3. The performance results are thoroughly reviewed and critically assessed in Section 4. The conclusion is presented in Section 5.

2. Related Work

The proposed method will be briefly compared to various algorithms in this section. The number of nodes, network architecture, IEEE standards, and simulation models for quality of service have all been contrasted and are reported in Table 1. The results for modern techniques [11–13] show the best network architecture based on metrics including throughput, jitter, and end-to-end latency, and deployed their models' using nodes from (3, 9, and 18), (20), and (2). However, the validation of their suggested techniques has only been carried out using the BSS network design. The influence that the spatial distribution of nodes (namely circular, random, and uniform) has on the efficiency of a network has been investigated for each of the six IEEE 802.11 protocols. Recent works, such as the reviews in [14–16], have not demonstrated this distinct area of study. On the other hand, some studies were solely focused on evaluating methods for IEEE 802.11, 11ac, and 11n technologies, such as [17–19]. Moreover, [20,21] examined IEEE technologies with fixed node counts of (16) and (10), respectively. A related model with a Uniform Random Ordered Policy (UROP) was used to attain an energy harvesting efficiency as in [22], which presents a resolution to the problem of scheduling data broadcasts to take place in the wireless sensor networks of energy production systems; it was shown that UROP accomplishes the best possible fairness performance under a relatively common energy harvesting procedure over an unlimited time scale.

Table 1. Outcomes comparing the proposed method to existing mixed-algorithm solutions.

Reference	Approach	Parameters for Measuring Quality of Service	The Number of Nodes in the Network	Architectural Components of a Network	Technology Developed by the IEEE	Modeling Simulation	Limitation
[20]	This article discussed how various performance assessment procedures might affect ad hoc networks and their effectiveness. Here, using NS2, the authors compare the performance of TCP, UDP, and SCTP over a range of metrics.	Jitter End-to-end delay Throughput Packet loss	10	IBSS	802.11	NS2	It used a limited number of nodes and only one IEEE standard technology in its architecture.
[18]	A wireless fiber architecture that combines a 5G WLAN with a 10G passive optical network (XGPON) was examined in this paper (IEEE 802.11ac). Both technologies have their advantages and disadvantages in regard to satisfying the quality of service requirements of paying customers.	Bandwidth Fairness End-to-End delay	8	BSS	802.11ac	OMNET++	The number of nodes that were utilized is not particularly high, and the network architecture and IEEE technology utilized were both singular.
[19]	In order to provide high QoS for various multimedia (video, voice, and FTP) services that customers need, the standard EDCA effective service differentiation structure is activated and evaluated (particularly critical or time-sensitive services).	Delay Throughput	3, 9, and 18	BSS	802.11n	OPNET	It only makes use of two QoS parameters, whereas the technology only made use of one.
[14]	In this study, an OPNET simulation was utilized to examine how different QoS methods affect the performance and capacity of a VoIP network. How high VoIP technology may go while still providing quality that meets standards was also explored.	Jitter Delay Throughput	5	ESS	802.11	OPNET	There was only one type of network architecture deployed, and the number of nodes was relatively low.
[13]	Taking into account all conceivable QoS methods, this research measures and assesses the behavior of web-based apps during a vertical handover between 802.16e and 802.11e technologies. OPNET Modeler was used to carry out the scenario evaluation. Used software included E-mail and web traffic, both of which were dynamic (HTTP + database).	HTTP load page delay Mail download and upload delay TCP delay DB query delay	2	BSS WiMAX	802.16e 802.11e	OPNET	The study only involves two nodes and two IEEE technologies.
[21]	In terms of network performance, the impact of the RTS and fragmentation thresholds was assessed. Additionally, different MAC access methods were used to assess the network's speed, and the findings were compared to industry norms.	Jitter End-to-end delay Throughput	10	IBSS	802.11e 802.11g	OPNET	This research used only IEEE 802.11e/g and a small network of 10 nodes.

Table 1. *Cont.*

Reference	Approach	Parameters for Measuring Quality of Service	The Number of Nodes in the Network	Architectural Components of a Network	Technology Developed by the IEEE	Modeling Simulation	Limitation
[20]	Exploration of how well the XG-PON and EDCA optimized network design deals with rapid growth in real-time traffic as a result of current global IP traffic distribution.	End-to-end delay Jitter Fairness Throughput	16	BSS	802.11n	NS3	There was only one technology and one type of BSS network architecture used in this research.
[15]	The robust performance of the OPNET-based communication IP network simulation model enabled the modeling of real-world network scenarios, the incorporation of performance specifications for the operation of existing equipment, and the provision of a versatile graphical environment and design for network communication.	Link data rate Throughput Delay	NA	ESS	802.11	OPNET	There was only one IEEE technology discussed, and the number of nodes used was not provided.
[16]	This article built simulations of different standard smart meter networks using evaluation metrics. Databases could be queried and files could be uploaded using both wired and wireless communications during typical data transmission and DDoS attacks on the network.	FTP request response from the server HTTP request received by the server	20	BSS	802.11	OPNET	The technology employed is old and outdated, and modern advancements are not even mentioned.
[12]	This technique evaluated the impact of jitter and delays in the network with regard to improving MAC layer QoS in a Wi-Fi downlink. According to the results of the simulations, the new classification of time-critical traffic access improved efficiency and led the way for the widespread deployment of time-sensitive networking and Wi-Fi systems in a variety of manufacturing settings.	Delay Jitter	20	BSS	802.11ac	Monte Carlo simulation	Two quality of service parameters and a single network topology were employed.
[11]	Review and evaluation of IEEE 802.11n random topology WLAN multimedia services are the focus of this research. The standard's impact on the network's output was explained by the optimized structure that included the necessary spatial stream of features at the MAC layer.	Attempts at retransmission and data loss Throughput Delay	3, 9, and 18	BSS	802.11n	OPNET	As well as only employing a single network architecture and a single IEEE 802.11 technology, the nodes used did not extend to medium or large networks.
Proposed study	Analyzed the best protocol and network architecture based on mixed application metrics for various IEEE 802.11 technologies.	Packet loss Jitter Throughput Delay	1–65	BSS ESS IBSS	802.11 802.11a 802.11b 802.11g 802.11e 802.11n	OPNET	

The WLAN 802.11 architecture is made up of various parts that work together to establish a connection to higher-level services. In IEEE 802.11 standards, waiting for medium access is one of the largest delays a node experiences. Unlike more modern IEEE 802.11 specifications, in which frames are transferred in bulk, traditional IEEE 802.11 specifications transfer frames individually, giving the node an opportunity to waste a considerable amount of time trying to reach the medium rather than actually transferring data. One simple approach to dealing with this issue is to deliver multiple frames together as a single aggregate frame [23]. The IEEE 802.11 MAC layer specifies two medium access coordination functions—the required DCF and the discretionary PCF. Asynchronous and synchronous transmissions are both possible in 802.11's access functions. All 802.11 stations are required to use DCF because of the asynchronous transmission it enables. The PCF offers a synchronous service that, in essence, implements polling-based access [24]. The primary goal of both of the updated IEEE 802.11e and IEEE 802.11n protocols was to boost the efficiency of the MAC layer when transmitting video data. IEEE 802.11e defines a new Distributed Coordination Function, known as EDCA, to minimize transmission latency for a group of high-priority video streams over a shared channel, while IEEE802.11n defines modern aggregation, block acknowledgement, and reverse direction improvements for high-throughput WLAN transmissions [25].

The Base Station Switch (BSS) is the central component of an 802.11 WLAN. A Base Station System (BSS) is a collection of base stations in a wireless network that are coordinated using either a Distributed Coordination Function (DCF) or a Points Coordination Function (PCF). Yet, the transmission medium degrades as a result of interference from neighboring stations sharing the same physical layer, making some stations seem "hidden" from others. Services such as data delivery, authentication, and confidentiality are all provided by a station in a wireless LAN. Ad hoc networking, or IEEE 802.11 IBSS, is the official name. All stations can have direct conversations with any other BSS station without resorting to AP transmission. An ESS is a collection of BSSs used for infrastructure purposes. Networks in the backbone must be constructed with APs that control the flow of data in transmission. The MAC Service Data Units (MSDUs) are transported via the Distribution System (DS), which is also the backbone of the wireless network and may be responsible for the installation of both wireless and wired networks. Digital Submarine (DS) signals carry data from one ESS access point to the next [26]. There are M primary users (transmitters), M secondary users (receivers), and K secondary receivers and channels in the work of [27], which deals with vehicle networks aided by cognitive radio. Moreover, a channel is allocated for primary user data transmission when a request for use is received. The receiver has a data backlog and does not know the channel state or the statistics of the channel's evolution; therefore, for each time slot, it picks a channel from K to M at random. Uniform Random Ordered Policy (UROP) is also introduced and shown to produce near-optimal throughput for a generic channel evolution process under the block fading assumption in this research.

Technology is improving the functionality of portable devices such as laptops and mobile phones, which allow users to access the internet and make calls when they are on the move. Self-organizing networks will make it easier for people to connect in the future by reducing the cost of communication [28] and simplifying the process of putting together and configuring a wireless network by eliminating the need for preexisting infrastructure. The use of an ad hoc network is suitable if you need to transfer a lot of data quickly from one device to another. There are few restrictions on where an ad hoc network can be set up. For this reason, they could be useful in a variety of settings, including commercial and non-profit enterprises, as well as for personal use at home. As a result, it is less complicated to use and costs less money for businesses [29].

Many issues arise because people are unable to properly configure a network's topology. A comparison between two extremes can show how similar or different they are. There is no established policy for this network [30].

A shift is occurring from the traditional desktop computing environment, in which workstations connect through shared servers on a single network, to one in which many different platforms communicate over many different networks [31]. Multiple network communication describes this scenario. There has been rapid acceleration in this shift.

It evolves and modifies itself to meet the requirements of mobile workers and their teams. The next generation of wireless communication systems will need to accommodate a large number of autonomous mobile users rapidly. In a MANET, all of the nodes in the network structure communicate with one another without any central authority. One alternative is for every node to serve as a router [32].

In contrast to the limitations previously described, this study demonstrates the implementation of a novel parametric assessment method capable of determining the optimal network configuration through the use of three distinct network architectures: one or more access points; an ESS/BSS and the non-availability of access points; and an ad hoc–IBSS architecture. The proposed method has been evaluated in accordance with the requirements of a total of six distinct IEEE standards for technical advancement, specifically 802.11, 11a, 11b, 11g, and 11n, for a total of five mixed-based applications with a range of different node sizes (1 to 65).

3. Network Prioritization for Smart Environments

The proposed network prioritization framework is a type of smart environment recommendation system that determines an appropriate WLAN standard or a combination of standards to best support a specific set of smart network applications in a specified environment. A network QoS modeling technique for smart services has been derived for the assessment of best-effort HTTP and FTP and the real-time performance of VoIP and VC services enabled via IEEE 802.11 protocols in order to discover optimal network architectures. Modeling, simulation, and experimentation are the three primary methods used in WLAN performance evaluation and network prioritization. It should be obvious that there are essential trade-offs, whereby it is truly crucial to each methodology to choose the right one for a given problem. Most of the time, the main goal is to enhance the key performance method and the strategic objectives that must be met. Instead of dissecting each strategy independently, we focus on the three trade-offs that are inevitable with any approach. When these concessions are carefully examined, an effective evaluation method whose implementation may be quite clear and obvious emerges.

Analytical or mathematical closed-form solution models are expected to provide a certain range of hypotheses for streamlining a system because the solution to the equations used to define the changes in the system is known. In addition, it is important to use a model faithfully by making sure that the model's accuracy and reliability are as crucial as the intended application demands. The simulation could be interpreted as a highly specific and fully automated model. Because they often only model subsets of the actual performance of the network, but are so detailed that they are essentially equivalent to experimental research, they fall into a middle ground between mathematical/analytical models and experiments. When comparing practical application to theoretical and experimental endeavors however, its utility as an evaluation tool for the former stands out. In recent years, IEEE 802.11 technologies have become more widely available and inexpensive, ushering in a more favorable epoch for experimental WLAN measurement. Recently, testbeds have become widespread and are being implemented in a wide range of settings. Considering that nothing is ever relevant to the actual program/scheme except the program/scheme on its own, it may be less common to challenge the fidelity of experiments. Time and money are the usual costs that are considered when evaluating different methods of performance review. Analytical modeling is best avoided in situations where quick answers are required because of the time and expertise required to build it. However, once a model is fully developed, it is usually much quicker at achieving performance than experiments or simulations. Furthermore, analytical models can be developed in completely open-source environments for next to no additional cost. Learning to use a simulation tool can take some time, but there are many free options for wireless simulation. While the price of hardware is comparable between analytical models and simulation configurations, the benefit of the latter is that the code that recreates the performance of the system has already been generated. However, depending on the complexity of the simulated network, the simulation runtime may be prohibitively long if parallelization is not an option. As the necessary facilities are likely to be quite pricey and the knowledge required to adequately design and implement experimental work takes a considerable amount of time to develop compared to mathematical models or simulation platforms, experimentation is traditionally the most costly technique.

When it comes to large-scale applications and scalability, simulation and analytical models take precedence over experimental work. In practice, only minor tweaks to the code are needed to simulate a network with hundreds of access points. It is important to remember that simulating and modeling on a massive scale will take more time and, probably, a lot of technical manpower. In the end, there are advantages and appropriate applications for all methods of performance evaluation. Network simulation and modeling are commonly used when a fast and low-cost result is required. Experimentation appears to always be the best option when it is critical to keep production linked and close to a practical wireless network. Additionally, all methods have one thing in common: they produce a number of useful and fortunate results.

Due to the diversity in evaluation methods, we have opted to construct a full suite of WLAN system simulations, which grants us great adaptability and allows us to scale the framework more effectively and cheaply. The works cited in this research offered a more comprehensive analysis of the system as a whole. As a result, rather than modeling the processes within each node separately, we decided to model and simulate the network as a group of nodes for the three different network configurations. Further, our method's distinctive feature is that it makes use of Riverbed's extensive standard model library to support a variety of network models, protocol configurations, and geographic distributions. Through its Rapid Configuration features, the Riverbed Academic Platform library incorporates the distribution patterns for three spatial distributions (circular, random, uniform) and Riverbed (OPNET), and automatically builds the necessary distribution from its C or C++ source code based on user requirements. To accomplish our goal of contributing an answer to the question, "What WLAN standard (or mix of standards) will result in the best overall performance for a given mix of applications in a given environment?", we chose a wireless protocol that meets user needs without any outside influence and developed a suite of system simulations. Furthermore, we made up a coefficient of importance for each QoS parameter used by each application. Five applications (VoIP, VC, HTTP, FTP, E-mail) were configured as mixing services and five mixed percentages were introduced that covered almost all distribution options for these services. Six IEEE technologies (11, 11a, 11b, 11g, 11e, 11n) were supported by OPNET academic licenses, with three network configurations (BSS, ESS, IBSS). All scenarios were run in all possible spatial distributions. The aforementioned conditions were tested across five distinct sets of nodes. Here, we want to highlight that we have presented a critical investigation of different network architectures that considers a range of communication protocols with the aim of enabling smart environments that focus on minimal jitter and higher throughput performance. The purpose and requirement of this study is to establish which network architecture is most suited for each of the five distinct mixed-use case studies that have been considered in this paper.

3.1. System Model and Preliminaries of Smart Environments

Wireless communication technologies require only a modest amount of cable infrastructure, making them an extremely efficient and cost-effective method for linking network nodes. Mobile networks are essential to the operation of a wide variety of applications, including C-ITS (Cooperative Intelligent Transportation Systems), automotive networks, precision farming with linked engines, and a large variety of functions available on smartphones. These technological advancements make it possible for applications to perform the purposes for which they were designed. The great majority of these applications depend on mobile nodes in order to achieve the maximum throughput that is possible for them. In order to obtain the greatest potential throughput, the communication equipment needs to be capable of achieving the highest feasible physical data rate. WLANs that are based on the IEEE 802.11 standard do away with the requirement for cables or mobility in public locations such as airports and workplaces [33]. WLANs are also essential due to the simplicity with which they may be installed and the rapidity with which they can transfer data. This section focuses on IEEE network infrastructure.

The streaming of live video, social networking, or the playing of online games can all benefit from Wi-Fi. High-quality video over WLANs [34] is still challenging to send due to bandwidth restrictions. Wireless communication has become an essential component of modern life as a result of rapid advancements in wireless technology and the increasing need for people to always be connected. In recent years, the amount of high-quality digital information that consumers have access to has grown, as has the method by which they consume it. The dominance of Wi-Fi in the market is due to a variety of causes, including a lack of competition. The standard's adaptability, affordability, and support for digital media are just a few of the many benefits it offers. As a result, network efficiency and usefulness have been hindered. Digital media delivery and streaming application growth have been spurred by the emergence of media platforms such as YouTube and Netflix among many others. Due to their impact on latency, jitter, and throughput for end users, these applications must be taken into consideration while designing networks.

The features of quality of service measurements are used to define performance metrics. As stated in Table 2, which describes the primary QoS expectations and standards for each application [35], the acceptable threshold for each QoS metric parameter may be found in the table (traffic to be carried by the bearer).

Table 2. Importance of QoS metrics for online applications.

Application	Importance (I) and Threshold (T)	Jitter (sec)	Packet Loss Rate (%)	Delay (sec)	Throughput (kbps)
E-mail	I	Very Low	Low	Low	Low
	T	0	10	1	30
HTTP	I	Very Low	Low	Medium	Low
	T	0	10	1	30
VC	I	High	Medium	High	High
	T	0.03	1	0.15	250
FTP	I	Very Low	High	Low	Medium
	T	0	5	1	45
VoIP	I	High	Low	High	Medium
	T	0.04	5	0.15	45

The following quality of service measures have a direct influence on the overall quality of applications:

- Latency: the amount of time measured in seconds that it takes for data or voice traffic to travel from node A to node B over the network.
- Jitter (sec) is a variation in latency that is caused by queuing.
- Throughput: the rate at which data packets are sent from one point to another during a specific amount of time, which is measured in bits per second.
- Packet loss: refers to the percentage of packets that are lost along the transmission channel after the user has already transmitted the packet across the network.
- An important coefficient, abbreviated as ICP, is assigned to each application parameter in accordance with the effect that the parameter will have on the data quality provided

by the service. The threshold values that are presented in Table 2 reflect the importance of every QoS parameter regarding the overall quality of each application. In order for these qualitative characteristics to be taken into account in a simulation, there must be a numerical representation of them (H = 1, M = 0.5, L = 0.01, and VL = 0).

3.2. Smart Environment Prioritization

Smart environment prioritization is used in this study to build and explore several smart application scenarios using an OPNET simulation platform called Riverbed Modeler 17.5 [36]. Thanks to Modeler's ease of use and scalability, it is now able to research communications networks, network equipment, business applications, services, and protocols. Technical companies who have been most successful in their R&D efforts have followed the approach shown below. OPNET was used to simulate the procedure, and the following two essential source inputs were considered. User and technical requirements (standards) can be set up in several ways. Here, you will find an explanation for each of these elements, as shown in Figure 1.

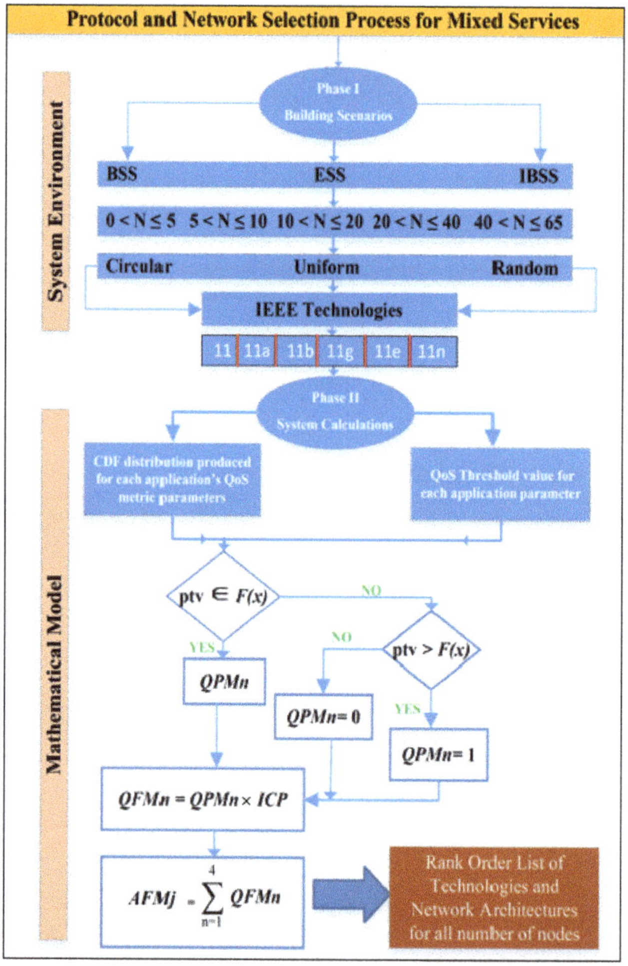

Figure 1. Flow diagram of the proposed algorithm in terms of both system environment and mathematical model.

Configurations for users (clients) can include a wide range of options, including but not limited to the following options:

- The total number of network nodes that must be present (where % is the total number of nodes in each application under consideration). The following research study used mixed applications:
    1. A mixture of 50% VoIP and 50% VC (real-time applications).
    2. E-mail (30%), FTP (30%), and HTTP (40%) (best-effort applications).
- Circular (oval), uniform (grid), or random topologies can be defined for these nodes using a spatial distribution.
- Technical Specifications.

It is possible to use the physical layer specifications to create a framework for many different design scenarios. Networks indicate communication between multiple wireless components in one of two ways: either without an access point (ad hoc) or with an access point present (Wi-Fi) (BSS and ESS), as shown in Figure 2. This network's nodes, which are divided into five distinct divisions, are essential (0–5, 6–10, 11–20, 21–40, and 41–65).

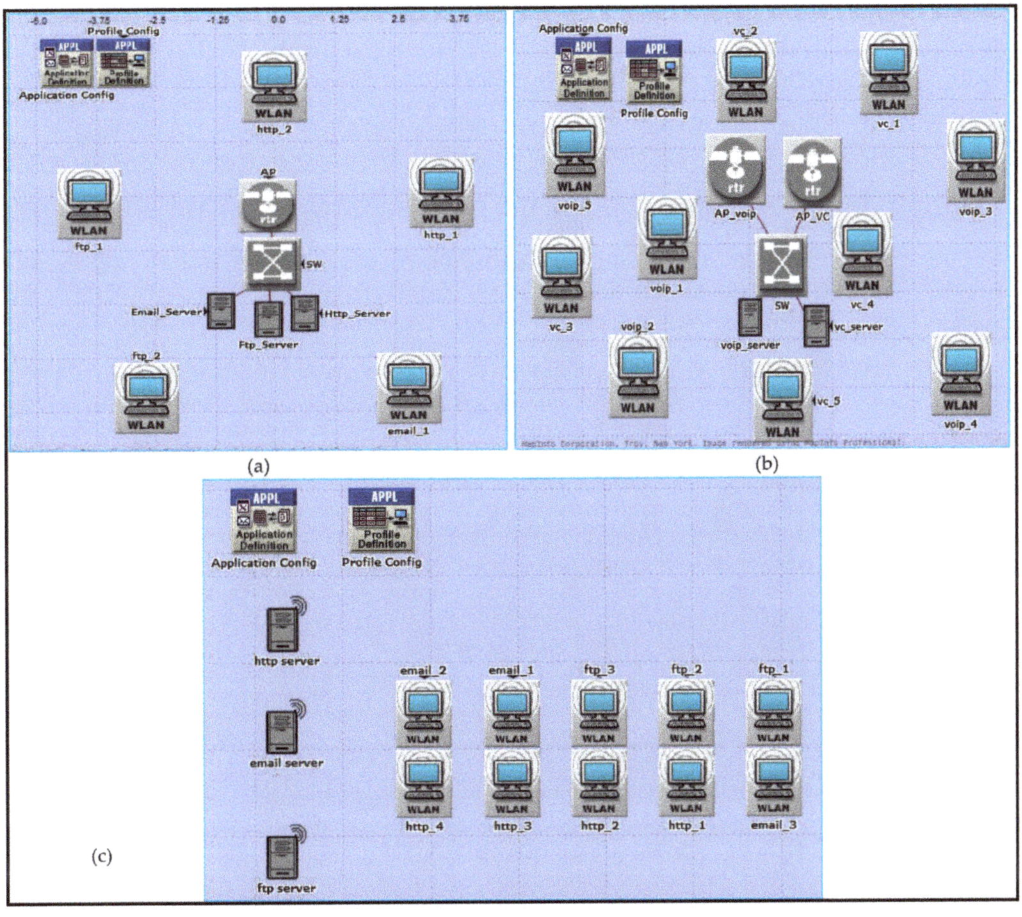

Figure 2. Design of the three network architectures across three spatial distributions for service mixing: (**a**) Basic Service Set (BSS), (**b**) Extended Service Set (ESS), and (**c**) Independent Basic Service Set (IBSS).

It is in accordance with the literature [20,37,38] that nodes are considered. To preserve network performance quality with these five groupings of nodes however, all observed outcomes are acceptable. These restrictions are due to the fixed capacity of bandwidth in a network; if the network has a large number of nodes, a modest amount of traffic can cause performance degradation. Because there are more nodes in the network, this is the case. This has only been tested in spaces between 2 and 3 m and 10 and 14 m because it is the normal size of a laboratory in a university, college, or school. An OPNET simulator was used to evaluate performance in a wide range of use cases for each application. There are other examples of these outcomes that are displayed in Figure 2a–c. The 802.11 (FHSS), 802.11a (OFDM), 802.11b (DSSS), 802.11g (OFDM), and 802.11n MAC layer technologies were used in this research (MIMO-OFDM). The 802.11e standard also enables contention-free bursts (CFBs) and defines quality of service (QoS) for 802.11. Because of CFBs, many frames can be sent at once if the transmission opportunity (TXOP) granted to a station is sufficient for this. Quality of service enabled access points (QAP) define hybrid coordinators, which feature an EDCA (Enhanced Distributed Channel Access) access mode. It is analogous to DCF, yet assigns varying weights to various services (such as DiffServ). IEEE 802.11e must be implemented in both the access point (AP) and the station (STA). The ability to work with STA devices that do not support QoS or 802.11e is also a benefit. VoIP traffic was configured for real-time applications with the following parameters during the simulation's 20 min runtime: G.711 encoding technique, one voice frame per voice packet, and interactive voice communication, which are all features of the G.711 standard. In addition, the frame inter-arrival time was 15 frames per second, and frame size was $128 \times 240/128 \times 240$ pixels (bytes). Files up to 50 KB can be transferred through FTP, whereas E-mail files can be up to 20 KB.

The customer is presented with a selection of options based on a variety of factors. There may be a trade-off between the expense and the speed of the vehicle. Since doing so may be prohibitively expensive for them, it is not true that customers will always opt for the fastest data rate available. A cost–performance comparison is what they are after so that they can find a service with the speeds they are willing to put up with for a price they can afford. It would be useless to provide a potential customer with a maximum data rate because it is merely a theoretical figure and frequently does not reflect the actual delivery rate. Since the theoretical data rate is so high, why do we not use it? The theoretical maximum speed of 802.11e is 54 Mbps; however, in practice, no one even comes near this [39]. Because the system handles both uplink and downlink applications, the minimum data rate shown in Table 3 is due to the fact that the theoretical upper bound of the system's uplink performance is very close to the minimum data rate mandated by the reference architecture, which is thus more likely to help users by being more realistic [40]. In light of the fact that the system manages both downlink and uplink application tasks, a minimum data rate is more likely to benefit consumers while also being more feasible.

Table 3. Data rates of IEEE 802.11 standards.

IEEE Standards	802.11	802.11a	802.11b	802.11g	802.11e	802.11n
Data rate (Mbps)	2	6	2	6	11	(base)/60 (max)

3.3. QoS Derivation for Smart Environments and Services

The mathematical model and system computations are depicted in the lower part of Figure 1, which represents Phase II. The CDF distribution and the QoS threshold values for each application are the mathematical inputs that are used by the algorithm. The results of the literature review are shown in Table 2 [41]. After each of the simulation scenarios was completed, the CDF distribution for the QoS metric parameters derived from OPNET was then created. The use of mathematical computations was required in order to determine whether or not a certain circumstance satisfied a number of crucial parameters for each application. The computations that were performed using this method will be discussed in the following parts, beginning with the results of each of the aforementioned projects.

QoS Performance Metric (QPM): the value that is generated by utilizing the parameter threshold value (PTV) as an application quality of service metric when defined in the cumulative distribution function (CDF), with distribution F(n) given by Equation (1), as illustrated in Figure 3, for each QoS performance criterion n.

$$QPM_n = F(ptv) \quad (1)$$

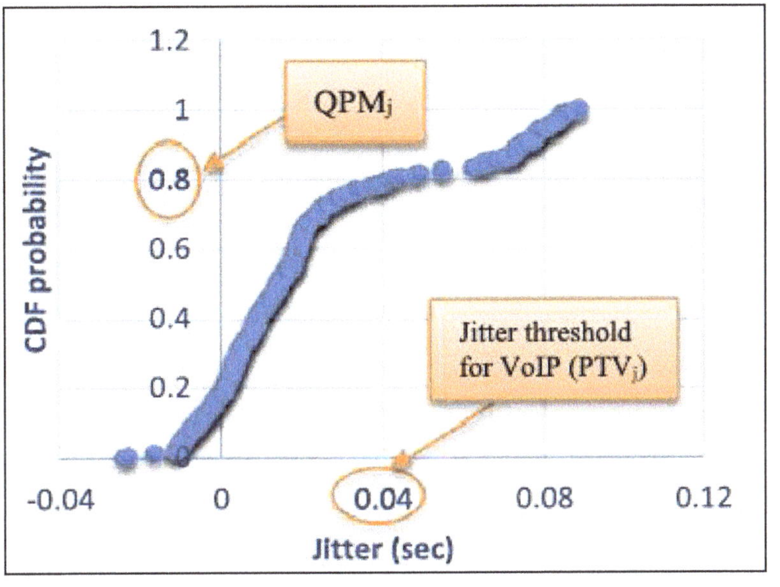

Figure 3. QPM for jitter.

The value created by applying a weighting to the QPM (given by importance) is known as the QoS Fitness Metric (QFM), and it is specified by Equation (2) for each QoS metric parameter.

$$QFM_n = QPM_n \times ICP \quad (2)$$

When the QFM is equal to 0.8 and weighted by 1, it generates 0.8, which is the performance measure for jitter for the same QoS parameter. The coefficient of importance is high (H = 1) for this parameter. As a result, the coefficient of importance (H = 1) is multiplied by 80% of adequacy.

The final phase involves accumulating all QFMs for n application QoS metrics (throughput, delay, packet loss, and jitter) for each IEEE 802.11 standard g, with M denoting the machine-specific percentages in the mixed services scenarios, as shown by Equation (3).

$$AFM_j = \sum_{n=1}^{4} QFM_n \quad (3)$$

Each of the three network designs will have a ranking for the six technologies based on the AFMs of IEEE 802.11 technology. Ideal network architecture performance is then established for every node grouping. Figure 1 depicts the mathematical formulas used to determine the AFM value for each IEEE MAC technology. OPNET Modeler's QoS metric settings, as well as the CDF distribution F(n) [41], will be provided and evaluated using the PTV in the following ways: For this metric parameter, the PTV has a CDF distribution that is identical to QPM if PTV F(n): ICP uses QPM as a weighting factor while generating

QFM. The AFM is then created by combining all QFMs and is used to categorize IEEE technologies, as seen here. This signifies that the QFM has expanded and the QPM value has reached 1 if PTV > F(n). A value of 0 for QPM and QFM will be the result if PTV > F(n). The preceding sections have described how to determine QoS metric parameters for all applications, except the one that accounts for packet loss. OPNET Modeler's packet loss parameter returns a Boolean value (0.0 or 1.0) that indicates whether or not a packet was accepted or rejected. However, a precise count of the number of packets lost is required for this investigation. Loss rate for an application packet is represented by ω_i on a node i. Equation (4) shows the proportion of discarded data packets (ki) to the total number of data packets (ρ_i) multiplied by 100%. A MATLAB tool was developed to calculate the percentage for all mixed applications. Packet loss percentages for hybrid applications and IEEE technologies are readily available within the OPNET Modeler.

$$\omega_i = (\,ki/\rho_i) \times 100\% \quad (4)$$

Utilizing the OPNET Modeler, traffic data are required in order to obtain information regarding the total number of packets that have been transmitted and received. The flowchart that was discussed before needs to be used to calculate the values of QPMs, QFMs, and AFMs. In order to do this, an exact packet loss ratio needs to be generated and presented in a CDF diagram. In order to determine which IEEE technologies are best suited for each application, each application's set of QoS metric parameters must be generated for all possible combinations of network architecture and the three possible locational distributions. The nodes in each of the three spatial distributions must then be divided into five equal groups. This was achieved by grouping the nodes into groups across all three spatial distributions. In light of this, it is generally accepted that the statistical correlations between parameters (which establish thresholds) need to be taken into consideration in order to guarantee that all applications included in the mix can be simulated. By adopting these methodologies and taking into account the individual statistics of the parameters (as opposed to the joint statistics), one can get a comparative gauge for overall performance that is both useful and informative.

All the scenarios were developed, configured, performed, and analyzed using OPNET Modeler. For a 50% VC implementation, you might set up 10 workstations in one of three different network topologies (IBSS, BSS, ESS) and one of three different spatial distributions (circular, uniform, random). All of the ten computers spread out over three places will be outfitted with the same set of six IEEE technologies (802.11, 11a, 11b, 11g, 11n, and 11e). The scenarios can then be performed and the results evaluated. Each of the following quality of service statistics has to have a cumulative distribution function (CDF) distribution generated for it: End-to-end delay in packet transmission (sec), jitters (sec), throughput (in bits per second), and the percentage of lost packets. Table 2 can be used to implement the algorithms and calculations used by the system.

Initially, it is necessary to define the QPM, which is the value produced in the CDF for VoIP by implementing the necessary threshold (QoS parameter) for each performance criterion. For each QoS parameter (H = 1, M = 0.5, L = 0.1, and VL = 0), the QFM value will be calculated using QPM weighting (as determined by importance). Additionally, each project will incorporate the six WLAN physical layer technologies into six distinct scenarios (802.11, 11a, 11b, 11g, 11n, and 11e). When the six scenarios have run for 20 min, the effects of each QoS parameter will be evaluated in the same way. The 802.11e scenario is used for the following computations:

- Jitter:

Table 2 demonstrates that a jitter value of 0.04 s is a VoIP service threshold where QoS application importance is high. The QPM is 1, as seen by the outcome in Figure 4. To calculate the QFM for jitter, we simply multiply 1 by the importance coefficient of 1 to obtain 1.

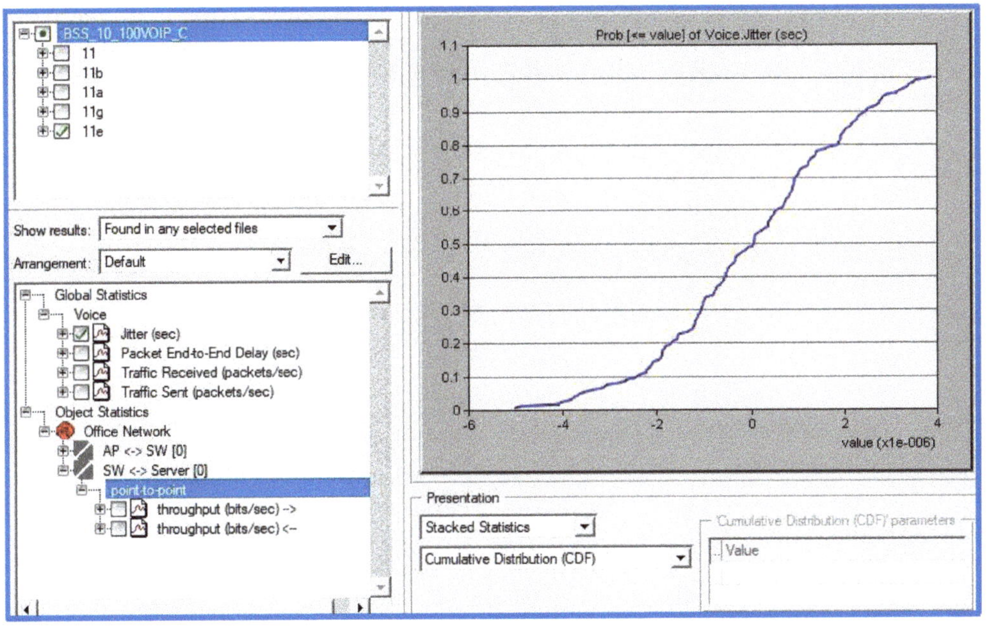

Figure 4. Jitter result of the scenario.

- Throughput:

As can be seen in Table 2, the VoIP throughput threshold value is 45 kbps when QoS application importance is set to medium. In Figure 5, we see that the QPM equals 0.052. To calculate the QFM, multiply 0.052 by its weight of 0.5, which is 0.0026 (0.5 is used because the throughput importance coefficient is medium (M = 0.5)).

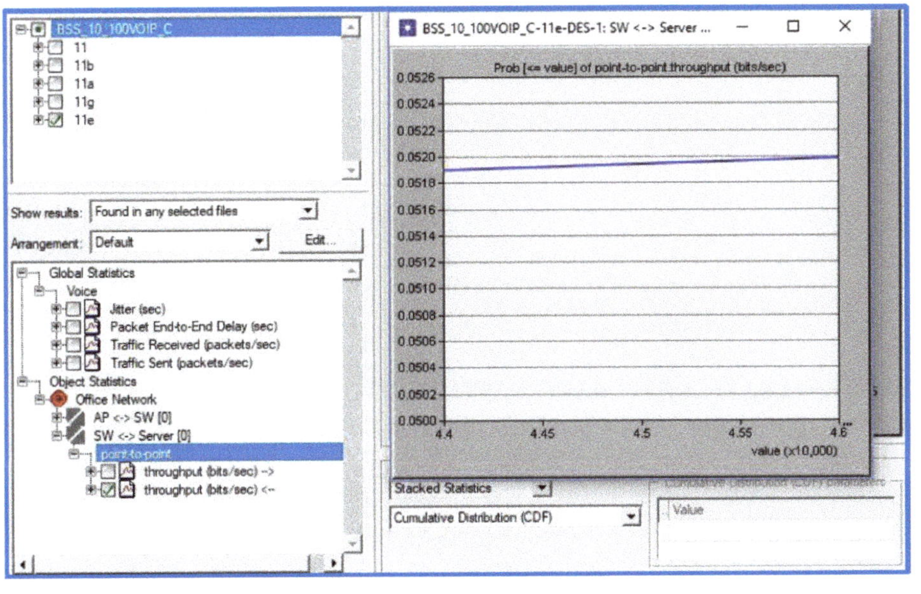

Figure 5. Throughput results of the scenario.

The same technique is to derive QoS data for various WLAN technologies (11, 11n, 11a, 11b, and 11g). The QFM values for each WLAN standard will be aggregated to determine the AFMs for all technologies. Based on WLAN SFM technologies, as indicated in Table 4, a ranking list of these six methods is presented. When ranking the six WLAN technologies, the same method is used for both the uniform and random distributions. All QoS values (QPMs, QFMs, SFMs, and AFMs) for all six technologies related to both applications (best-effort and 50% VC) in IBSS and ESS networks across all three spatial distributions were calculated using the same system algorithms and applied to the BSS and ESS networks to determine the best performing WLAN technology (or technologies) across these two network configurations.

Table 4. Settings and parameters used in typical simulations.

		System Settings	
1	Profile start time (sec)	60	
2	Simulation time (min)	20	
3	Value Per Statistic	200	
4	IP Routing	EIGRP Enable	
5	VC	Parameters	Values
		Frame Interarrival Time Information	10–15 frames/sec
		Symbolic Destination Name	Video Destination
		Frame Size Information (bytes)	128 × 120/128 × 240 pixels
		Type of service (TOS)	Interactive multimedia
6	VoIP	Parameters	Values
		Voice frame per packet	1
		Application	Voice
		Codec	G.711
		Compression and Decompression delay	0.02 sec
		Types of service (TOS)	Interactive voice
7	HTTP	Parameters	Values
		HTTP Specification	HTTP 1.1
		Page Interval Time (sec)	Exponential (60)
		Types of service (TOS)	Best Effort
8	FTP	Parameters	Values
		Command Mix (Get/Total)	50%
		Inter-Request Time (sec)	Exponential (360)
		File Size (bytes)	50,000
		Types of service (TOS)	Best Effort
9	Email	Parameters	Values
		Send Interarrival Time (sec)	Exponential (360)
		Receive Interarrival Time (sec)	Exponential (360)
		E-Mail Size (bytes)	20,000
		Symbolic Server Name	Email Server
		Types of service (TOS)	Best Effort

Table 4 summarizes the various settings that must be used in the OPNET Modeler to develop scenarios with mixed applications.

We also analyzed how different spatial distributions (topologies) would affect the performance of each WLAN technology; however, instead of settling on one, we provided a range of options, such as uniform, circular, and random topologies. If a school or university needed to set up a lab, we figured that the nodes would be distributed in one of these three topologies. If a business requires a meeting or videoconferencing space with many computers, the same procedure will be followed to accommodate its needs. The devices have been distributed in every feasible way, including in a circle, uniformly, and randomly. Using Rapid Configuration, the Riverbed Platform library can generate a user-specified distribution from its C or C++ source code, as seen in Figure 6.

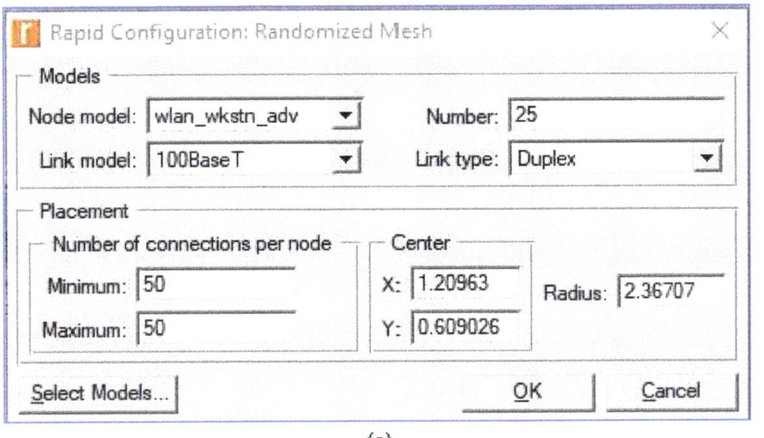

(**a**)

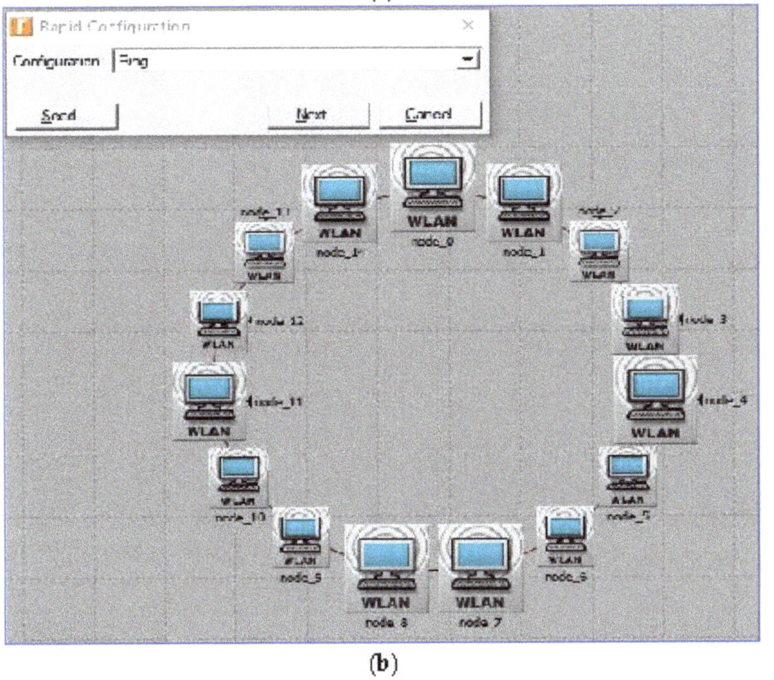

(**b**)

Figure 6. *Cont.*

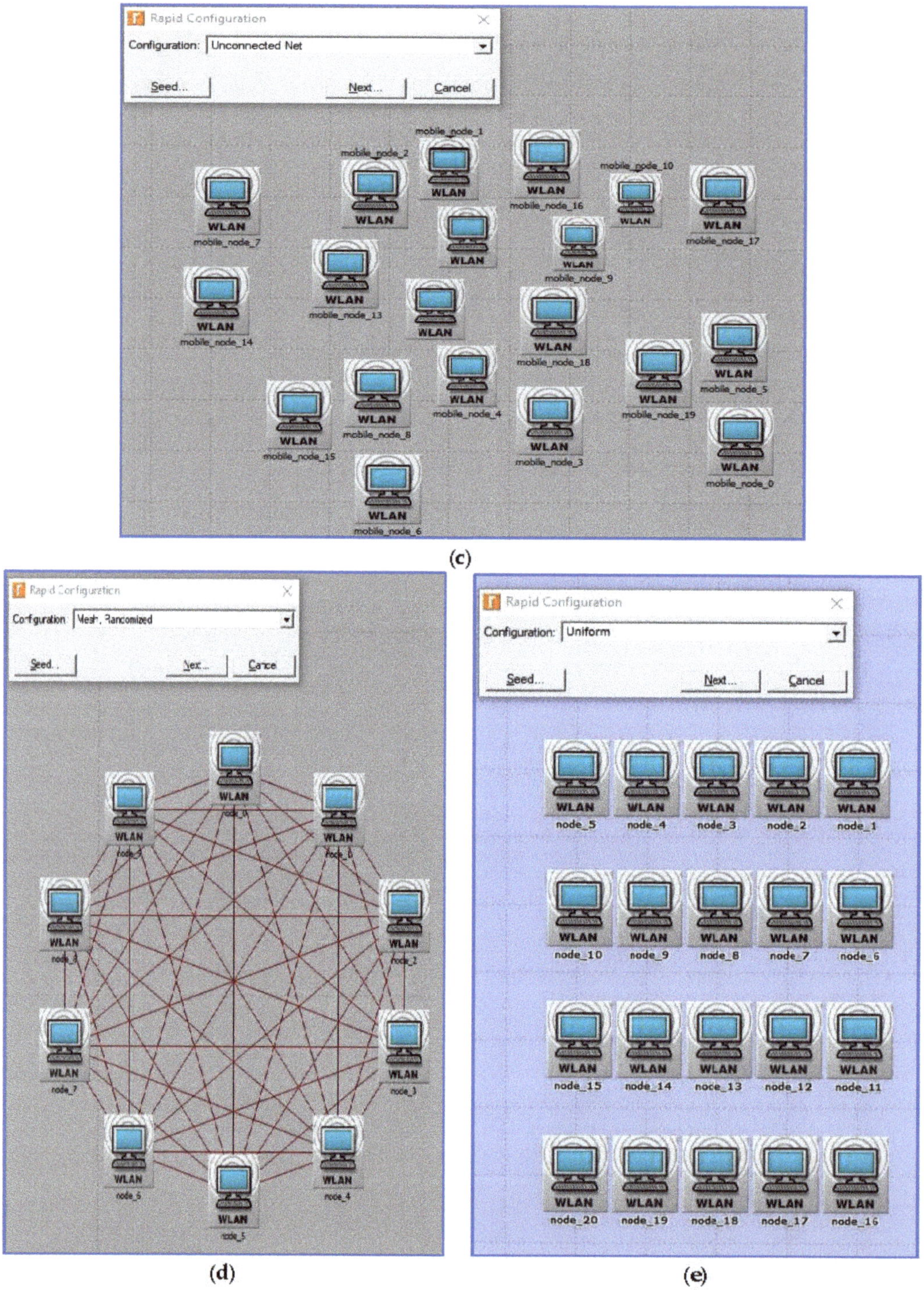

Figure 6. (**a**) Riverbed Rapid Configuration dialog box; (**b**) circular (ring) topology; (**c**) unconnected net (random) topology; (**d**) randomized mesh topology; (**e**) uniform topology (Riverbed, 2017).

4. Performance Evaluation

In this section, we will examine the outcomes by making use of a wide array of application strategies. For each best-effort application (HTTP, FTP, and E-mail), as well as real-time applications (VoIP and Video Conferencing), IEEE 802.11 technologies were assessed in terms of three different spatial distributions: circular, random, and uniform distributions. In order to be more particular, we looked at how well the applications performed in circular, random, and uniform settings. The purpose of this study is to establish which network architecture is most suited for each of the five distinct mixed-use case studies discussed in this paper. This was accomplished by gathering information from a variety of sources. There are six possible technology rankings for IEEE 802.11, namely 802.11a, 11b, 11g, 11e, and 11n, each of which is designed for a certain combination of applications operating in a particular environment. According to the rankings of IEEE 802.11 technologies, the algorithm that has been suggested here would have the highest quality services in addition to the best overall network efficiency. In light of these attributes, the algorithm that has been provided would result in the highest possible level of service quality and the most effective overall network. Two distinct situations involving mixed applications were assessed and analyzed regarding a variety of parameters, which encompassed everything from location to the number of nodes to the architecture of the network as a whole. OPNET was used as an implementation tool, in particular Riverbed Modeler 17.5 [42].

Here, we want to clarify that in our minimal jitter- and greater throughput-centric critical investigation of a range of networking technologies for smart environments, two types of applications were used: best-effort applications and real-time applications. Three different network architectures were considered with a range of communication protocols for making prioritization decisions that focused on minimal jitter and greater throughput. Standard protocol settings and predefined similar traffic loads were applied in the implementation for better clarity in terms of jitter and throughput performance in the considered network scenarios and communication protocols:

1. Real-time applications consisting of 50% VoIP and 50% VC (Voice over Internet Protocol and Video Conferencing): All scenarios in this case study were solely focused on investigating and analyzing the mixtures in real-time applications. This phrase has been abbreviated to "50VC" for the sake of readability.
2. The best-effort applications should comprise 40% HTTP, 30% FTP, and 30% E-mail. The best-effort application mixtures tested here form the basis of all configurations used in this case study. "Best-effort" has been condensed in order to make the sentence more readable. A flowchart and a bar chart are included in the format of the results for every set of nodes and every mixed application, respectively. A flowchart has been used to determine the number of nodes, the network topologies, the spatial distributions, and the WLAN technologies that are being utilized.

The results obtained in mixed applications are referred to as the Scenario Fitness Metrics (SFMs). Because the outcomes depend on the presence of the access point, the tables of outcomes are displayed (interpreted), which will be demonstrated later for each application in two flowcharts: the IBSS chart and the generic flowchart. The very last thing that needs to be accomplished is that the technology that has the optimum performance for each individual case study (network configuration) needs to be highlighted, as well as the optimal choice for each group of nodes:

- If there is at least one access point available, the suggested method is implemented in the manner depicted in Figure 1, and the flowchart results are displayed in Figure 7a, and subsequently. This is true for the BSS architecture layer, as well as the ESS architecture layer.
- The method proposed in Figure 1, as well as the algorithms shown in Figure 7b, and subsequently for the findings of the IBSS network, can be utilized even if the network is set up without an access point.

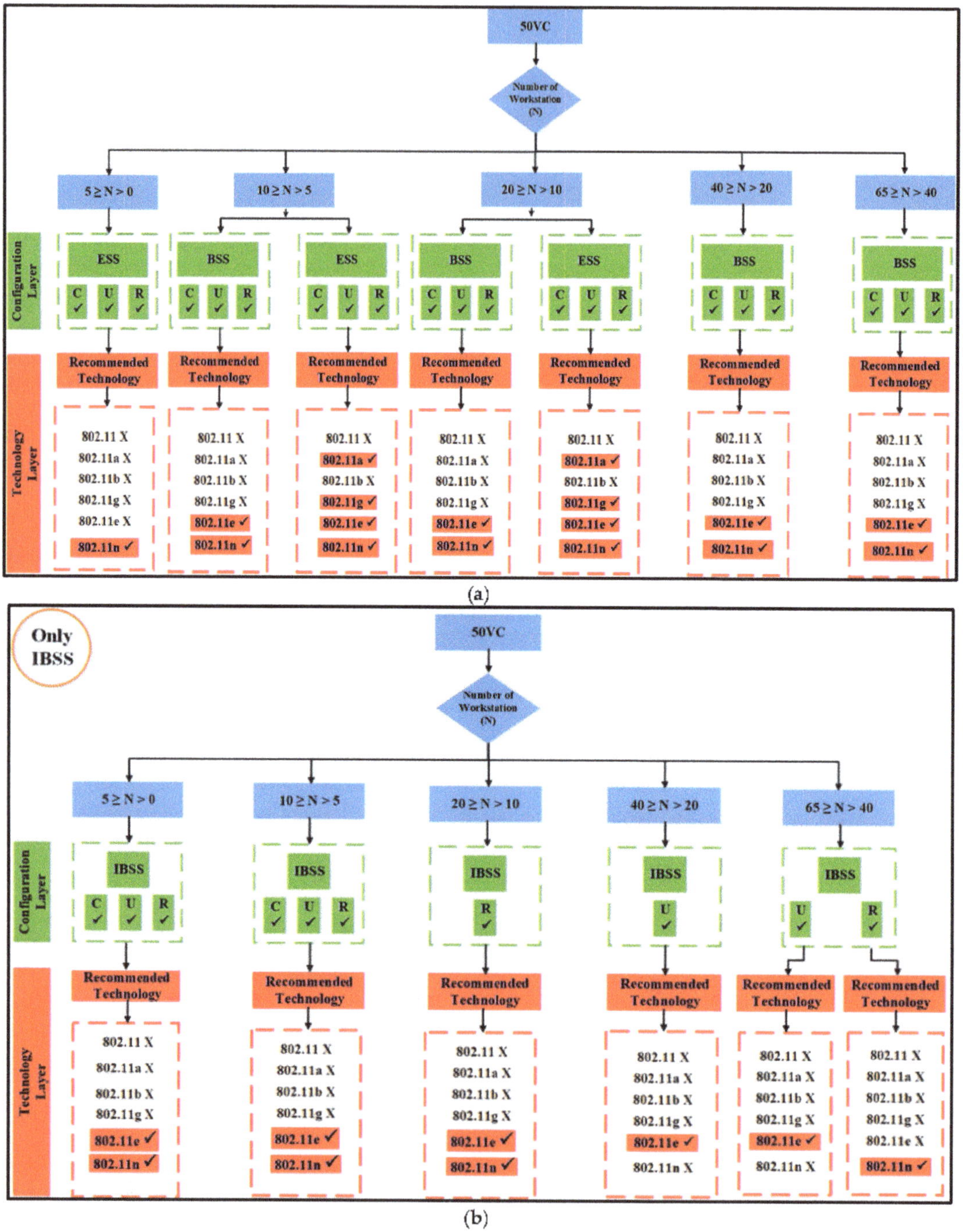

Figure 7. The proposed 50% VC algorithm. (**a**) BSS and ESS; (**b**) only IBSS.

4.1. A Mixture of Applications Working in Real Time (50% VoIP and 50% VC)

The algorithms for both set of results are displayed below in Figures 8–10 for all six IEEE 802.11 technologies across all 65 nodes in the case study with three different geographical deployments involving 50% VoIP and 50% VC. These figures cover the entirety of the case study and apply to all six IEEE 802.11 technologies. In addition, a case study that was conducted consisted of calls that were split evenly between VoIP and traditional videoconferencing (C, U, R). Both technology 11e and technology 11n produced nearly optimal performance for all spatial distributions, which is consistent with the case of IBSS, with the exception that the uniform and random distributions are dominant for medium and larger nodes. Both of these technologies are consistent with the case of IBSS (20–65). The research indicates that the performance of ESS and BSS for each of the five groups of nodes is virtually the same. In addition, the findings indicate that both approaches generate performance that is nearly optimal across the board for spatial distributions.

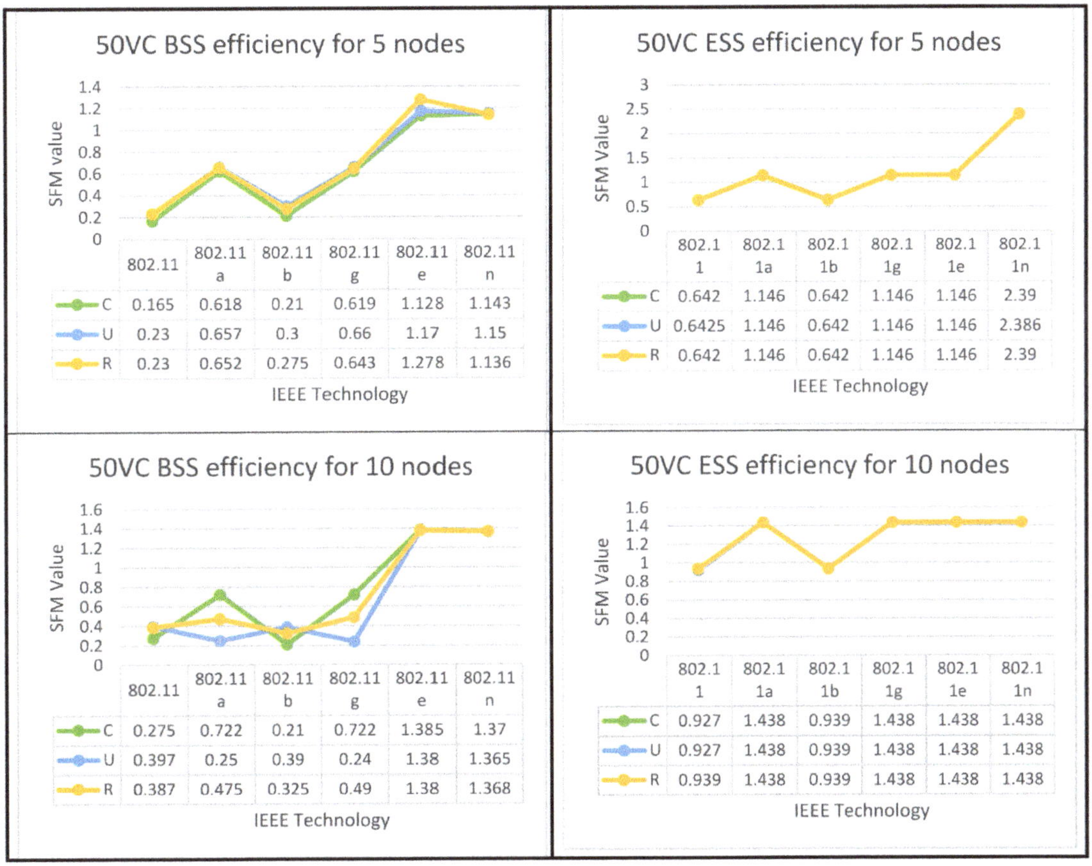

Figure 8. *Cont.*

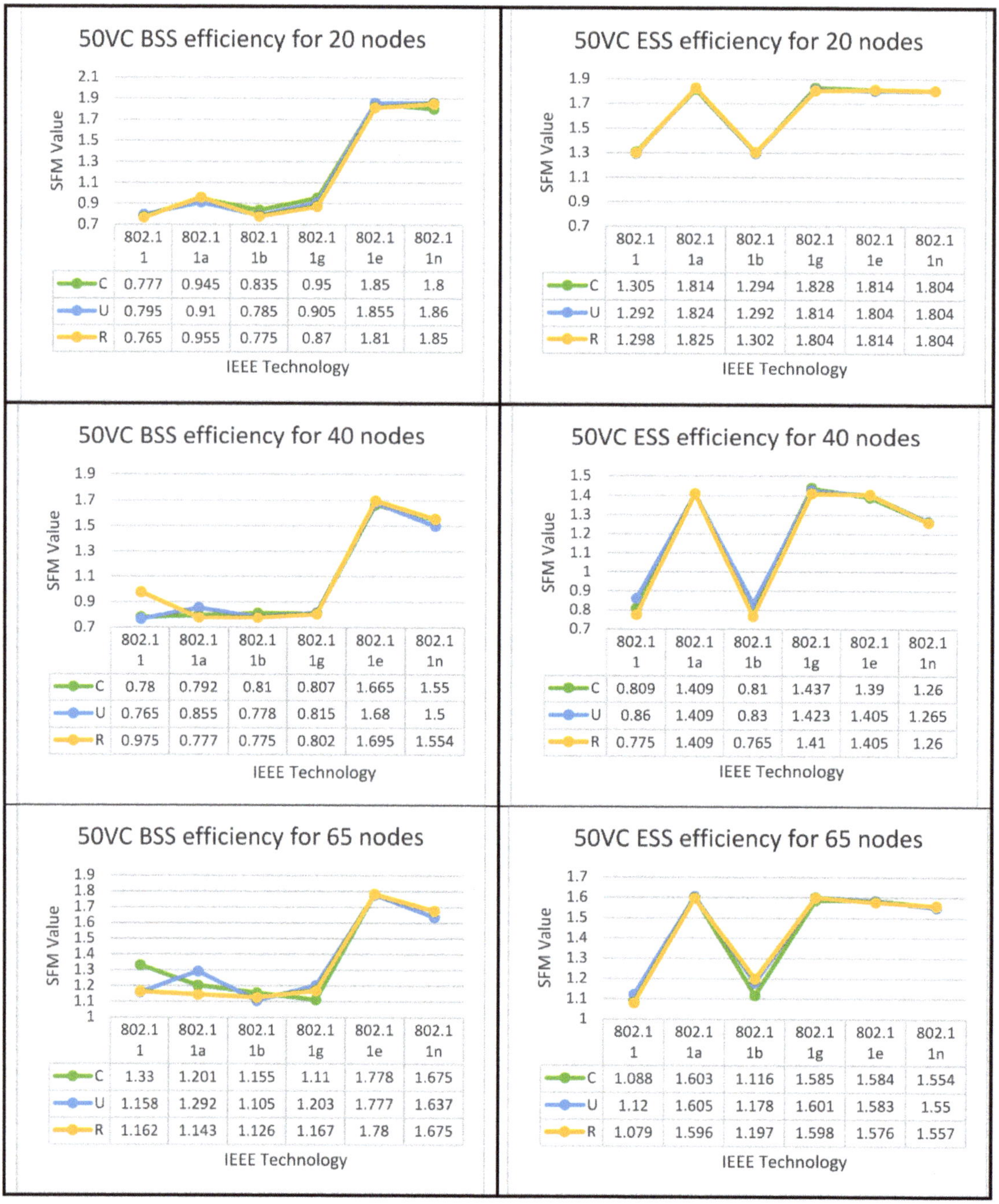

Figure 8. BSS & ESS Performance Optimization for (50%VC algorithm).

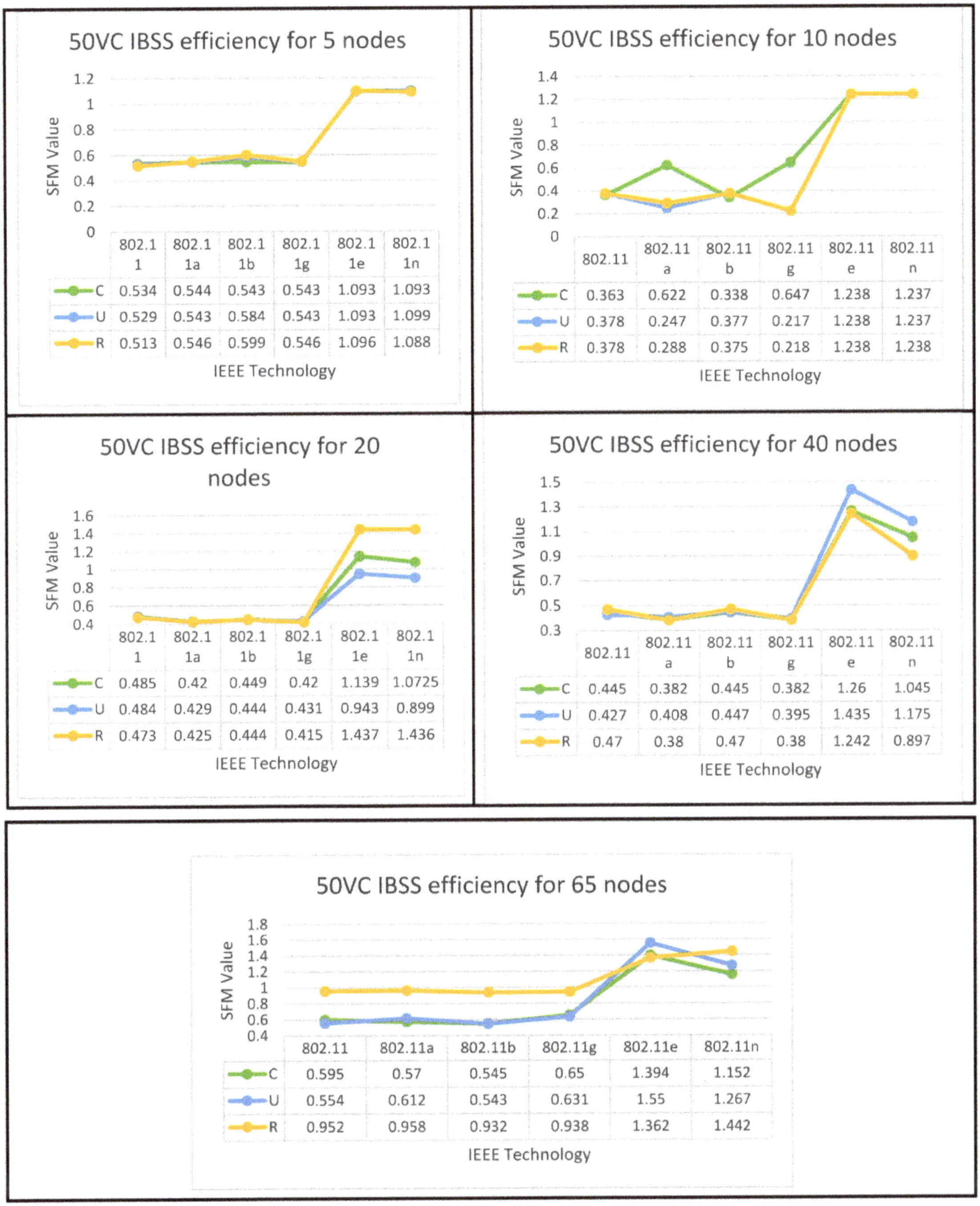

Figure 9. IBSS Performance Optimization for (50%VC algorithm).

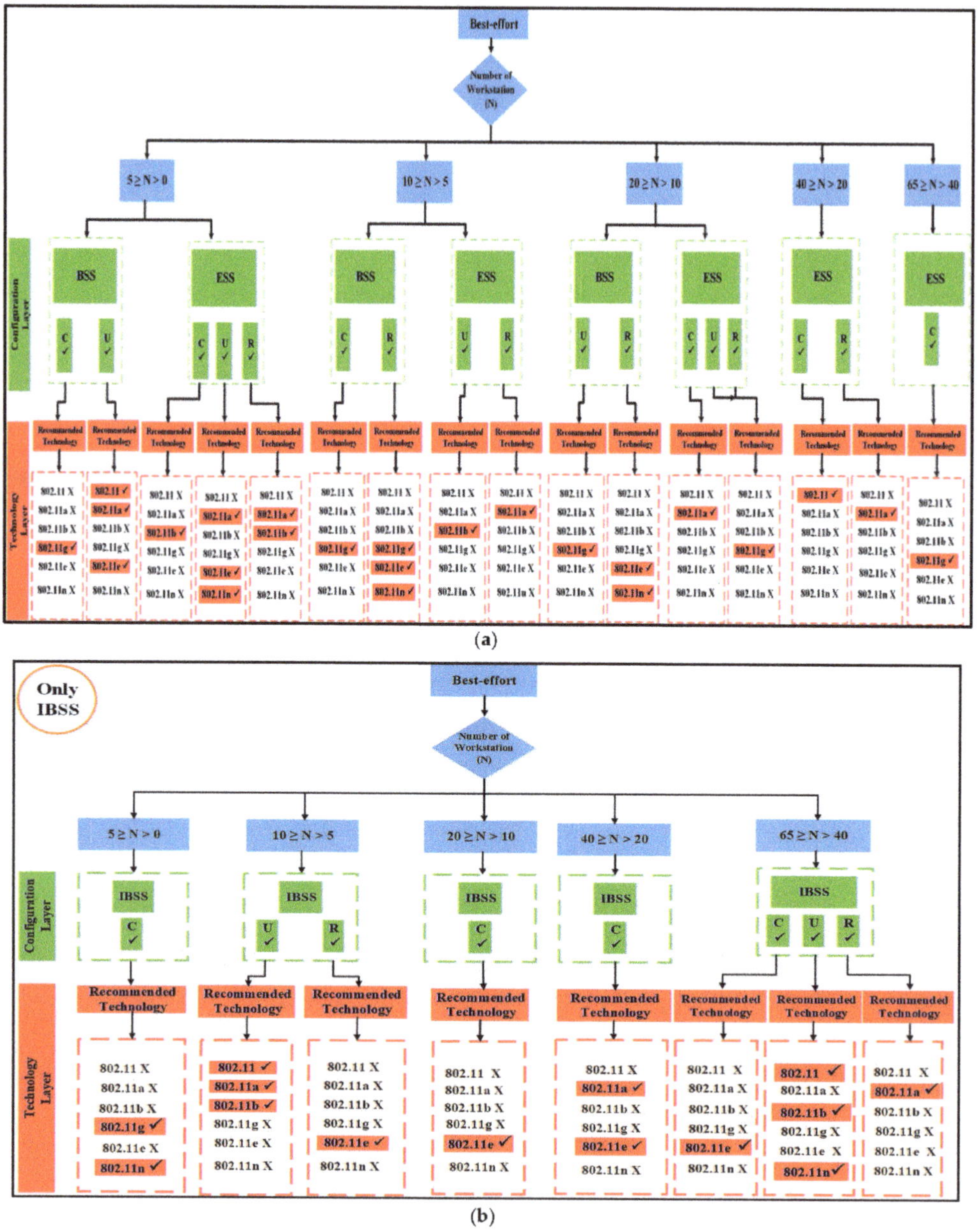

Figure 10. Proposed best-effort algorithm. (**a**) BSS and ESS; (**b**) only IBSS.

IBSS reaches its maximum performance potential on both 802.11e and 11n when the client builds a network with a limited number of nodes. This holds true for both of these standards, where 5 ≥ N > 0. It turns out that the only technologies that perform adequately in the second and third groups of space distribution are IEEE 802.11e and 11n. This is the case for both of these groups.

Following the flowchart for the IBSS led to the discovery of this information. 802.11n will deliver the best performance in the fifth category if the configuration is randomized;

however, if the setup is uniform, 802.11e will be the best option for categories four and five in the IBSS results. Figure 9 provides a visual representation of this point.

4.2. A Mixture of Applications Made with Best Effort (40% HTTP, 30% E-Mail, and 30% FTP)

The results of both algorithms are depicted in Figures 10–12 for all nodes that participated in the best-effort case study and used all six WLANs. Three spatial configurations using IEEE 802.11 standards are examined. As can be observed in Figures 10a and 11, both network topologies, BSS and ESS, are able to function in small and medium-sized networks (1–20), with 11g, 11e, and 11n technologies providing the highest level of performance. In the first group, where $5 \geq N > 0$, there is a variety of options provided by BSS and ESS. Only a circular network layout benefits from the use of IEEE 802.11g technology in BSS architecture. As an added note, IEEE 802.11, 11a, and 11e technologies only function properly in a standard setting. ESS, on the other hand, provides a number of alternatives. The only situations where IEEE 802.11b is considered the best option are those with a circular or random distribution of problems. Figures 11 and 12 show that the best user output is achieved with a circular configuration of IEEE 802.11g and 11n technologies, respectively, while the best performance is achieved with a uniform configuration of IEEE 802.11e and 11n technologies. In the second range, $10 \geq N > 5$, BSS and ESS have a wide variety of options to choose from. For BSS architecture, IEEE 802.11g technology performs best with a circular network layout.

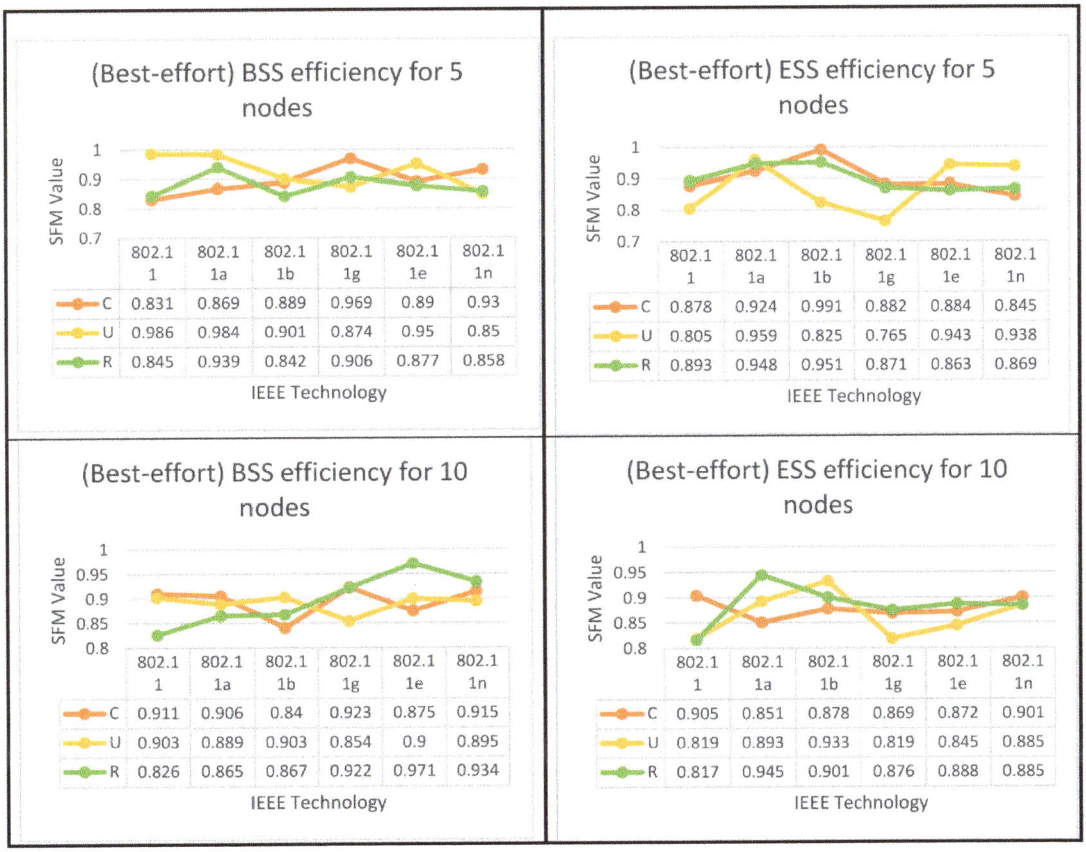

Figure 11. *Cont.*

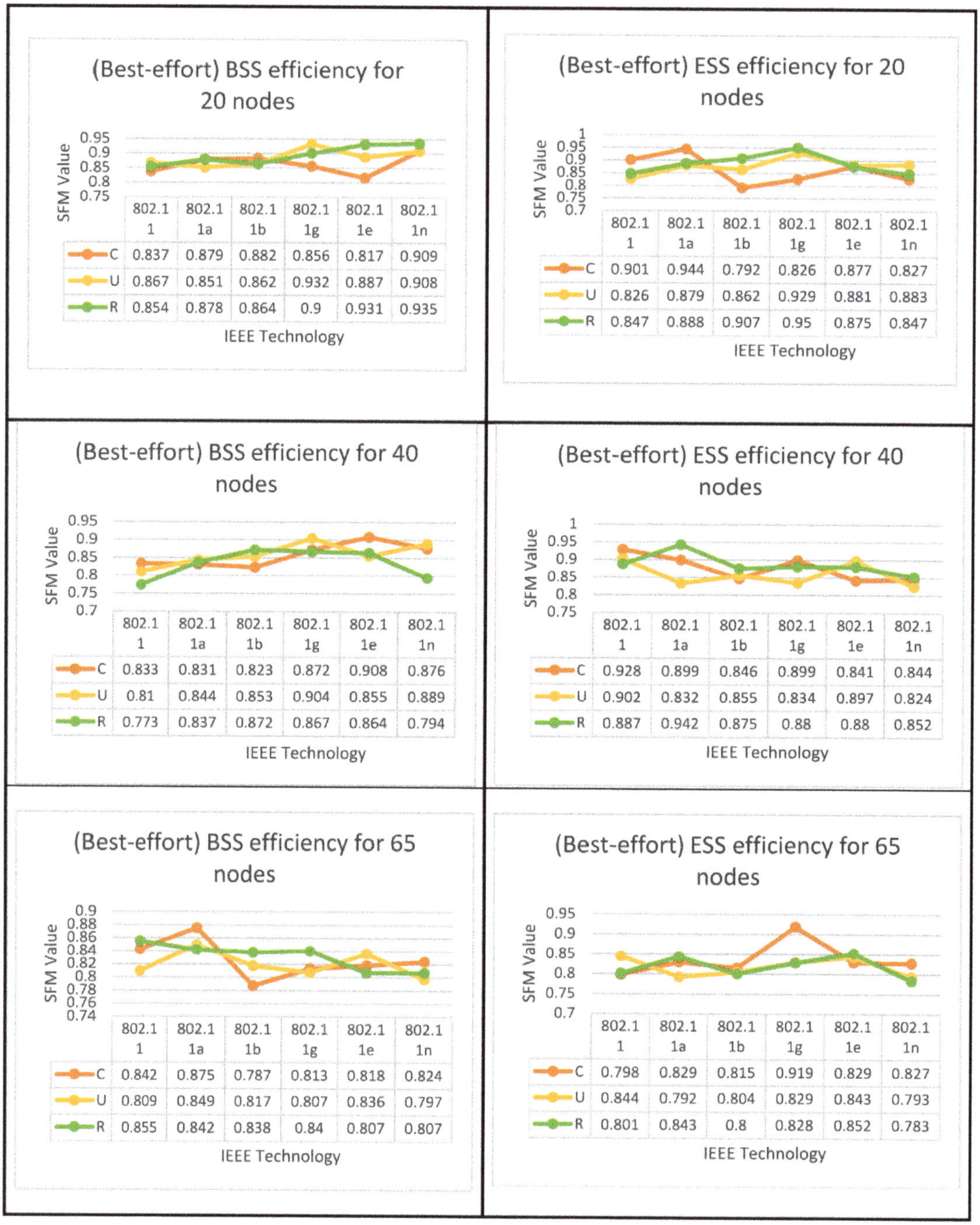

Figure 11. BSS and ESS performance optimization for the best-effort algorithm.

Figure 12. IBSS performance optimization for best-effort applications.

In addition, the performance of the IEEE 802.11g, 11e, and 11n technologies was outstanding, even when the configuration was performed in a haphazard manner. On the other hand, ESS provides users with a variety of choices from which to select. It has been demonstrated that the optimal choice is IEEE 802.11b when the network is configured in a consistent manner. Furthermore, as can be seen in Figure 11, the optimal configuration for

IEEE 802.11a is one in which the settings are completely arbitrary, which results in the best performance. According to the findings of the IBSS network, the optimal results for the client can be achieved with a circular deployment of 802.11g and 11n, which is depicted as an example in Figure 12.

In the third group, where $20 \geq N > 10$, both BSS and ESS offer a wide range of different choices to their customers. Only when the network is constructed consistently do IEEE 802.11g technologies perform to their full potential in BSS architectures. In addition, the optimal performance of IEEE 802.11e and 11n technologies can only be achieved when the configuration is performed in a randomized fashion. Nonetheless, the ESS architecture provides a variety of options to select from. Only in a network that is a completely closed loop will IEEE 802.11a become the superior choice. In addition, the performance of IEEE 802.11g is excellent, regardless of whether the configuration is uniform or arbitrary, whereas the results of the IBSS architecture show that IEEE 802.11e has the best performance when it comes to circular distribution.

According to the standard flowchart, the ESS architecture achieves its highest level of performance in the fourth and fifth categories. The client has the option of choosing between two different sets of nodes for the fourth set based on the data that are presented in Figure 11. The 802.11 standard is the way to go if you only plan on arranging your network in a circular fashion as part of your deployment.

The other alternative is to employ 802.11a technology with a haphazard setup. Using 802.11g technology in a circular configuration is ideal for the fifth set of nodes. Figure 12 shows that if only a circular network is designed for an IBSS, either 802.11a or 11e would be the best technology to use. The 802.11e standard is recommended for use in the fifth category if the network is to be set up in a circular configuration, while the 802.11, 11b, or 11n standards are more appropriate for use in a network set up in a uniform fashion. However, if 802.11a is configured arbitrarily, it is the best choice.

4.3. Mixed Service Scenario Layer Configuration

This section describes how to configure the configuration and technology layers of a system using the system's algorithms and their corresponding mathematical models (equations), with examples of how this is achieved in the context of mixed-service use cases. A university has requested a networking lab, which will have 40 machines due to the 40 students enrolled. The university estimates that, at any given time, traffic will be split between best-effort services, with 40% allocated to HTTP, 30% allocated to E-mail browsing, and 30% allocated to file sharing and transfer. The university makes the call after determining how much time will be spent on the internet for things such as HTTP traffic, E-mail, and file transfer. The optimal IEEE technology and network architecture, as well as the optimal spatial distribution of these 40 machines, should be provided by our algorithm and mathematical equations.

Forty machines will initially be set up in 54 configurations due to the fact that there are three major projects to be built for each of the three main network configurations (BSS, ESS, IBSS), and all of these must take into account the spatial distributions of the workstations (circular, uniform, random). For each possible configuration of spatial distributions for the six WLAN technologies, six separate scenarios are developed (802.11, 11a, 11b, 11g, 11e, and 11n). In addition, there are four applications represented here, with 40% of nodes running HTTP (16 machines), 30% running FTP (12 machines), and 30% running E-mail (12 machines). Figure 13 provides a graphical representation of the projects involved in the BSS, ESS, and IBSS scenarios.

In the second phase, depicted at the bottom of Figure 1, the system calculations and mathematical model are displayed. The CDF distribution for each application's QoS metric and the QoS threshold values for those applications are used as inputs for the algorithm's calculations. In Table 2, we can see the relative threshold values for each mixed-use application and the relative qualitative importance of each QoS parameter. Once the OPNET simulation scenarios have been run, a CDF distribution is generated for these QoS metric parameters. For each mixed application, we will calculate the underlying mathematical equations to understand how a given scenario has met the required performance metrics. This algorithm's computations are explained, and the results of the three projects are analyzed using the following equations.

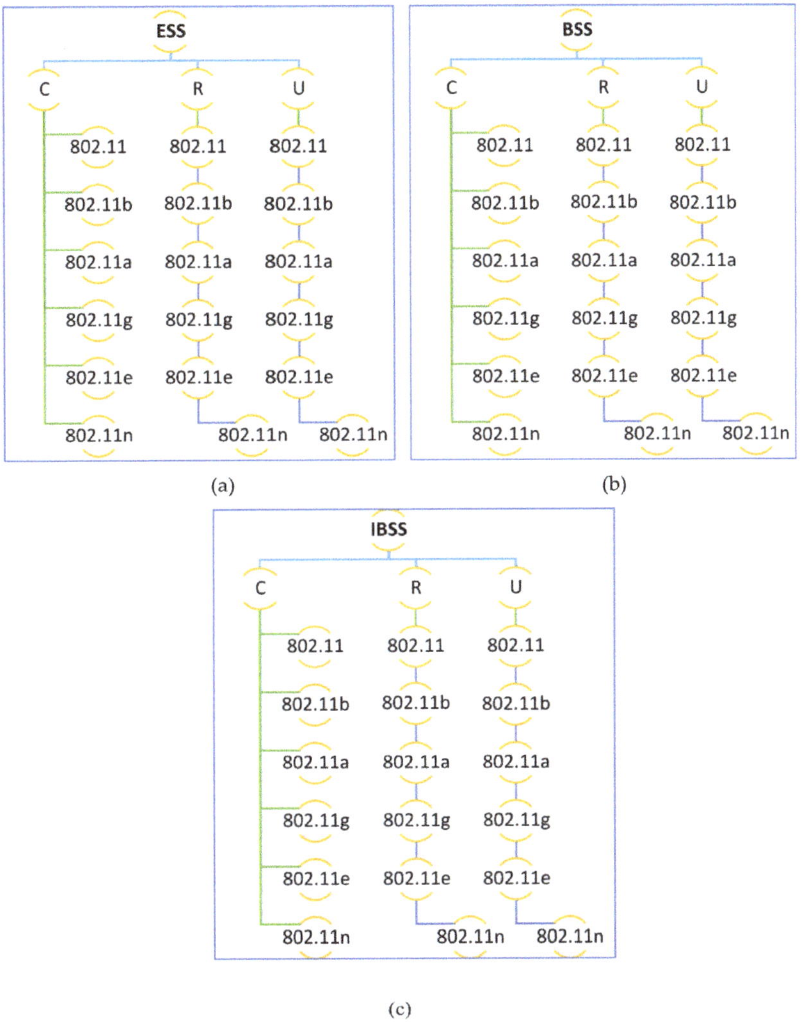

Figure 13. Project scenarios for all three network configurations: (**a**) BSS, (**b**) ESS, and (**c**) IBSS.

Table 5 reveals that among the three mixed services, HTTP accounts for 40%, FTP for 30%, and E-mail for 30%. Quantitative QoS parameters n for each type of mixed service will be calculated. First, the QoS parameters for HTTP 40% are used to calculate the QPM for both metric parameters across all six IEEE technologies (g). Next, the QFM is calculated across all six technologies using Equation (2). These two equations will be used to determine the QPM and QFM if throughput is used as TH, and the implementation of these equations and the methods for determining these quantities are described in detail in the sub-section above.

Table 5. Service mix (40% HTTP, 30% FTP, and 30% E-mail).

Application	Technology	802.11	802.11b	802.11a	802.11g	802.11e	802.11n
Email 30%	PR	QFM_{PR}	QFM_{PR}	QFM_{PR}	QFM_{PR}	QFM_{PR}	QFM_{PR}
	TH	QFM_{TH} QFM_{TH}	QFM_{TH} QFM_{TH}	QFM_{TH} QFM_{TH}	QFM_{TH} QFM_{TH}	QFM_{TH} QFM_{TH}	QFM_{TH} QFM_{TH}
	PL	QFM_{PL}	QFM_{PL}	QFM_{PL}	QFM_{PL}	QFM_{PL}	QFM_{PL}
	1	Sum $(QFM_J + QFM_D + QFM_{TH} + QFM_{PL}) \times 40\%$	AFM_{11b}	AFM_{11a}	AFM_{11g}	AFM_{11e}	AFM_{11n}
HTTP 40%	PR	QFM_{PR}	QFM_{PR}	QFM_{PR}	QFM_{PR}	QFM_{PR}	QFM_{PR}
	TH	QFM_{TH} QFM_{TH}	QFM_{TH} QFM_{TH}	QFM_{TH} QFM_{TH}	QFM_{TH} QFM_{TH}	QFM_{TH} QFM_{TH}	QFM_{TH} QFM_{TH}
	PL	QFM_{PL}	QFM_{PL}	QFM_{PL}	QFM_{PL}	QFM_{PL}	QFM_{PL}
	2	Sum $(QFM_{PR} + QFM_{TH} + QFM_{PL}) \times 40\%$	AFM_{11b}	AFM_{11a}	AFM_{11g}	AFM_{11e}	AFM_{11n}
FTP 30%	DR	QFM_{DR}	QFM_{DR}	QFM_{DR}	QFM_{DR}	QFM_{DR}	QFM_{DR}
	TH	QFM_{TH} QFM_{TH}	QFM_{TH} QFM_{TH}	QFM_{TH} QFM_{TH}	QFM_{TH} QFM_{TH}	QFM_{TH} QFM_{TH}	QFM_{TH} QFM_{TH}
	PL	QFM_{PL}	QFM_{PL}	QFM_{PL}	QFM_{PL}	QFM_{PL}	QFM_{PL}
	3	Sum $(QFM_{DR} + QFM_{TH} + QFM_{PL}) \times 30\%$	AFM_{11b}	AFM_{11a}	AFM_{11g}	AFM_{11e}	AFM_{11n}
SFM		1 + 2 + 3	SFM_{11b}	SFM_{11a}	SFM_{11g}	SFM_{11e}	SFM_{11n}
Rank		IEEE technology	IEEE technology	IEEE technology	IEEE technology	IEEE technology	IEEE technology

The percentage of packets lost (ω_i) by an application on a given node i as of Equation (4) shows that, for the percentage of dropped data packets ki divided by the total number of dropped packets ρ_i multiplied by 100%, a code has been written in MATLAB to determine this percentage for each of the three mixed-use cases, as shown in Figure 14. This technique is integrated with the OPNET Modeler to generate an accurate packet loss percentage for each unique combination of applications and IEEE technologies. The AFM for individual services and the SFM for combined services will be determined using Equation (3) and (5).

$$SFM_g = \sum_{n=1}^{4} AFM_n \tag{5}$$

```
filename='C:\ BSS_10_100VOIP_C-11e-DES-
1__Voice.xlsx';

A = xlsread(filename,'A:A');

B = xlsread(filename,'B:B');

C = xlsread(filename,'C:C');

for i=1:1:199;

nr(i)=(B(i+1)+B(i))./2. X (A(i+1)-A(i));

ns(i)=(C(i+1)+C(i))./2. X (A(i+1)-A(i));

pl(i)=(ns(i)-nr(i))./(ns(i)) X 100;

end

[f,x1] = ecdf(pl)

plot(x1,f)
```

Figure 14. Evaluating packet loss through the utilization of MATLAB.

The WLAN SFM values are used to rank these six technologies for the circular distribution of mixed services in BSS scenarios. When ranking the six WLAN technologies, the same method will be used for both the uniform and random distributions. For the ESS and IBSS network topologies, the system will apply algorithms and perform calculations to identify the top-performing WLAN technology in each topology and generate the QPMs, QFMs, AFMs, and SFMs for all QoS parameters for all six technologies with respect to each mixed service in this scenario's projects across all three spatial distributions.

Table 6 displays the relative rankings of all WLAN technologies for each of the three geographical distributions of the best-effort mixed services.

Table 6. Outcomes for best-effort mixed BSS, IBSS, and ESS services across 40 workstations.

Application / Technology	BSS			IBSS			ESS		
	C	U	R	C	U	R	C	U	R
802.11	SFM_{11}	SFM_{11}	SFM_{11}	SFM_{11}	SFM_{11}	SFM_{11}	SFM_{11}	SFM_{11}	SFM_{11}
802.11a	SFM_{11a}	SFM_{11a}	SFM_{11a}	SFM_{11a}	SFM_{11a}	SFM_{11a}	SFM_{11a}	SFM_{11a}	SFM_{11a}
802.11b	SFM_{11b}	SFM_{11b}	SFM_{11b}	SFM_{11b}	SFM_{11b}	SFM_{11b}	SFM_{11b}	SFM_{11b}	SFM_{11b}
802.11g	SFM_{11g}	SFM_{11g}	SFM_{11g}	SFM_{11g}	SFM_{11g}	SFM_{11g}	SFM_{11g}	SFM_{11g}	SFM_{11g}
802.11e	SFM_{11e}	SFM_{11e}	SFM_{11e}	SFM_{11e}	SFM_{11e}	SFM_{11e}	SFM_{11e}	SFM_{11e}	SFM_{11e}
802.11n	SFM_{11n}	SFM_{11n}	SFM_{11n}	SFM_{11n}	SFM_{11n}	SFM_{11n}	SFM_{11n}	SFM_{11n}	SFM_{11n}

5. Conclusions

We have expanded our previous work to the implementation of Internet applications as a stand-alone service that includes mixed applications. Different services are executed and configured at different rates depending on the circumstances. Additionally, new nodes and IEEE technologies were included. In this body of work, a novel approach to analyzing network performance was developed. The results of 50% VC applications show that it is only preferable to use the BSS network with a high number of workstations/nodes for all three spatial distributions. This is due to high packet loss and delays that might appear in the network owing to the increase in the number of workstations. Interestingly, ESS networks are suitable for best-effort services for a large number of nodes. Furthermore, IBSS networks worked efficiently with 802.11e and 802.11n technologies for almost all selected numbers of nodes for 50% VC when they were configured randomly or uniformly in medium-sized networks, whereas for best-effort services the performance of different technologies was based on the size of the network, with the circular distribution shown to be the dominant distribution.

The next step for this study is to apply the same methods developed here to other real-time and best-effort qualities, as well as upcoming ones, in order to determine the most appropriate IEEE standard for use with different kinds of application services. As a result of the increased network size and the incorporation of new network protocols, we intend to propose and implement new applications for the resulting service mix. The Internet of Things (IoT), which has expanded from simple machine-to-machine (M2M) communication, is just one example of the many promising directions in which researchers could go next. M2M communications are useful for businesses because they allow computers to be connected to the cloud, devices to be controlled remotely, and data to be collected. The connectivity enabled by M2M communication is the backbone of the Internet of Things. To continue, it is expected that Machine Learning (ML) algorithms and Artificial Intelligence (AI) will serve as the backbone of many different kinds of technologies and applications. Answers and forecasts from ML depend on the data available from the network. Finally, wireless networks have become increasingly popular, with the release of Wi-Fi 6 (IEEE 802.11ax) and Wi-Fi 7 (IEEE 802.11be) attracting an ever-increasing number of users. As a result, digital information consumption across all mediums has increased dramatically.

Author Contributions: Conceptualization, A.M.A.; formal analysis, M.R.H.; investigation, A.M.A.; methodology, A.M.A.; resources, A.a.-Q., A.H., K.A.-Q. and M.A.; supervision, M.R.H., M.A., O.K. and J.L.; validation, A.A.-K.; writing, A.M.A.; review and editing, M.A., O.K. and J.L. All authors have read and agreed to the published version of the manuscript.

Funding: This research is supported by Communications and the Computer Engineering Department, Faculty of Engineering, Al-Ahliyya Amman University, Amman 19328, Jordan and the Computer Science Department, Faculty of Information Technology, Zarqa University, Zarqa 13110 Jordan.

Institutional Review Board Statement: Not applicable.

Informed Consent Statement: Not applicable.

Data Availability Statement: Research data will be available on individual requests to the corresponding author considering collaboration possibilities with the researcher or research team and with restrictions that the data will be used only for further research in the related literature progress.

Acknowledgments: This research is scientifically supported by the research team in Nottingham Trent University, UK.

Conflicts of Interest: The authors declare no conflict of interest.

References

1. Sendra, S.; Fernandez, P.; Turro, C.; Lloret, J. IEEE 802.11 a/b/g/n Indoor Coverage and Performance Comparison. In Proceedings of the 6th International Conference on Wireless and Mobile Communications, Valencia, Spain, 20–25 September 2010; pp. 185–190.
2. Garcia, M.; Martinez, C.; Tomas, J.; Lloret, J. Wireless Sensors self-location in an Indoor WLAN environment. In Proceedings of the International Conference on Sensor Technologies and Applications (SENSORCOMM 2007), Valencia, Spain, 14–20 October 2007.
3. Tramarin, F.; Vitturi, S.; Luvisotto, M.; Zanella, A. On the use of IEEE 802.11 n for industrial communications. *IEEE Trans. Ind. Inform.* **2016**, *12*, 1877–1886. [CrossRef]
4. Lopez-Aguilera, E.; Garcia-Villegas, E.; Casademont, J. Evaluation of IEEE 802.11 coexistence in WLAN deployments. *Wirel. Netw.* **2019**, *25*, 87–104. [CrossRef]
5. Gao, Y.; Sun, X.; Dai, L. Sum Rate Optimization of Multi-Standard IEEE 802.11 WLANs. *IEEE Trans. Commun.* **2018**, *67*, 3055–3068. [CrossRef]
6. Mohd Ali, A.; Dhimish, M.; Alsmadi, M.; Mather, P. Algorithmic Identification of the Best WLAN Protocol and Network Architecture for Internet-Based Applications. *J. Inf. Knowl. Manag.* **2020**, *19*, 2040011. [CrossRef]
7. Mohd Ali, A.; Dhimish, M.; Mather, P. WLAN Protocol and Network Architecture Selection for Real-time Applications. *Int. J. Adv. Comput. Eng. Netw. (IJACEN)* **2019**, *7*, 8–14.
8. Mohd Ali, A.; Dhimish, M.; Glover, I. WLAN Protocol and Network Architecture Identification for Service Mix Applications. *Int. J. Adv. Comput. Eng. Netw. (IJACEN)* **2020**, *8*, 24–30.
9. Mohd Ali, A.; Dhimish, M.; Mather, P. An Algorithmic Approach to Identify the Optimum Network Architecture and WLAN Protocol for VoIP Application. *Wirel. Pers. Commun.* **2021**, *119*, 3013–3035. [CrossRef]
10. Coronado, E.; Villalón, J.; Garrido, A. Improvements to Multimedia Content Delivery over IEEE 802.11 Networks. In Proceedings of the IEEE/IFIP Network Operations and Management Symposium, Budapest, Hungary, 20–24 April 2020; IEEE: Piscataway, NJ, USA, 2020. [CrossRef]
11. Farej, Z.; Jasim, M. Performance evaluation of the IEEE 802.11n random topology WLAN with QoS application. *Int. J. Electr. Comput. Eng.* **2020**, *10*, 1924. [CrossRef]
12. Genc, E.; Del Carpio, L. Wi-Fi QoS Enhancements for Downlink Operations in Industrial Automation Using TSN. In Proceedings of the 15th IEEE International Workshop on Factory Communication Systems (WFCS), Sundsvall, Sweden, 27–29 May 2019; IEEE: Piscataway, NJ, USA, 2019. [CrossRef]
13. Khiat, A.; Bahnasse, A.; EL Khail, M.; Bakkoury, J. Wi-Fi and WiMax QoS Performance Analysis on High-Level Traffic using OPNET Modeler. *Pertanika J. Sci. Technol.* **2017**, *25*, 1343–1356.
14. Refaet, A.; Ahmed, M.; Aish, Q.; Jasim, A. VoIP Performance Evaluation and Capacity Estimation Using different QoS Mechanisms. In Proceedings of the 3rd International Conference on Sustainable Engineering Techniques (ICSET 2020), Munich, Germany, 2–3 July 2020. [CrossRef]
15. Cao, C.; Zuo, Y.; Zhang, F. Research on Comprehensive Performance Simulation of Communication IP Network Based on OPNET. In Proceedings of the International Conference on Intelligent Transportation, Big Data & Smart City (ICITBS), Xiamen, China, 25–26 January 2018; IEEE: Piscataway, NJ, USA, 2018. [CrossRef]
16. Bhatt, T.; Kotwal, C.; Chaubey, N. Implementing AMI Network using Riverbed OPNET Modeler for DDoS attack. *Int. J. Comput. Sci. Eng.* **2019**, *7*, 569–574. [CrossRef]
17. Vivekananda, G.; Chenna Reddy, P. Performance Evaluation of TCP, UDP, and SCTP in Manets. *ARPN J. Eng. Appl. Sci.* **2018**, *13*, 3087–3092.

18. Mohammadani, K.; Aslam Butt, R.; Memon, K.; Hassan, F.; Majeed, A.; Kumar, R. Highest Cost First-Based QoS Mapping Scheme for Fiber Wireless Architecture. *Photonics* **2020**, *7*, 114. [CrossRef]
19. Farej, Z.; Jasim, M. Investigation on the Performance of the IEE802.11n Based Wireless Networks for Multimedia Services. In Proceedings of the 2nd International Conference for Engineering, Technology and Sciences of Al-Kitab (ICETS), Karkuk, Iraq, 4–6 December 2018; IEEE: Piscataway, NJ, USA, 2018. [CrossRef]
20. Kaur, R.; Gupta, A.; Srivastava, A.; Chatterjee, B.; Mitra, A.; Ramamurthy, B.; Bohara, V. Resource Allocation and QoS Guarantees for Real World IP Traffic in Integrated XG-PON and IEEE 802. 11e EDCA Networks. *IEEE Access* **2020**, *8*, 124883. [CrossRef]
21. Refaet, A.; Amed, M.; Abed, W.; Aish, Q. WLAN performance evaluation in different wireless access techniques (DCF, PCF, HCF). *Period. Eng. Nat. Sci.* **2020**, *8*, 1297–1308. [CrossRef]
22. Gul, O.M. Achieving Near-Optimal Fairness in Energy Harvesting Wireless Sensor Networks. In Proceedings of the 2019 IEEE Symposium on Computers and Communications (ISCC), Barcelona, Spain, 29 June–3 July 2019; pp. 1–6. [CrossRef]
23. Nosheen, S.; Khan, J. High Throughput and QoE Fairness Algorithms for HD Video Transmission over IEEE802.11ac Networks. In Proceedings of the International Conference on Computing, Networking and Communications (ICNC), Big Island, HI, USA, 17–20 February 2020; IEEE: Piscataway, NJ, USA, 2020. [CrossRef]
24. Wen, H.; Xiaofeng, T.; Defeng, R. The analysis of IEEE 802.11 PCF protocol based on LEO satellite Wi-Fi. *MATEC Web Conf.* **2018**, *189*, 04014. [CrossRef]
25. AL-Maqri, M.; Alrshah, M.; Othman, M. Review on QoS Provisioning Approaches for Supporting Video Traffic in IEEE802.11e: Challenges and Issues. *IEEE Access* **2018**, *6*, 55202–55219. [CrossRef]
26. Ergenç, D.; Onur, E. Plane-separated routing in ad-hoc networks. *Wirel. Netw.* **2022**, *28*, 331–353. [CrossRef]
27. Gul, O.M.; Kantarci, B. Near optimal scheduling for opportunistic spectrum access over block fading channels in cognitive radio assisted vehicular network. *Veh. Commun.* **2022**, *37*, 100500. [CrossRef]
28. Zhang, S.; Li, X.; Liu, Y. Analysis of scheduling delay and throughput of multiple radio multiple access protocols in wireless ad hoc networks. In *Advances in Guidance, Navigation and Control*; Springer: New York, NY, USA, 2022; pp. 5419–5428.
29. Fatima, M.; Khursheed, A. Heterogeneous Ad hoc Network Management: An Overview. In *Cloud Computing Enabled Big-Data Analytics in Wireless Ad-Hoc Networks*; CRC Press: Boca Raton, FL, USA, 2022; p. 103.
30. Rani, P.; Verma, S.; Kaur, N.; Wozniak, M.; Shafi, J.; Ijaz, M.F. Robust and secure data transmission using artificial intelligence techniques in ad-hoc networks. *Sensors* **2022**, *22*, 251. [CrossRef]
31. Sharma, B.; Vaid, R. A comprehensive study on vulnerabilities and attacks in multicast routing over mobile ad hoc network. In *Cyber Security and Digital Forensics*; Springer: New York, NY, USA, 2022; pp. 253–264.
32. Qasim, O.A.; Noori, M.S.; Dabag, A.; Ahmad, M.L. Performance evaluation of ad-hoc on-demand distance vector protocol in highway environment in VANET with MATLAB. *Telkomnika* **2022**, *20*, 194–200. [CrossRef]
33. Sammour, I.; Chalhoub, G. Evaluation of Rate Adaptation Algorithms in IEEE 802.11 Networks. *Electronics* **2020**, *9*, 1436. [CrossRef]
34. Chang, C.-Y.; Yen, H.-C.; Lin, C.-C.; Deng, D.-J. QoS/QoE support for H. 264/AVC video stream in IEEE 802.11 ac WLANs. *IEEE Syst. J.* **2015**, *11*, 2546–2555. [CrossRef]
35. Zaidi, T.; Nand Dwivedi, N. Voice Packet Performance Estimation through Step Network Using OPNET. In Proceedings of the 3rd International Conference on Computing, Communication and Security (ICCCS), Kathmandu, Nepal, 25–27 October 2018; IEEE: Piscataway, NJ, USA, 2018. [CrossRef]
36. Riverbed. Retrieved from Riverbed Web Site. 2020. Available online: https://www.riverbed.com/gb/index.html (accessed on 16 November 2022).
37. Hassan, W.; Idrus, S.; King, H.; Ahmed, S.; Faulkner, M. Idle sense with transmission priority in fibre-wireless networks. *IET Commun.* **2020**, *14*, 1428–1437. [CrossRef]
38. Dhawankar, P.; Kumar, A.; Crespi, N.; Busawon, K.; Qureshi, K.N.; Javed, I.T.; Prakash, S.; Kaiwartya, O. *Next-Generation Indoor Wireless Systems: Compatibility and Migration Case Study*; IEEE Access: Piscataway, NJ, USA, 2021; Volume 9, pp. 156915–156929.
39. Kelly, G. Forbes. Retrieved from Forbes Web Site. December 2014. Available online: https://www.forbes.com/sites/gordonkelly/2014/12/30/802-11ac-vs-802-11n-wifi-whats-the-difference/#2061f1073957 (accessed on 16 November 2022).
40. Aagela, H.; Holmes, V.; Dhimish, M.; Wilson, D. Impact of video streaming quality on bandwidth in humanoid robot NAO connected to the cloud. In Proceedings of the Second International Conference on Internet of Things, Data and Cloud Computing, Cambridge, UK, 22–23 March 2017; p. 134.
41. Yates, R.D.; Goodman, D.J. *Probability and Stochastic Processes: A Friendly Introduction for Electrical and Computer Engineers*; John Wiley & Sons: New York, NY, USA, 2014.
42. Lloret, J.; López, J.J.; Turró, C.; Flores, S. A fast design model for indoor radio coverage in the 2.4 GHz wireless LAN. In Proceedings of the 1st International Symposium on Wireless Communication Systems, Mauritius, 20–22 September 2004; pp. 408–412.

Disclaimer/Publisher's Note: The statements, opinions and data contained in all publications are solely those of the individual author(s) and contributor(s) and not of MDPI and/or the editor(s). MDPI and/or the editor(s) disclaim responsibility for any injury to people or property resulting from any ideas, methods, instructions or products referred to in the content.

sensors

MDPI

Article

Evaluation of Machine Leaning Algorithms for Streets Traffic Prediction: A Smart Home Use Case

Xinyao Feng [1], Ehsan Ahvar [2,†] and Gyu Myoung Lee [3,*]

1 Learning, Data and Robotics Laboratory, ESIEA Graduate Engineering School, 75005 Paris, France; feng@et.esiea.fr
2 Nokia, 91300 Massy, France; ehsan.ahvar@nokia.com
3 School of Computer Science and Mathematics, Liverpool John Moores University, Liverpool L3 3AF, UK
* Correspondence: g.m.lee@ljmu.ac.uk
† Ehsan Ahvar recently joined Nokia. This work was done when the author was at ESIEA.

Abstract: This paper defines a smart home use case to automatically adjust home temperature and/or hot water. The main objective is to reduce the energy consumption of cooling, heating and hot water systems in smart homes. To this end, the residents set a temperature (i.e., X degree Celsius) for home and/or hot water. When the residents leave homes (e.g., for work), they turn off the cooling or heating devices. A few minutes before arriving at their residences, the cooling or heating devices start working automatically to adjust the home or water temperature according to the residents' preference (i.e., X degree Celsius). This can help reduce the energy consumption of these devices. To estimate the arrival time of the residents (i.e., drivers), this paper uses a machine learning-based street traffic prediction system. Unlike many related works that use machine learning for tracking and predicting residents' behaviors inside their homes, this paper focuses on predicting resident behavior outside their home (i.e., arrival time as a context) to reduce the energy consumption of smart homes. One main objective of this paper is to find the most appropriate machine learning and neural network-based (MLNN) algorithm that can be integrated into the street traffic prediction system. To evaluate the performance of several MLNN algorithms, we utilize an Uber's dataset for the city of San Francisco and complete the missing values by applying an imputation algorithm. The prediction system can also be used as a route recommender to offer the quickest route for drivers.

Keywords: machine learning; deep learning; recommendation system; energy consumption; smart home; neural network; performance evaluation

Citation: Feng, X.; Ahvar, E.; Lee, G.M. Evaluation of Machine Leaning Algorithms for Streets Traffic Prediction: A Smart Home Use Case. *Sensors* **2023**, *23*, 2174. https://doi.org/10.3390/s23042174

Academic Editor: Daniele Cenni

Received: 6 December 2022
Revised: 6 February 2023
Accepted: 7 February 2023
Published: 15 February 2023

Copyright: © 2023 by the authors. Licensee MDPI, Basel, Switzerland. This article is an open access article distributed under the terms and conditions of the Creative Commons Attribution (CC BY) license (https://creativecommons.org/licenses/by/4.0/).

1. Introduction

Smart homes can make our lives more comfortable and exciting. Security, user privacy and energy consumption are considered as the three main challenges in smart homes.

In this paper, we focus on the energy consumption issue. Heating, ventilation, and air conditioning account for around half of building energy consumption in the U.S. and between 10 and 20% of total energy consumption in developed countries [1].

On the other hand, the number of vehicles in large-size smart cities has sharply increased in recent years. In large-size cities, traveling duration between two special locations can be varied depending on the time of day and even the day of week (i.e., we can have different traffic situations). This has made the traffic one of the smart city challenges. A short-term prediction for the street traffic situation of routes is not only essential for drivers (e.g., to save time and automotive fuel), but can also be used as useful information for various use cases and applications.

Putting all pieces together, we propose a smart home use case to automatically adjust home temperature and hot water temperature. Here, the street traffic prediction system predicts the arrival time of a home resident (i.e., who is driving) and, just a few minutes before the arrival, the cooling or heating devices will be turned on to reduce the energy consumption of the devices.

The system predicts the street traffic situation using a machine learning technique (model). In general, machine learning and neural network-based (MLNN) algorithms play an important role on the performance of machine learning-based systems.

However, the performance of an MLNN algorithm may depend on the application and/or the utilized dataset. While an MLNN algorithm can show a great performance for one application, it may not necessarily provide this level of performance for another application or even another dataset.

For this reason, we evaluate the performance of several famous MLNN algorithms to find which one is more appropriate for our application (i.e., traffic prediction). We evaluate the performance of the following MLNN algorithms: K-nearest neighbors (KNN), decision tree (DT), random forest (RF), support vector machines (SVMs), multilayer perceptron (MLP), long short-term memory (LSTM), single layer perceptron (SLP) and categorical naive Bayes (CNB).

In addition to MLNN algorithms, the utilized datasets play an important role in the street traffic prediction process and results. Even if the systems are efficient, certain limitations can occur because of the uncertainty in traffic-related data [2].

While some datasets are available for the traffic situation of several highways/freeways or some special streets of different cities, finding comprehensive and complete datasets covering all small and large-size streets of a city is still a big issue. As a solution, in this paper, we utilize an Uber dataset (which covers a large number of small streets) for the city of San Francisco and then complete the missing values by applying an imputation algorithm.

We use Scikit-Learn [3] and Pytorch [4] to implement the algorithms and consider the following evaluation metrics: Precision, F1-Score, and Accuracy.

We summarize our contributions as follows:

- Propose a use case for smart home to adjust the home temperature and hot water temperature by controlling the heating and cooling devices with the main goal of reducing the energy consumption of the devices;
- Propose a dual-objective machine learning-based streets traffic prediction system. It estimates a resident (i.e., a driver) arrival time and sends it (i.e., as a context) to the context-aware smart home application. It can be also used as a route recommender to guide the quickest routes for drivers;
- Implementing several famous MLNN algorithms for the streets' traffic prediction;
- Analyzing and comparing the performance of the implemented MLNN algorithms and introducing the most appropriate one for the problem of streets traffic prediction.

The rest of the paper is organized as follows. The related work is presented in Section 2. Section 3 introduces our use case. An overview of the implemented and analyzed algorithms in this paper is presented in Section 4. Section 5 evaluates the performance of several famous MLNN algorithms to find the most appropriate one for our application. We will have a discussion and conclusion in Sections 6 and 7, respectively.

2. Related Work

2.1. Machine Learning for Smart Homes

Machine learning has been used for various applications in smart homes. For example, Taiwo et al. [5] presented a system to control home appliances, monitor environmental factors, and detect movement in the home and its surroundings. They utilized a deep learning model for motion recognition and classification based on the detected movement patterns. Using a deep learning model, they designed a system for intruder detection. Based on the walking pattern, a human detected by the surveillance camera is categorized as an intruder or home residence.

Filipe et al. [6] presented a voice-activated smart home controller. They proposed an architecture that focuses on the use of Online Learning to develop a smart home controller capable of controlling multiple connected devices according to the resident's preferences and habits.

Fahim et al. [7] proposed ApplianceNet to detect daily life activities in smart homes. It is based on the energy consumption patterns of home appliances attached to smart plugs. They use a multi-layer, feed-forward neural network to classify the home appliances.

Machine learning can also help smart homes reduce energy consumption.

Fakhar et al. [8] recently provided a survey of smart home energy conservation techniques. They identified various critical features in energy conservation techniques (e.g., user and appliance energy profiling) to perform a comparative analysis among various techniques. They also presented various energy conservation techniques and provided a statistical analysis of the existing literature.

Kim et al. [9] proposed a temperature controller based on machine learning. The system learns the life patterns and desired temperature of individual residents (according to each situation) and allows the learning results to be reflected in the temperature control (i.e., cooling/heat devices).

All the aforementioned machine learning-based works have mainly focused on tracking and predicting the home resident's mobility patterns or behavior inside homes. In contrast, our work uses machine learning to predict the arrival time of a resident based on the predicted (route) traffic situation.

2.2. Streets Traffic Prediction

Several approaches have been proposed to predict the traffic situations of streets and highways. Street traffic prediction approaches can be divided into two main categories: parametric (e.g., exponential smoothing, the Kalman filtering [10] and ARIMA [11]) and non-parametric (e.g., machine learning-based approaches).

The parametric approaches estimate the traffic situation based on the strong theoretical assumption. The studies show that non-parametric approaches outperform parametric solutions because of their ability to deal with a large number of parameters and big data [2,12].

Jose Braz et al. [13] developed and compared three deep learning models for forecasting the traffic flow in the Barra and Costa Nova regions. They divided their dataset into training, validation, and testing sets (i.e., the holding out method).

Kim et al. [14] presented a structural RNN architecture that combines the road network map with the traffic speed data to predict the future traffic speed. They used a traffic speed dataset from the case studies of the SETA EU project [15].

In contrast to the mentioned works, we utilize the walk forward validation. For the time series data analysis, the walk forward validation can offer a better result than holdout and k-fold cross-validation [16].

In addition, one common challenge for the existing related work is that the traffic road datasets under their study only cover several inter-city highways or the main avenues of a city or a smaller number of streets which are most of the time sparsely located. When we want to predict the traffic situation of every requested route (i.e., in source–destination pair format) in a city, these datasets are not very useful. To solve this problem, we used an Uber dataset for the city of San Francisco and completed the missing values by applying an imputation algorithm (see details in Section 5).

3. Use Case

In this paper, we present a use case to automatically adjust one's home temperature (and/or hot water temperature) according to the desire temperature which has been already set by the resident. The main objective is to reduce the energy consumption of the cooling or heating devices in smart homes. When the residents are not at home, the cooling or heating devices should be off. A few minutes before a resident returns, the devices start working to adjust the requested temperature. To this end, it is necessary to know the estimated arrival time of the residents. This use case is more useful for the resident who spends a considerable duration of time outside their home. Notice that calculating the needed time to adapt the home temperature degree (or water temperature) according to a resident request is possible. This mainly depends on several factors such as the home size, number and power of the cooling and heating devices at home, outside temperature and requested temperature degree. However, this is out of scope of our work (i.e., this paper).

3.1. Smart City

We consider a smart city including a number of intersections. Here, an intersection is a place or point where two or more streets cross each other. A street (or a part of that) which is located between two intersections is considered as a segment (see Figure 1).

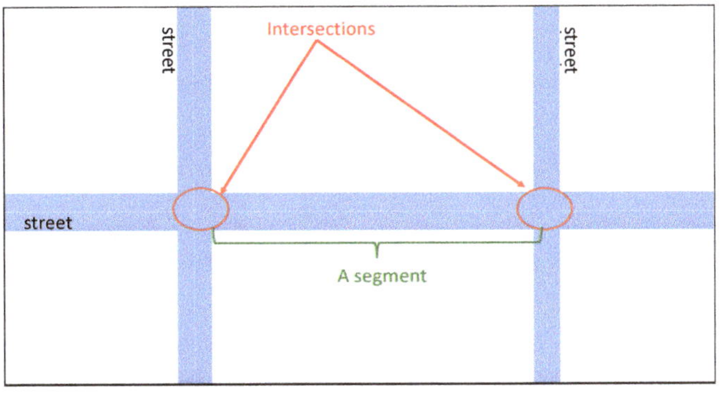

Figure 1. Intersection and segment concepts.

3.2. Scenario

We present the use case considering the following scenario. Assume that a driver is now somewhere in a large-size city and they decide to return home. We estimate the arrival time of the driver and send it to the related home application (i.e., here, the application of adjusting home temperature) to adjust the home temperature based on the arriving time of the driver.

We assume that every driver has a list of source/destination pairs and selects an appropriate source/destination pair from the list. Here, the source is the closest intersection to the current location of the driver and the destination is the closest intersection to their home.

We propose a system that receives a driver request (i.e., including the leaving time and their selected source/destination pair) and recommends the best routes to the driver and the estimated arrival time to the smart home application.

As Figure 2 shows, we define 12 steps to better present the use case and working mechanism of the system. In step 1, a driver (i.e., a home resident which is driving somewhere in the city now) sends a route request to the system including a source/destination pair and leaving time.

When the system (i.e., central controller) receives the request, it selects the four shortest routes between the requested source and destination with the help of the route finder component (i.e., steps 2 and 3). In fact, the route finder component creates a graph where the intersections are its vertices and the segments are its edges. Having this graph, it can find the four shortest routes between every requested source and the destination using a modified version of Dijkstra's algorithm (or any other shortest route/path finder algorithm). Every selected route consists of a set of segments. For every segment of a selected route, we have historical data (introduced in Section 5.1). The system applies an MLNN algorithm to the historical data of every segment of selected routes one-by-one to predict the traffic situation of them for the requested time interval (i.e., steps 4 and 5).

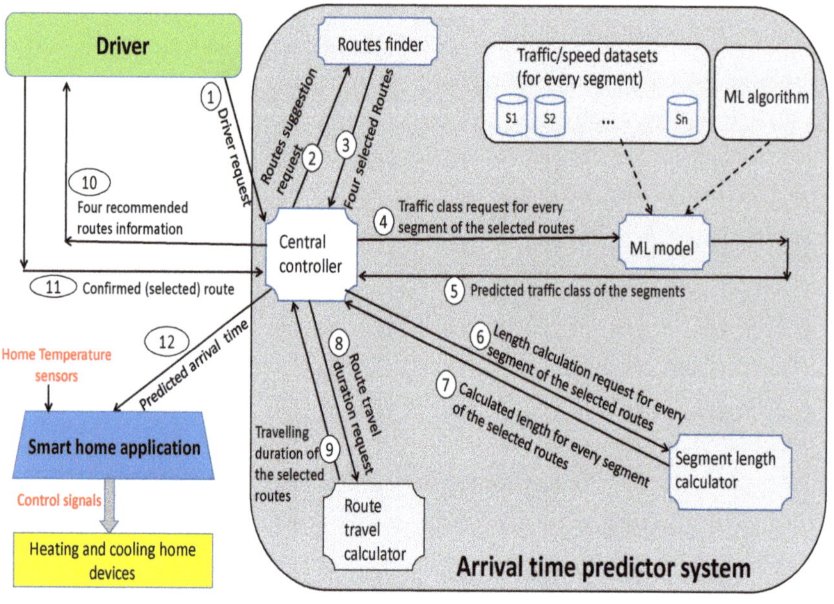

Figure 2. Working mechanism of the proposed system.

To this end, our system considers three levels of (average) car movement speeds for segments called L1 (heavy traffic), L2 (normal traffic), and L3 (low traffic) considering two thresholds (i.e., F1 and F2) as follows:

- The average speed less than or equal to F1 is defined as heavy traffic or level L1;
- The average speed between F1 and F2 is defined as normal traffic or level L2;
- The average speed equal or above F2 is defined as low traffic or level L3.

Here, the classification technique for traffic prediction gives us the opportunity to send simple and easy-to-understand information to drivers (i.e., sending traffic information of every segment as low, normal or heavy).

In the next steps (i.e., steps 6 and 7), the system calculates the length of every segment of the selected routes utilizing the segment length calculator component. Recall that the system has the geographical information of intersections and segments.

In steps 8 and 9, having the length and traffic class (i.e., low, normal, or heavy) information of the segments of every selected route, the system estimates the traveling duration of the four selected routes.

In step 10, the system suggests the selected routes, the traveling duration of every selected route and the information related to segments of every selected route (i.e., traffic level) to the requested driver. In step 11, the driver selects one of the four suggested routes and informs the system about it. Finally, our system sends the estimated arrival time of the driver to the smart home application (i.e., as an external context) in step 12. Here, we assume that the system can communicate to both drivers and smart home applications. The smart home application knows the current temperature of the home (i.e., using the home temperature sensors). After receiving the estimated arrival time, if necessary, the application sends appropriate control signals to turn the cooling or heating devices on.

Recall that this paper mainly focuses on the benefits of combining smart home use cases and the street traffic prediction system (i.e., here, the arrival time predictor system) rather than going into the details of the smart home applications. As already mentioned, this traffic prediction system can also help a home residence (i.e., when driving) to select the most appropriate route. The traffic prediction system (i.e., the arrival time predictor system) can be installed on the Cloud and provide the service to several drivers.

4. Machine Learning and Neural Network Algorithms

As we already mentioned, we cannot find an MLNN algorithm that works perfectly for all types of scenarios, applications and datasets. In this case, we need to study and evaluate several well-known MLNN algorithms to find which one can be more appropriate for our work. This section briefly presents MLNN algorithms that we analyze in this paper.

4.1. K-Nearest Neighbors (KNN)

KNN is a simple and lazy learning algorithm that can be used for both classification and regression. It stores all the available data and classifies a new data point based on a similarity measure (e.g., Euclidean distance function). In other words, it does a simple majority vote of its k-nearest neighbors for classifying a new data. Detecting an optimal number of neighbors for a new data plays an important role in the performance of KNN [17].

4.2. Decision Tree (DT) and Random Forest (RF)

Similar to KNN, DT can be utilized for both classification and regression tasks. It uses a hierarchical tree-like structure where it sets different conditions on the branches. Iterative Dichotomiser 3 (ID3), C4.5, and Classification and Regression Trees (CART) are well known DT algorithms. RF is another well-known machine learning algorithm. It is an ensemble classification technique consisting of many DTs. In other words, the forest is an ensemble of independent DTs in a way that every tree sets conditional features differently. After receiving a sample, the root node sends it to all the sub-trees. In the next step, each sub-tree predicts the class label for the sample. Finally, the class in the majority is given to that sample [17–19].

4.3. Support Vector Machines (SVMs)

SVM is considered a supervised learning algorithm to solve classification and regression prediction problems based on statistical learning theory [20]. In fact, it maps data points to high-dimensional vectors. For data points in an n-dimensional space, a $(n-1)$-dimensional hyperplane is considered as a classifier [21].

4.4. Categorical Naive Bayes (CNB)

Naive Bayes is a simple algorithm for classification dependent on Bayes' theorem with a supposition of freedom among predictors. In CNB, it is naively considered that the features are independent [21–23].

4.5. Single-Layer Perceptron (SLP)

SLP is a binary classifier consisting of a node (called artificial neuron or perceptron) with a set of inputs and an output. SLP considers a weight for every input of the perceptron. It then multiplies the inputs with their respective weights. After that, the results of the products are added together. The calculated value is then compared with a threshold value to detect the (output) label [21].

4.6. Multilayer Perceptron (MLP)

MLP is a neural network algorithm. It can use several perceptrons where these perceptrons are distributed in at least three layers (i.e., one input layer, one or more hidden layers, and one output layer). MLP is usually more powerful and complex than SLP [21,24].

4.7. Long Short-Term Memory (LSTM)

LSTM is considered an improved version of the recurrent neural network (RNN) [25]. It is proposed to expand the RNN memory. This memory can keep information over an arbitrary length of time. Three gates (i.e., input, output, and forget gates) control the information flow into and out of the neuron's memory [26].

5. Performance Evaluation

In this section, we first present the utilized dataset and the way of preparing it. After that, we present technical details for the implementation including the utilized tools, hyper-parameters, utilized features, validation technique and evaluation metrics. Finally, the third part describes how we implemented MLNN algorithms by the introducing main utilized Python functions.

5.1. Dataset Description

We utilized an Uber traffic dataset for San Francisco [27]. We selected a zone covering a total number of 439 segments in San Francisco. Traffic observations were recorded in approximately 1-hour intervals for each street segment during 18 h per day (i.e., from 6 AM to 24) for 5 weeks. The real dataset contains around 197,496 records. It then becomes equal to 276,570 records after calculating and filling its missing values (i.e., by applying the imputation algorithm. Details in Section 5.1.1).

Every row in the dataset refers to the situation of a segment at a specific timestamp. The columns correspond to the following content:

- date: the date (year–month–day) of the record;
- hour: the time (in hours) of the record (local city time);
- segment_id: shows ID of the corresponding street segment;
- osm_start_node_id: starting node (intersection) of the segment, ID of the node according to OpenStreetMap;
- osm_end_node_id: ending node of the segment, ID of the node according to OpenStreetMap;
- mean_speed: average speed of Uber vehicles during the time interval in this street segment in km/h;
- traffic_class: shows traffic situation of the segment (i.e., low, normal, heavy) during the time interval.

5.1.1. Data Cleaning

We found that the obtained dataset contains several missing lines (data) where the observations are not regularly and continuously recorded every 1 h for each segment during the considered 5 weeks.

In order to solve this problem, we defined a two-step solution: step (1) dataset size reduction (to remove unnecessary lines/records); step (2) dataset completion (to add necessary/utilized missing data).

- step 1—dataset size reduction: we found that a majority of the missing values were from the night. It is normal. Because Uber does not have a lot of services during the night (i.e., between 00:00 and 6:00). On the other hand, we know that we usually do not have heavy traffic between 00:00 and 6:00 and, therefore, it is not necessary to use the traffic recommendation system during these hours. As a result, we reduced the dataset size. Instead of considering observations for 24 h per day, we consider a dataset file keeping observations for 18 h per day (i.e., recorded observations between 6:00 and 24:00). This could remove a large number of missing values from our dataset (approximately 7.44 percent);
- step 2—dataset completion: in order to find appropriate values for the missing parts and complete the dataset, we use the Bayesian temporal matrix factorization (BTMF) model recently proposed in [28].

As an example, Figure 3 shows a part of the initial dataset for segment 0 (on 5 January 2020, from 6:00 to 24:00) with two missing values. Figure 4 shows that two missing values were calculated in step 2 (i.e., using BTFM).

year	month	day	hour	osm_start_node_id	osm_end_node_id	speed_mph_mean	segment_id
2020	1	5	6	65287111	65317959	14.203	0
2020	1	5	9	65287111	65317959	10.849	0
2020	1	5	10	65287111	65317959	9.497	0
2020	1	5	11	65287111	65317959	9.698	0
2020	1	5	12	65287111	65317959	10.341	0
2020	1	5	13	65287111	65317959	9.148	0
2020	1	5	14	65287111	65317959	8.226	0
2020	1	5	15	65287111	65317959	8.279	0
2020	1	5	16	65287111	65317959	10.075	0
2020	1	5	17	65287111	65317959	8.844	0
2020	1	5	18	65287111	65317959	9.48	0
2020	1	5	19	65287111	65317959	8.837	0
2020	1	5	20	65287111	65317959	14.807	0
2020	1	5	21	65287111	65317959	12.822	0
2020	1	5	22	65287111	65317959	10.698	0
2020	1	5	23	65287111	65317959	19.287	0

Figure 3. An illustration of the initial dataset for segment 0 (on 5 January 2020, from 6:00 to 24:00) with two missing values.

year	month	day	hour	osm_start_node_id	osm_end_node_id	speed_mph_mean	segment_id
2020	1	5	6	65287111	65317959	14.203	0
2020	1	5	7	65287111	65317959	15.42	0
2020	1	5	8	65287111	65317959	13.736	0
2020	1	5	9	65287111	65317959	10.849	0
2020	1	5	10	65287111	65317959	9.497	0
2020	1	5	11	65287111	65317959	9.698	0
2020	1	5	12	65287111	65317959	10.341	0
2020	1	5	13	65287111	65317959	9.148	0
2020	1	5	14	65287111	65317959	8.226	0
2020	1	5	15	65287111	65317959	8.279	0
2020	1	5	16	65287111	65317959	10.075	0
2020	1	5	17	65287111	65317959	8.844	0
2020	1	5	18	65287111	65317959	9.48	0
2020	1	5	19	65287111	65317959	8.837	0
2020	1	5	20	65287111	65317959	14.807	0
2020	1	5	21	65287111	65317959	12.822	0
2020	1	5	22	65287111	65317959	10.698	0
2020	1	5	23	65287111	65317959	19.287	0

Figure 4. An illustration of the final dataset for segment 0 (5 January 2020, from 6:00 to 24:00) where the two missing values (shown in red color) were calculated using BTFM.

5.1.2. Dataset Individualization

We separated our dataset into 439 sub-datasets where each new dataset contains traffic observations for only one individual segment. The dataset separation helped us reduce the complexity and processing time and improve the performance. Recall that, according to the use case under study (see Section 3), we can predict the traffic situation for every segment of a selected route separately.

5.2. *Implementation Technical Details*

5.2.1. Tools and Software

We utilized the Scikit-learn library [3] to implement the classic machine learning algorithms and PyTorch [4] for the deep learning algorithm.

5.2.2. Input Features

Considering the dataset information and requirements of our use case, the following input features are defined and extracted:

- segment_id: shows ID of the corresponding segment;
- observation time: keeps the time (hour) of each observation;
- traffic level/class: is calculated based on the average speed vehicles in every segment. For defining the three traffic levels/classes, F1 and F2 threshold values are considered 10 and 30, respectively. The values are selected after doing several tests with different values. In this case, the average speed:
  - Less than or equal to 10 km/h is considered as level L1 (heavy traffic);

- Between 10 and 30 is considered as level L2 or normal traffic;
- Equal or above 30 is considered as level L3 or low traffic.

5.2.3. Hyper-Parameters

Hyper-parameters play an important role in performance of the created models.

We use the Scikit-learn's RandomSearch cross-validation [29] to find appropriate values for hyper-parameters.

Recall that CNB does not have hyper-parameters to tune.

One week (i.e., 20%) of the dataset is used for tuning the hyper-parameters. The obtained hyper-parameter values are shown in Tables 1 and 2.

Table 1. Obtained values for hyper-parameters—Part I.

KNN		DT		RF	
Parameter	**Value**	**Parameter**	**Value**	**Parameter**	**Value**
n_neighbors	14	max_depth	8	max_depth	19
metric	manhattan	criterion	entropy	criterion	gini
weights	distance	min_samples_leaf	3	min_samples_leaf	4
				n_estimators	270

Table 2. Obtained values for hyper-parameters—Part II.

SVM		SLP		MLP		LSTM	
Par.	**Val.**	**Par.**	**Val.**	**Par.**	**Val.**	**Par.**	**Val.**
kernel	rbf	loss	perceptron	hidden_size	130	input_size	2
C	100	alpha	100.0	solver	adam	hidden_size	20
gamma	1	penalty	l2	activation	logistic	layer_dim	1
-	-	learning_rate	adaptive	learning_rate	invescaling	-	-
-	-	eta0	0.001	-	-	-	-

5.2.4. Validation Technique

We know that our data (dataset) constitutes a time series. As some usual techniques such as the k-fold cross-validation technique do not work perfectly for time series datasets [16], we utilize the walk-forward validation technique (i.e., a technique preserving the order of data).

We separate our dataset into five equal units or partitions (i.e., a unit is equal to one week). The units are chronologically ordered. We considered the last four consecutive weeks of the dataset (i.e., 80% of the dataset) for training and testing our models. In each execution, we consider all the available data before the last unit as the training set and the last unit itself as our testing set.

As shown in Figure 5, for the first execution (i.e., Execution 1), we consider the W1 unit for training and W2 for testing. We then consider W1 and W2 units together for training and W3 for testing in the second execution. Finally, for the last execution (i.e., Execution 3), W1, W2, and W3 units are considered for training and W4 for testing.

Finally, the model scores (e.g., accuracy) are calculated as the average results of all executions [16,30].

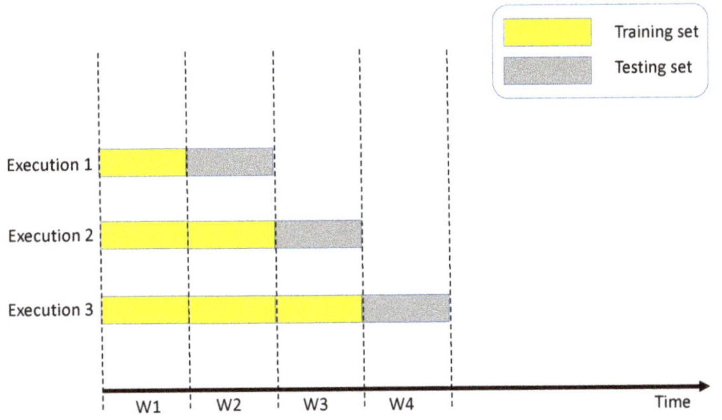

Figure 5. The walk-forward working mechanism for 5 weeks (units).

5.2.5. Evaluation Metrics

In this section, we evaluate the performance of the following MLNN algorithms: KNN, DT, RF, SVMs, MLP, LSTM, SLP and CNB.

In order to evaluate the aforementioned MLNN algorithms, we consider the accuracy, precision and F1-score metrics as well as the execution time of algorithms (i.e., including both training and testing times).

5.3. Implementation and Evaluation

We used Scikit-learn built-in machine learning classifiers (e.g., DecisionTreeClassifier for DT algorithm) to implement algorithms (except LSTM). The random search function available in Scikit-learn was utilized to detect appropriate values for hyper-parameters.

For every algorithm, we calculated its training/testing execution time by measuring the starting time and ending time of the execution. To do this, we utilized the available "time" function. As we used the walk forward methodology where we executed an algorithm several times (with different training and testing sizes), we calculated the execution time of an algorithm as an average of all execution times of that algorithm.

Notice that we utilized the one-vs.-one method for using binary classification algorithms such as SLP for our three-class classification.

Table 3 shows the average scores of the algorithms (models) and their training and testing times (in seconds) given 1 h predictions.

Table 3. Average cross-validation scores of models and their training and testing times.

Metrics	CNB	DT	KNN	MLP	RF	SLP	SVM	LSTM
Accuracy	90.09	89.81	89.20	87.34	90.53	83.89	89.75	81.55
Precision	90.58	90.10	89.38	89.09	90.99	83.43	89.64	93.17
F1_score	88.31	88.72	88.25	84.03	89.10	82.93	88.72	77.10
Training	0.0013	0.0012	0.0011	0.2334	1.0843	0.0013	0.0021	0.7882
Testing	0.0008	0.0010	0.0016	0.0016	0.0172	0.0006	0.0017	0.0006

By analyzing Table 3, we extracted the best algorithms in Table 4. We can observe that RF could outperform other algorithms in accuracy and F1_score. It is also the second best algorithm in terms of precision. However, for the training and testing times, it is one of the worst algorithms. This is because it is an ensemble learning algorithm.

CNB can be considered the second best algorithms (i.e., ranked 2 in accuracy and testing time and ranked 3 in precision, training time and F1_score).

Table 4. The best algorithms according to different metrics.

Accuracy	F1_Score	Precision	Training	Testing
RF(90.53)	RF(89.10)	LSTM(93.17)	KNN(0.0011)	LSTM, SLP(0.0006)
CNB(90.09)	DT/SVM(88.72)	RF(90.99)	DT(0.0012)	CNB(0.0008)
DT(89.81)	CNB(88.31)	CNB(90.58)	CNB, SLP(0.0013)	DT(0.0010)

DT shows promising results. It is ranked among the top three algorithms in accuracy, F1_score, training time and testing time. Its precision is equal to 90.10 which makes it the fourth algorithm in the table.

We can see that SVM and KNN are ranked in the middle of the table. While the results of SVM are slightly better than KNN in terms of accuracy and precision, KNN has a slightly better results for testing and training times.

The neural network-based algorithms (i.e., SLP, MLP and LSTM) have the worst results in accuracy and F1_score. While the training time for MLP and LSTM, as two complex algorithms, is rather long, SLP has a rather short training time.

Putting all the pieces together, RF has the best cross-validation scores. However, as the results show, a complex algorithm such as RF is time- and energy-consuming. As reducing energy consumption is the main objective of our work, our priority is to select and use a simpler and quicker algorithm (having a performance close to that of RF) such as CNB or even DT for using in our use case.

6. Discussion

Energy efficiency (cost) is an important factor for smart home applications. Many studies have been performed to improve the energy consumption of smart home applications using machine learning. However, they generally focus on tracking and predicting resident behavior in smart homes and use it as a context for applications. In contrast, our work shows that machine learning can also help smart home applications improve energy consumption by considering and predicting a resident's behavior outside a smart home (i.e., as an external context for smart home systems and applications). The proposed street traffic prediction system can be also used as a route recommender (i.e., to save time and automotive fuel) when the residents are not home or as an assistant in other use cases and applications.

To find the most suitable MLNN algorithm for traffic prediction, we evaluated the performance of a set of well-known MLNN algorithms (i.e., KNN, DT, RF, SVM, MLP, LSTM and SLP) based on (a modified version of) a real Uber parking dataset in San Francisco. Although the RF algorithm shows the best performance, its training is very time- and energy-consuming. On the other hand, we can see that the performance of some simple algorithms such as CNB and DT is very close to RF. In this situation, it can be a good idea to use a simple algorithm with a lower execution time.

For traffic prediction using machine learning, there are still many open research problems and challenges that have not been fully investigated. Finding appropriate solutions to consider abnormal conditions (e.g., extreme weather or temporary traffic control) is one of the current challenges. Benchmarking traffic prediction, the interpretability of models, real-time prediction and choosing an optimal network architecture are additional issues in using machine learning for traffic prediction [31].

7. Conclusions

We proposed a smart home use case (i.e., temperature adjustment by controlling the heating and cooling systems automatically). The main goal was to reduce energy consumption of the cooling and heating systems. A machine learning-based traffic prediction system was designed for the use case to estimate residents' arrival time based on the predicted traffic situation. It can also recommend residents what appropriate routes to take when they are in the city and want to return home. Unlike many works that use machine learning to track and predict the residents behavior inside their homes, we used machine learning to predict the residents behavior (i.e., arrival time) when they are outside their smart home (i.e., driving). We believe that predicting residents behavior outside smart homes can also help to improve energy consumption of some smart home applications. We also found that some simple algorithms (e.g., CNB and DT) can provide a very close performance to RF performance with a better execution time.

Author Contributions: Conceptualization, E.A.; Methodology, E.A.; Validation, X.F.; Formal analysis, X.F.; Investigation, X.F.; Writing—original draft, X.F. and E.A.; Writing—review & editing, E.A and G.M.L.; Supervision, E.A. and G.M.L. All authors have read and agreed to the published version of the manuscript.

Funding: This research received no external funding.

Conflicts of Interest: The authors declare no conflict of interest.

References

1. Gupta, A.; Badr, Y.; Negahban, A.; Qiu, R.G. Energy-efficient heating control for smart buildings with deep reinforcement learning. *J. Build. Eng.* **2021**, *34*, 101739. [CrossRef]
2. George, S.; Santra, A.K. Traffic Prediction Using Multifaceted Techniques: A Survey. *Wireless Pers. Commun.* **2020**, *115*, 1047–1106. [CrossRef]
3. Pedregosa, F.; Varoquaux, G.; Gramfort, A.; Michel, V.; Thirion, B.; Grisel, O.; Blondel, M.; Prettenhofer, P.; Weiss, R.; Dubourg, V.; et al. Scikit-learn: Machine Learning in Python. *J. Mach. Learn. Res.* **2011**, *12*, 2825–2830.
4. Paszke, A.; Gross, S.; Massa, F.; Lerer, A.; Bradbury, J.; Chanan, G.; Killeen, T.; Lin, Z.; Gimelshein, N.; Antiga, L.; et al. PyTorch: An Imperative Style, High-Performance Deep Learning Library. *Adv. Neural Inf. Process. Syst.* **2019**, *32*, 8024–8035.
5. Taiwo, O.; Ezugwu, A.E.; Oyelade, O.N.; Almutairi, M.S. Enhanced Intelligent Smart Home Control and Security System Based on Deep Learning Model. *Wirel. Commun. Mob. Comput.* **2022**, *2022*, 961. [CrossRef]
6. Filipe, L.; Peres, R.S.; Tavares, R.M. Voice-Activated Smart Home Controller Using Machine Learning. *IEEE Access* **2021**, *9*, 66852–66863. [CrossRef]
7. Fahim, M.; Kazm, S.M.A.; Khattak, A.M. ApplianceNet: A Neural Network Based Framework to Recognize Daily Life Activities and Behavior in Smart Home Using Smart Plugs. *Neural Comput. Appl.* **2022**, *34*, 12749–12763. [CrossRef]
8. Fakhar, M.Z.; Yalcin, E.; Bilge, A. A survey of smart home energy conservation techniques. *Expert Syst. Appl.* **2023**, *213*, 118974. [CrossRef]
9. Kim, S.; Kim, J.-H.; Yun, J.; Lee, S.H. Machine Learning-based Temperature Control for Smart Home Environment. In Proceedings of the EEECS2017, Dubrovnik, Croatia, 4–6 July 2017; pp. 35–39.
10. Okutani, I.; Stephanedes, Y.J. Dynamic forecasting of traffic volume through Kalman filtering theory. *Transp. Res. Part B* **1984**, *18*, 1–11. [CrossRef]
11. Ahmed, M.S.; Cook, A.R. Analysis of freeway traffic time-series data by using Box-Jenkins techniques. *Transp. Res. Rec.* **1979**, *722*, 1–9.
12. Navarro-Espinoza, A.; López-Bonilla, O.R.; García-Guerrero; E, E.; Tlelo-Cuautle, E.; López-Mancilla, D.; Hernández-Mejía, C.; Inzunza-González, E. Traffic Flow Prediction for Smart Traffic Lights Using Machine Learning Algorithms. *Technologies* **2022**, *10*, 5. [CrossRef]
13. Braz, F.J.; Ferreira, J.; Gonçalves, F.; Weege, K.; Almeida, J.; Baldo, F.; Gonçalves, P. Road Traffic Forecast Based on Meteorological Information through Deep Learning Methods. *Sensors* **2022**, *22*, 4485. [CrossRef]
14. Kim, Y.; Wang, P.; Mihaylova, L. Structural Recurrent Neural Network for Traffic Speed Prediction. In Proceedings of the ICASSP 2019—2019 IEEE International Conference on Acoustics, Speech and Signal Processing (ICASSP), Brighton, UK, 12–17 May 2019; pp. 5207–5211. [CrossRef]
15. SETA EU Project, A Ubiquitous Data and Service Ecosystem for Better Metropolitan Mobility, Horizon 2020 Program. 2016. Available online: http://setamobility.weebly.com/ (accessed on 2 July 2016).
16. Falessi, D.; Huang, J.; Narayana, L.; Thai, J. F.; Turhan, B. On the need of preserving order of data when validating within-project defect classifiers. *Empir. Softw. Eng.* **2020**, *25*, 4805–4830. [CrossRef]
17. Sarker, I.H. Machine Learning: Algorithms, Real-World Applications and Research Directions. *SN Comput. Sci.* **2021** *2*, 160. [CrossRef]
18. Awan, F.M.; Saleem, Y.; Minerva, R.; Crespi, N. A Comparative Analysis of Machine/Deep Learning Models for Parking Space Availability Prediction. *Sensors* **2020**, *20*, 322. [CrossRef]
19. Belavagi, M.C.; Muniyal, B. Performance Evaluation of Supervised Machine Learning Algorithms for Intrusion Detection, *Twelfth Int. Conf. Commun. Netw. ICCN* **2016**, *89*, 117–123.
20. Feng, H.; Wang, W.; Chen, B.; Zhang, X. Evaluation on Frozen Shellfish Quality by Blockchain Based Multi-Sensors Monitoring and SVM Algorithm During Cold Storage. *IEEE Access* **2020**, *8*, 54361–54370. [CrossRef]

21. Verbraeken, J.; Wolting, M.; Katzy, J.; Kloppenburg, J.; Verbelen, T.; Rellermeyer, J.S. A Survey on Distributed Machine Learning. *ACM Comput. Surv.* **2020**, *53*, 1–33. [CrossRef]
22. Kaviani1, K.; Dhotre, S. Short Survey on Naive Bayes Algorithm. *Int. J. Adv. Eng. Res. Dev.* **2017**, *4*, 607–611.
23. Abed, M.; İbrıkçı, T. Comparison between Machine Learning Algorithms in the Predicting the Onset of Diabetes. In Proceedings of the 2019 International Artificial Intelligence and Data Processing Symposium (IDAP), Malatya, Turkey, 21–22 September 2019; pp. 1–5. [CrossRef]
24. Marius, P.; Balas, V.E.; Perescu-Popescu, L.; Mastorakis, N.E. Multilayer perceptron and neural networks. *WSEAS Trans. Circuits Syst.* **2009**, *8*, 579–588.
25. Zeng, C.; Ma, C.; Wang, K.; Cui, Z. Parking Occupancy Prediction Method Based on Multi Factors and Stacked GRU-LSTM. *IEEE Access* **2022**, *10*, 47361–47370. [CrossRef]
26. Ozdemir, T.; Taher, F.; Ayinde, B.O.; Zurada, J.M.; Tuzun Ozmen, O. Comparison of Feedforward Perceptron Network with LSTM for Solar Cell Radiation Prediction. *Appl. Sci.* **2022**, *12*, 4463. [CrossRef]
27. Uber Technologies, Inc. Uber Movement. Available online: https://movement.uber.com (accessed on 2 July 2022).
28. Chen, X.; Sun, L. Bayesian Temporal Factorization for Multidimensional Time Series Prediction. *IEEE Trans. Pattern Anal. Mach. Intell.* **2021**, *44*, 4659–4673. [CrossRef] [PubMed]
29. Random-Search in Scikit Learn. Available online: https://scikit-learn.org/stable/modules/generated/sklearn.model_selection.RandomizedSearchCV.html (accessed on 2 July 2022).
30. Dia, I.; Ahvar, E.; Lee, G.M. Performance Evaluation of Machine Learning and Neural Network-Based Algorithms for Predicting Segment Availability in AIoT-Based Smart Parking. *Network* **2022**, *2*, 225–238. [CrossRef]
31. Yin, X.; Wu, G.; Wei, J.; Shen, Y.; Qi, H.; Yin, B. Deep Learning on Traffic Prediction: Methods, Analysis and Future Directions. *IEEE Trans. Intell. Transp. Syst.* **2021**, *23*, 4927–4943. [CrossRef]

Disclaimer/Publisher's Note: The statements, opinions and data contained in all publications are solely those of the individual author(s) and contributor(s) and not of MDPI and/or the editor(s). MDPI and/or the editor(s) disclaim responsibility for any injury to people or property resulting from any ideas, methods, instructions or products referred to in the content.

sensors

MDPI

Article

Comparison of Two Paradigms Based on Stimulation with Images in a Spelling Brain–Computer Interface

Ricardo Ron-Angevin [1,*], Álvaro Fernández-Rodríguez [1], Clara Dupont [2], Jeanne Maigrot [2], Juliette Meunier [2], Hugo Tavard [2], Véronique Lespinet-Najib [3] and Jean-Marc André [3]

1 Departamento de Tecnología Electrónica, Universidad de Málaga, 29071 Malaga, Spain
2 ENSC, Bordeaux INP, 33400 Bordeaux, France
3 Laboratoire IMS, CNRS UMR 5218, Cognitive Team, Bordeaux INP-ENSC, 33400 Talence, France
* Correspondence: rron@uma.es

Abstract: A P300-based speller can be used to control a home automation system via brain activity. Evaluation of the visual stimuli used in a P300-based speller is a common topic in the field of brain–computer interfaces (BCIs). The aim of the present work is to compare, using the usability approach, two types of stimuli that have provided high performance in previous studies. Twelve participants controlled a BCI under two conditions, which varied in terms of the type of stimulus employed: a red famous face surrounded by a white rectangle (RFW) and a range of neutral pictures (NPs). The usability approach included variables related to effectiveness (accuracy and information transfer rate), efficiency (stress and fatigue), and satisfaction (pleasantness and System Usability Scale and Affect Grid questionnaires). The results indicated that there were no significant differences in effectiveness, but the system that used NPs was reported as significantly more pleasant. Hence, since satisfaction variables should also be considered in systems that potential users are likely to employ regularly, the use of different NPs may be a more suitable option than the use of a single RFW for the development of a home automation system based on a visual P300-based speller.

Keywords: brain–computer interface (BCI); event-related potential (ERP); row-column paradigm (RCP); stimulus; face; picture

Citation: Ron-Angevin, R.; Fernández-Rodríguez, Á.; Dupont, C.; Maigrot, J.; Meunier, J.; Tavard, H.; Lespinet-Najib, V.; André, J.-M. Comparison of Two Paradigms Based on Stimulation with Images in a Spelling Brain–Computer Interface. *Sensors* **2023**, *23*, 1304. https://doi.org/10.3390/s23031304

Academic Editor: Daniele Cenni

Received: 15 December 2022
Revised: 11 January 2023
Accepted: 20 January 2023
Published: 23 January 2023

Copyright: © 2023 by the authors. Licensee MDPI, Basel, Switzerland. This article is an open access article distributed under the terms and conditions of the Creative Commons Attribution (CC BY) license (https://creativecommons.org/licenses/by/4.0/).

1. Introduction

The brain–computer interface (BCI) is a technology that uses brain activity to create a gateway between the brain and the external environment. This technology may be especially important for users with severe motor impairments such as amyotrophic lateral sclerosis (ALS) [1]. The most widely studied devices employing BCI technology are spellers, which enable communication through the selection of characters from a display. The most frequent control signal used by BCI spellers is brain activity recorded via electroencephalography (EEG) [2], as the main advantages of this method are that it captures brain signals in a non-invasive way, with a suitable temporal resolution [3]. Previous BCI spellers based on EEG have used specific signals, such as event-related potentials (ERPs), steady-state visual evoked potentials (SSVEPs), or sensorimotor rhythms (SMRs) (e.g., Zhang et al. [4], Nakanishi et al. [5], and Cao et al. [6], respectively). The EEG signals most commonly used by BCI spellers are the ERPs, and the present work therefore focuses on this signal. Of the ERPs, such as the P100, N200, or N400 components, the most important is the P300, and for this reason, ERP-based spellers are typically referred to as P300 spellers. P300 is a positive potential with a latency of roughly 300 ms, and it can be elicited in response to task-relevant stimuli [7].

Farwell and Donchin [8] designed the precursor to the P300 speller and were the first to develop this technology. Their speller was composed of a 6 × 6 matrix containing 36 characters (26 letters and 10 digits). In this paradigm, rows and columns are flashed (i.e., highlighted from grey to white) one by one, and this paradigm is therefore called the row-column paradigm (RCP). To select a character, the user pays attention to the flashing of a specific target character, as this acts as the task-relevant stimulus that elicits the P300 potential. Once the P300 has elicited a specific row and column, the BCI is able to determine the user's target character.

Improving the usability of the P300 speller is an important goal of the BCI community (see work by Pasqualotto et al. [9] and Guy et al. [10], for example). According to the Organization for Standardization, the definition of usability is composed of three dimensions: effectiveness, efficiency, and satisfaction [11]. Effectiveness represents the completeness and accuracy with which users can complete their objectives. To measure this parameter, error rates or the quality of the solution may be used. Efficiency relates to the resources employed to achieve goals, and indicators of efficiency include the task completion time and task learning time. The last dimension of usability is satisfaction, which reflects the user's attitude toward and comfort with the system. An attitude rating scale can be used to measure satisfaction, such as the System Usability Scale (SUS) [12]. These three measures should be studied separately, as they are independent aspects of usability [13].

In order to improve the performance of the P300 speller, several visual aspects associated with the presented stimuli have been studied. These visual factors have included the color (e.g., Ryan et al. [14]), the size (e.g., Kellicut-Jones & Sellers [15]), and the use of alternative stimuli, such as faces (e.g., Kaufmann et al. [16]). Kaufmann et al. [16] showed that the use of famous faces as stimuli that appear above the letters can improve performance compared to the use of classical grey-to-white flashing, since this type of stimulus can induce the apparition of several ERPs, notably N170, P300, and N400. Next, Q. Li et al. [17] showed that the use of famous faces colored in green as stimuli improved the performance versus naturally colored faces. It therefore appears that the use of artificially colored faces could provide performance improvements for the P300 speller. Following this, S. Li et al. [18] evaluated the use of red, green, and blue faces, and showed that the use of red faces as stimuli led to superior performance. Finally, Zhang et al. [4] combined the use of a famous face colored in red with a surrounding square of a specific color (white, blue, or red, depending on the experimental conditions). They showed that red faces surrounded by a white rectangle gave the best performance and are thus the best current visual stimulus for use in a P300 speller.

In addition to faces, alternative stimuli have been studied in regard to the control of a P300 speller. Fernández-Rodríguez et al. [19] investigated the emotional properties of pictures and their effects on BCI performance and found that the presentation of pictures above letters, regardless of their emotional content, resulted in better performance compared to the classic use of gray-to-white flashing letters. Unlike the pattern with familiar faces devised by Zhang et al. [4], different images were used for each letter in the paradigm presented by Fernández-Rodríguez et al. [19]. This feature could be useful in the design of P300 spellers, as pictograms could be used to improve and facilitate communication [20].

The aim of the present study is to compare two different paradigms that have already been proven to improve the performance of a P300 speller versus the use of standard gray-to-white flashing letters: (i) one inspired by Zhang et al. [4], in which a single red-colored famous face is displayed surrounded by a white rectangle, and (ii) one based on the scheme of Fernández-Rodríguez et al. [19], in which various neutral pictures are displayed. A preliminary study by Ron-Angevin et al. [21] (N = 4) found that both paradigms may be similar in terms of effectiveness. It was therefore considered appropriate to carry out a more comprehensive assessment of this comparison. In particular, the sample size was increased, and a more complete approach to usability was used, i.e., with measures of effectiveness, efficiency, and satisfaction. The main benefit of the usability approach is that it allows a more comprehensive assessment to be carried out by considering certain dimensions that may be of particular relevance for practical real-life scenarios in which users actually would want to use these interfaces. Moreover, our study permitted us to evaluate the issues of usability and user-friendliness, as suggested by Zhang et al. [4].

2. Methods

2.1. Participants

The study initially involved 14 healthy French-speaking students from the École Nationale Supérieure de Cognitique de Bordeaux—where the experiment was carried out—although two of these were discarded due to unusable EEG calibration data. The remaining sample contained 12 participants (aged 21.2 ± 1.5, four male and eight female, referred to here as S1 to S12). All of them had normal or corrected-to-normal vision, and they provided written consent. The study was approved by the Ethics Committee of the University of Malaga and met the ethical standards of the Helsinki Declaration. All participants were of legal age. According to self-reports, none of the participants had any history of neurological or psychiatric illness or were taking any medication regularly that could affect the results of the experiment. Before the experiment began, all subjects were informed of the experimental protocol and were able to stop it at any time.

2.2. Data Acquisition and Signal Processing

The EEG was recorded via eight electrodes: Fz, Cz, Pz, Oz, P3, P4, PO7, and PO8. All channels were referenced to the right earlobe, using FPz as ground. The signal was modulated by a 16-channel g.USBamp amplifier, which gave a bandwidth from 0.5 to 100 Hz with a sensitivity of 500 µV. A notch filter at 50 Hz was applied, and the EEG was digitized with a sampling frequency of 256 Hz. All aspects of EEG data collection and processing were controlled by BCI2000 software [22]. A stepwise linear discriminant analysis (SWLDA) of the EEG data was performed to obtain the weights for the P300 classifier, to calculate the accuracy, and to enable online spelling.

2.3. Spelling Paradigms

The software used to design the layout of the spelling paradigms was the UMA-BCI speller [23], which serves as a user-friendly frontend for BCI2000. Two paradigms were considered in the present study: (i) one involving red faces surrounded by a white rectangle (RFW) and (ii) one with neutral pictures (NP). Each paradigm was evaluated using a 5 × 5 matrix (showing the letters A–Y) using the RCP. The choice of this grid size was made to reduce the number of flashes and to increase the writing speed. As a result, the letter Z did not form part of this study, and none of the written words contained this letter. The background color was black, and the letters on which the stimulus appeared (i.e., faces or pictures, depending on the paradigm) were white. For both paradigms, a stimulus onset asynchrony (SOA) of 281.25 ms and an inter-stimulus interval (ISI) of 93.75 ms were used, meaning that each stimulus was presented for 187.5 ms. The stimulus size for both conditions was the same: 4.1 × 2.7 cm (2.35 × 1.55° at ~100 cm). In the following, the specific characteristics of the stimuli used in each of the paradigms will be described.

In the RFW paradigm, which was based on the proposal by Zhang et al. [4], the stimulus overlaying the white letters was a semi-transparent image of the face of Barack Obama, colored in red with a white surrounding rectangle (Figure 1a). In the NP paradigm, which was inspired by Fernández-Rodríguez et al. [19], each character of the matrix was flashed using a different neutral picture (Figure 1b). In order to ensure that the pictures were neutral, they were obtained from the International Affective Picture System (IAPS) [24]. Pictures with the lowest score for arousal level and that had the aforementioned size (i.e., images that filled all the space without black padding) were chosen.

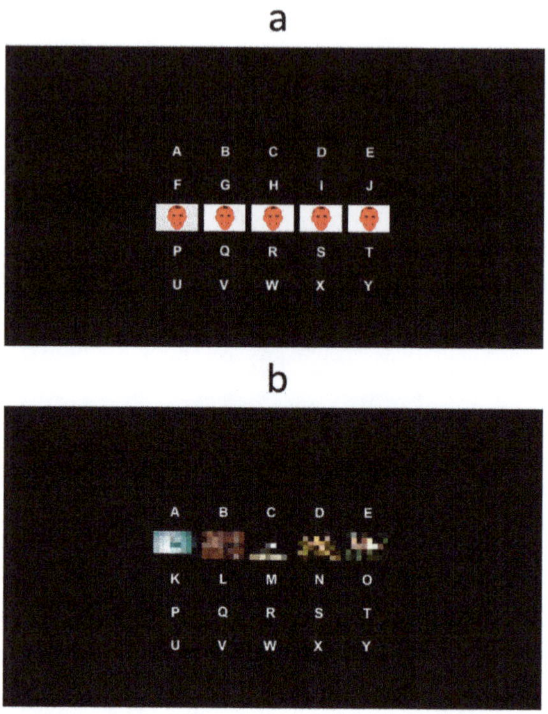

Figure 1. Two spelling paradigms were tested in the experiment, which varied in terms of the flashing stimulus used: (**a**) famous faces shown in red with a white rectangle (RFW), based on [4]; and (**b**) neutral pictures (NP), based on [19]. For copyright reasons, the neutral pictures have been pixelated here.

2.4. Procedure

This experiment was carried out in an isolated room. Participants sat a distance of approximately 1 m from the computer screen. Instructions were given in both written and verbal form, and each experiment was performed in a single session through an intrasubject design, during which the same participant used both paradigms (Figure 2). Also, in order to prevent the effect of extraneous variables that could influence the results—such as fatigue or learning—the order in which the paradigms were presented was counterbalanced across the participants.

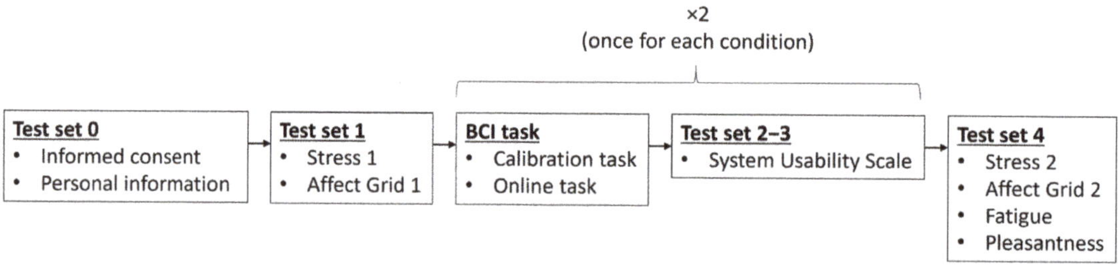

Figure 2. Sequence of questionnaires and tasks completed by each participant during the experimental session.

Immediately after explaining the experiment to the participant, obtaining written consent, and collecting the questionnaire used to gather personal information (such as gender, age, and technological experience) (Test Set 0), a questionnaire with one item related to stress level and the Affect Grid questionnaire were administered (Test Set 1). These elements will be described in detail in Section 2.5. Once these initial questionnaires had been completed, the BCI task could begin.

The evaluation of each paradigm consisted of two parts: (i) a calibration phase, in which the system was adapted to the user; and (ii) a copy-spelling phase, in which the participant wrote four words of four letters each. The subject was free to ask for a pause after each word, in addition to the mandatory pause between the two paradigms. Initially, during the calibration phase, participants were asked to select 12 characters, divided into three words of four letters, in each condition. The number of sequences (i.e., the number of times each row and column were highlighted) was fixed to 10, meaning that each letter was overlaid by a face or a picture (depending on the paradigm) 20 times. The three specific French words used for calibration were: "*FEUX*" (fire), "*CHAT*" (cat), and "*PURE*" (pure). No feedback on the correctness of the spelling was provided to the participants, since the purpose of this phase was to create the classifier. At the end of the calibration phase, a SWLDA was performed to obtain the weights of the P300 classifier. Following this, the copy-spelling phase involved the participant spelling four specific words: "*ABRI*" (shelter), "*LUNE*" (moon), "*YOGA*" (yoga), and "*CHEF*" (boss). The number of sequences used during this phase to select characters was that in which the user obtained the second consecutive best accuracy in the calibration task. In cases where the maximum accuracy was not repeated consecutively or there was only one, the first best sequence was selected. During this phase, each time a participant selected a character, it was displayed at the top of the screen. An incorrect letter could be selected, but the subject was warned that he/she had to continue with the next one.

After completing the copy-spelling phase using each paradigm, the user was asked to complete the SUS (Test Sets 2 and 3). Finally, after finishing all the BCI tasks, the user was again asked to answer the item related to stress level (to measure the change produced by the BCI session) and the Affect Grid, and, for the first time, an ad hoc item concerning fatigue and two items related to the level of pleasantness for each paradigm (Test Set 4).

2.5. Evaluation

As described in the introduction, our goal was to evaluate the usability and the performance of two different conditions: famous faces shown in red on a white background (RFW) and neutral pictures (NP). In order to carry out this evaluation, we built our analysis on three dimensions: (i) effectiveness, (ii) efficiency, and (iii) satisfaction.

2.5.1. Effectiveness

Effectiveness denotes the performance with which the user succeeds in the task. Three specific variables were considered in the evaluation of effectiveness: (i) the number of sequences used; (ii) the accuracy; and (iii) the information transfer rate (ITR, bit/min), based on the formula presented by Wolpaw et al. [25]. The ITR is an objective measure used to determine the communication speed of the system, and is expressed as:

$$\text{ITR} = \frac{\log_2 N + P \log_2 P + (1-P)\log_2 \frac{1-P}{N-1}}{T}$$

where P denotes the classification accuracy, N denotes the number of available characters in the interface (25 in this experiment), and T denotes the time needed to perform a selection.

Since the performance of the system depends on the recorded EEG signal, it would be interesting to study the grand average of the ERP waveforms (μV). More specifically, the larger the differences between the amplitudes for the target and non-target stimuli in a given paradigm, the easier it will be for the classifier to discriminate between them. Hence, the two conditions (NP and RFW) were compared using the amplitude difference as a variable (μV, amplitude of the target stimulus minus the amplitude of the non-target stimulus, from 0 to 800 ms). To perform this analysis, a baseline ranging from −200 to 0 ms and low-pass filtered at 30 Hz was used for each electrode.

2.5.2. Efficiency

Efficiency is related to the number of resources needed by the participant to achieve the required task. Two specific variables related to this dimension were collected at specific times during the experiment: (i) stress and (ii) fatigue. The stress level was measured before and after the BCI tasks using the following item (during Test Sets 1 and 4): "Evaluate your current stress level". The users' responses were graded on a scale of one to five, with five indicating the highest stress level. Fatigue was measured through two different ad hoc items after performing all the BCI tasks (during Test Set 4). First, users were asked about their level of fatigue: "To what extent did you find this experiment tiring?" The response to this item was on a scale of one to five, with five representing the highest level of fatigue. The second item related to fatigue aimed to find out which paradigm had been the most exhausting for the user: "Which paradigm would you say was the most exhausting?" There were three possible answers to this item: RFW, NP, or equivalent.

2.5.3. Satisfaction

Satisfaction is related to the participants' attitudes on the system, i.e., their perceived comfort and their well-being. These two notions were explored using three questionnaires: (i) the SUS (Test Sets 2 and 3), (ii) the Affect Grid (during Test Set 4), and (iii) an ad hoc item related to pleasantness (during Test Set 4).

The SUS is commonly used to collect a user's subjective rating of the usability of a product in a quick and easy format [12]. In the present study, participants were asked to answer the questionnaire after completing the copy-spelling task in both paradigms. This survey contained questions about the convenience, the complexity, and the integration of the functionalities in the interface. The answers permitted us to calculate a global score (of between zero and 100). Bangor et al. [26] explained how to interpret a SUS score of this type. In order to clarify the meaning of this score, seven terms are used: worst imaginable, poor, acceptable, good, excellent, and best imaginable. The minimal score for acceptance is 50.9; however, the probability of acceptance for the system is low when scores fall between this minimal score and about 63. Above 63, the probability is rather higher, and above 71.4, the system's usability is judged to be good and commonly accepted. The best usability is obtained for a score of above 90.9.

The Affect Grid was developed to quickly assess the current mood of a subject [27]. This questionnaire was administered twice during the experiment, that is, before and after the BCI tasks (Figure 2). Therefore, the purpose of this questionnaire was not so much to establish a comparison between the two paradigms used (RFW versus NP) but between before and after the performance of the BCI tasks in order to assess the effect it could have on the user's condition. The matrix employed by the Affect Grid should be read along two axes (Figure 3). The abscissa refers to the pleasure level (from unpleasant to pleasant feelings), while the ordinate represents the arousal level (from sleepiness to arousal). Participants were asked to choose the number that best matched their current state. The four corners of the grid correspond to specific feelings: stress, excitement, depression, and relaxation. As explained by Russell et al. [27], two parameters can be extracted from the Affect Grid: the pleasure score (P) and the arousal score (A). The P score corresponds to the column of the number chosen by the participant (between one and nine, counting from the left), while the A score corresponds to the row (between one and nine, counting from the bottom). For example, if a participant estimates his or her mood as corresponding to the number 24, the P and A scores are equal to six and seven, respectively.

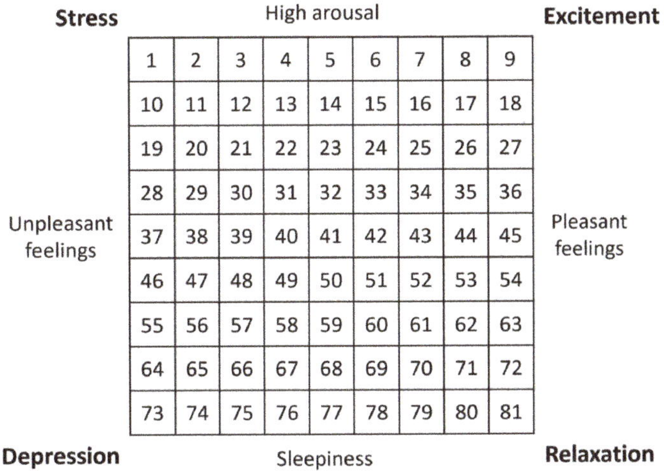

Figure 3. The matrix employed in the Affect Grid questionnaire to indicate the participant's affective state using two dimensions (pleasantness and arousal).

Finally, an ad hoc item was used to find out how pleasant the participant found the use of the system under each of the paradigms ("How pleasant was the keyboard to use?"). This item was answered after completing all the BCI tasks and was measured on a scale of one to five points, with five indicating maximum pleasure.

2.6. Statistical Analysis

Several statistical analyses were carried out. The aims were to study the potential differences between the two conditions (RFW and NP) and to examine the number of sequences, accuracy, ITR, ERP waveform, the most tiring condition, SUS score, and pleasantness score. There was also one variable (the user's stress level) that was not compared between conditions, but it was measured at the beginning and end of the experimental session to evaluate whether it was influenced by the overall execution of the experiment.

With the exception of the ERP waveform analysis, the rest of the variables were analyzed using SPSS software (v24) [28]. Before proceeding with each analysis, the assumption of normality was checked before proceeding to a Student's *t*-test for repeated samples or a Wilcoxon test (except for the most tiring condition variable), depending on whether the criterion was met or not, respectively. The variable that explored which condition had been more tiring was evaluated through a binomial test of the proportion of users who chose one condition over the other. For the ERP waveform, statistical analyses were carried out using EEGLAB (v2022.0) [29], where the options to use parametric statistics and the false discovery rate (FDR) as a correction method for multiple comparisons were selected.

3. Results

3.1. Effectiveness

3.1.1. BCI Performance

Table 1 shows the number of sequences used, the accuracy, and the ITR for each participant and condition. The average number of sequences used during the copy-spelling phase was 4.83 ± 2.04 for the RFW, and 3.58 ± 1.38 for NP. The average accuracy was 97.42 ± 4.23% for RFW, and 94.83 ± 5.95% for NP. Finally, the average ITR was 22.94 ± 11 bit/min for RFW, and 27.92 ± 10.47 bit/min for NP. No significant differences were found in relation to any of the variables reported: number of sequences ($Z = 1.621$; $p = 0.105$), accuracy ($Z = 1.089$; $p = 0.276$), or ITR ($Z = 0.941$; $p = 0.347$).

Table 1. Results for each participant, and average (±standard deviation), in the copy-spelling phase in terms of number of sequences used, accuracy (%), and ITR (information transfer rate, bit/min) for each condition (red faces surrounded by a white rectangle (RFW) and neutral pictures (NP)).

User	Number of Sequences		Accuracy (%)		ITR (bit/min)	
	RFW	NP	RFW	NP	RFW	NP
S1	3	7	100	100	33.02	14.15
S2	6	2	87	100	12.41	49.53
S3	2	3	100	100	49.53	33.02
S4	5	5	100	81	19.81	13.11
S5	10	4	100	94	9.91	21.55
S6	4	3	94	94	21.55	28.74
S7	5	4	100	100	19.81	24.77
S8	5	3	94	94	17.24	28.74
S9	5	3	100	94	19.81	28.74
S10	3	4	100	100	33.02	24.77
S11	6	2	94	94	14.37	43.11
S12	4	3	100	87	24.77	24.82
Average	4.83 ± 2.04	3.58 ± 1.38	97.42 ± 4.23	94.83 ± 5.95	22.94 ± 11	27.92 ± 10.47

3.1.2. ERP Waveform

The ERP waveforms found were within the expected range for a visual ERP-based BCI under the RCP condition. The main component to note was a strong positivity of around 350–600 ms, which was generalized along the scalp surface (Figure 4). Some channels around the parietooccipital region (P4, PO7, PO8, and Oz) also showed positivity (~200 ms), which was immediately followed by a pronounced negativity (~300 ms).

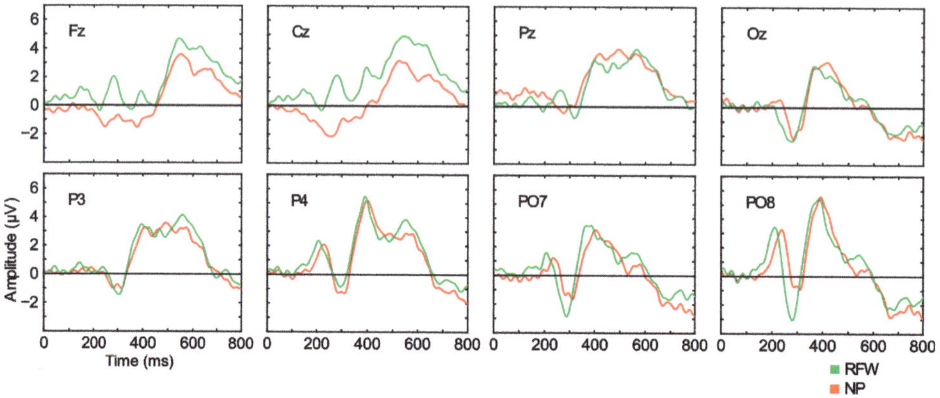

Figure 4. Grand average event-related potential (ERP) waveforms for the amplitude differences (μV) between the target and non-target stimuli for each condition (red faces surrounded by a white rectangle (RFW) and neutral pictures (NP)) and channel (Fz, Cz, Pz, Oz, P3, P4, PO7, and PO8), represented over the time interval from 0 to 800 ms. No significant differences were found between the target and non-target stimuli for any of the channels.

A comparison of the conditions showed no significant differences in any of the registered channels. These results can help us to understand the lack of significant differences in the BCI performance; if the EEG signal does not differ significantly between conditions (RFW and NP), it is difficult for the BCI performance to do so. Nevertheless, a tendency can be observed in the frontal channels (Fz and Cz) of RFW towards higher positivity throughout the temporal interval studied compared to NP, as well as a stronger negative peak in PO7 and PO8 for RFW. The interpretation of the possible components and a comparison with the related literature will be discussed in a later section.

3.2. Efficiency

3.2.1. Stress

From the item that measured the participant's current stress level, we saw that the average stress score before performing any of the BCI tasks was equal to 1.83 ± 1.19 points, whereas this score was equal to 1.58 ± 0.67 points after completing all the BCI tasks. There were no significant differences between the stress levels before and after the experience ($Z = 1.089$; $p = 0.276$).

3.2.2. Fatigue

From the item measuring how tiring the participant found the experiment, we saw that they scored it with an average of 3.42 ± 1 points, on a scale from one to five. From the item asking participants to select the most exhausting condition, the following results were obtained: eight participants chose the RFW condition, three chose the NP condition, and one found the two conditions to be equivalent. A binomial test indicated that there was no statistically significant difference between the proportion of participants who chose one paradigm as the most tiring (eight out of 11 for RFW and three out of 11 for NP; $Z = 1.206$; $p = 0.227$).

3.3. Satisfaction

3.3.1. System Usability Scale

The average scores obtained on the SUS after completing the copy-spelling phase for each paradigm were 66.04 ± 12.27 points for RFW and 66.04 ± 12.18 points for NP. Hence, no significant differences were found between the two conditions ($Z = 0.655$; $p = 0.512$).

3.3.2. Affect Grid

Data on each participant's state of mind, gathered via the Affect Grid before and after all BCI tasks, were analyzed. More precisely, the pleasure and arousal scores were investigated. Figure 5 shows that before the experiment (red points), users' states were scattered across different regions of the grid; however, after the experiment (blue points), their states were all within the same region, which was related to a state of relaxation (shown with a blue dashed outline in Figure 5).

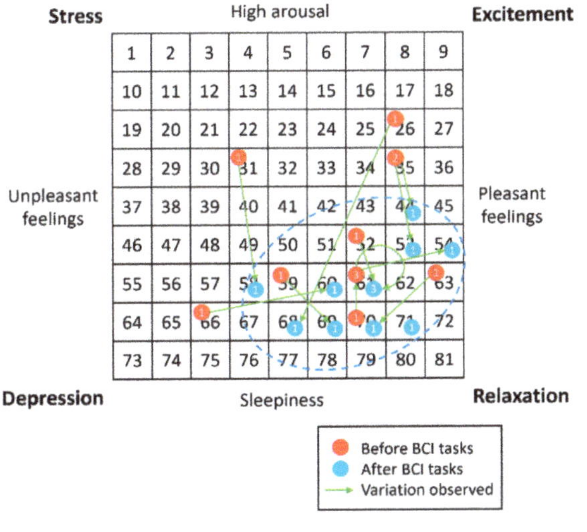

Figure 5. Results for each participant on the Affect Grid questionnaire, before and after the brain–computer interface (BCI) tasks. The number inside each circle refers to the number of participants who obtained that score.

3.3.3. Pleasantness

After completing all the BCI tasks, participants rated how pleasant the use of RFW and NP had been. On average, RFW scored 2.67 $\pm$ 0.89 points, while NP scored 3.92 $\pm$ 0.79. In this case, significant differences between conditions were found ($Z = 2.319$; $p = 0.02$). It can therefore be stated that NP was found to be more pleasant to use than RFW.

4. Discussion

The aim of this work was to compare two paradigms that had been previously studied, referred to here as RFW and NP. Zhang et al. [4] showed that flashing red faces with a white background led to better results than other studies using famous faces. However, Fernández-Rodríguez et al. [19] examined the valence of pictures and their performance on the P300 speller and proved that the use of pictures, both neutral and with a valence, increased the performance versus the standard flashing of letters from grey to white. Hence, a comparison of both paradigms for the first time might offer valuable insights.

In addition to this comparison, it is important to study the usability of this device, since this has not been conducted in previous studies. The inclusion of usability concerns in the development of a P300 speller could be valuable for research on specific potential users (e.g., ALS patients). In this study, three dimensions of usability were analyzed: (i) effectiveness, (ii) efficiency, and (iii) satisfaction. Our study provides a comparison of RFW and NP based on these dimensions. The following paragraphs discuss the results for each dimension.

4.1. Effectiveness

4.1.1. BCI Performance

The present work confirmed the preliminary results reported by Ron-Angevin et al. [21], in that there was no significant difference between the two paradigms in terms of performance. Thus, performance is not a factor that affects the decision on which type of stimulus is more appropriate for the control of a visual P300 speller. In the following, the three variables used to evaluate performance will be discussed: the number of sequences, accuracy, and ITR.

The number of sequences chosen for the copy-spelling phase corresponds to the number of sequences needed to guarantee adequate accuracy, based on the performance of the classifier in the calibration task. Although the results from each paradigm were similar, fewer sequences were necessary to achieve an accuracy of 100% in the NP experiment than in the RFW one. Thus, more sequences were used in the RFW condition than in NP in the copy-spelling phase (4.83 and 3.58 sequences, respectively). The number of sequences could affect other variables related to the writing time, such as the ITR. Hence, as long as the accuracy remains acceptable, a decrease in the number of sequences used should be considered positive, as it will reduce the time required for each selection and will speed up the control of the interface, which could also improve the user experience.

The results for the accuracy also did not indicate that one paradigm was better than the other, with a difference of only 1.5% in favor of RFW versus NP (97.42% and 94.83%, respectively). The question of whether this increase in the case of the RFW condition compensates for the fact that the user needed to use on average 1.25 sequences more than in the NP condition (4.83 and 3.58, respectively) therefore needs to be answered, and we attempt to do this using the following metric.

The amount of information transmitted in a given period of time impacts the quality of communication, so a high ITR is necessary to improve the performance of BCIs. This variable did not show significant differences between conditions. However, it should be noted that the ITR of NP was 21.71% (4.98 bit/min) higher than that of RFW. Hence, although it was not significant, NP had a somewhat higher ITR than RFW (27.92 and 22.94 bit/min, respectively).

Our results for the effectiveness are compared with those of Zhang et al. [4] and Fernández-Rodríguez et al. [19] in Table 2. Since we used the stimuli proposed in both these studies, it would be interesting to compare the accuracy and ITR for the three schemes. The results reported by Zhang et al. [4] and our results were almost the same for the accuracy (96.94% and 97.42%, respectively); we therefore note that their results were replicated and that a RFW paradigm is indeed able to provide high levels of accuracy. Since we found that there were no significant differences between the RFW and NP paradigms, it can also be concluded that the NP paradigm had a high classification accuracy. However, the results obtained by Fernández-Rodríguez et al. [19] and the present study for the NP paradigm were quite different, in terms of both accuracy and ITR (Table 2). This could be explained by the differences between the studies, since the conditions differed in certain respects such as the number of possible selectable elements and the size of the interface, parameters that could influence the performance of a P300 speller [30,31].

Table 2. Comparison of average performance for the conditions considered here and the results of other studies.

Work	Accuracy (%)	ITR (bit/min)
Present study: red faces with a white rectangle (RFW)	97.42	22.94
Present study: neutral pictures (NP)	94.83	27.92
Zhang et al. [4]	96.94	-
Fernández-Rodríguez et al. [19]	99.17	42.6

4.1.2. ERP Waveform

As mentioned in the Results section, the ERP waveforms found in this study were in line with those in the literature. However, it should be noted that researchers focusing on BCIs aim to optimize system performance and maximize the user experience, and these devices are not intended to provide theoretical answers about ERP components; neither the presentation paradigm nor the protocol were designed for this purpose. This means that the interpretations provided below must be considered carefully. The positivity found at around 350–600 ms, which was generalized across the entire scalp surface, was probably a P300 signal, the component that gives its name to this type of BCI system (i.e., a P300 speller). This positivity has been also reported at similar temporal intervals in several other studies (e.g., by Kellicut-Jones and Sellers [15] and M. Li et al. [32]). The negativity found for the occipital areas (PO7, PO8, and Oz) at around 280 ms could be an N170 signal (e.g., Kaufmann et al. [16], Q. Li et al. [17], and Lu et al. [33]). The N170 has been associated with image processing, and particularly with face processing (e.g., Eimer [34] and Tian et al. [35]). This would make sense considering that this component was larger for RFW than for NP (although this was not a significant effect).

4.2. Efficiency

4.2.1. Stress

Regarding the items related to stress, it should be noted that most of subjects (nine out of 12) had the same level of stress before and after the BCI tasks. Of the three participants whose stress scores varied, one (participant S2) became more stressed after the experiment (from one to two points) and the two others (participants S3 and S11) were more stressed before the experiment (with stress levels of five and three points, respectively) and became more relaxed afterwards (three and one points, respectively). However, the least stressed subjects did not achieve better results in terms of performance. Despite some variations, we conclude that this experiment had no effect on stress, as demonstrated by the statistical analysis ($p = 0.276$). The average stress scores before and after the experiment were 1.83 and 1.58 points, respectively.

4.2.2. Fatigue

The average level of fatigue was higher than three points (3.42 ± 1) on a scale from one to five, which could indicate that participants found the experiment tiring. Participants

S1, S3, S10, S11, and S12 felt very tired (four or five points out of five). These results are important, since the use of BCI systems may be challenging and require a lot of effort, especially for their target population (i.e., people with severely impaired motor skills). Thus, reducing the impact on fatigue must be a priority in future studies. If one of the two paradigms had to be chosen, the NP paradigm should be considered, even though the differences were not significant, since 66.67% of the participants found it less exhausting than the RFW paradigm.

4.3. *Satisfaction*

4.3.1. System Usability Scale

According to Bangor et al. [26], the scores obtained from the SUS allow us to establish a series of thresholds to classify the usability of the system ("worst imaginable", "poor", "acceptable", "good", "excellent", and "best imaginable"). In the present study, the two paradigms showed similar scores (66.04 points for both conditions). In particular, since both of these scores were above 50.9 points, each of the two conditions can be labeled as acceptable (the threshold for labeling the system as "good" is 71.4). There is still a wide margin for improvement to increase the usability of these systems, and, as stated above, future studies should make efforts in this direction, rather than only focusing on the dimensions related to performance.

4.3.2. Affect Grid

As shown in Figure 5, after the experiment, the results for all subjects were gathered in the same region. This region corresponds to pleasant feelings (high pleasure), sleepiness (low arousal), or even relaxation (high pleasure with low arousal) for some participants, meaning that in general, participants tended towards a state of well-being at the end of the experiment. Although most participants were more tired after the experiment, we also noted that some of them felt better (moving from a depressed state to a lightly relaxed one). For some, there was a reduction in their well-being, but they were still in a neutral/relaxed state after the experiment. However, the state of the users generally improved (from stress to relaxation). It is possible that this improvement towards a relaxed state was due to the initial excitement of the participant when entering an experimental scenario, and that as he/she became habituated to the context and acquired confidence in using the system, this state gradually improved. In short, it seems that using the system at least did not worsen the user's initial state.

4.3.3. Pleasantness

The NP paradigm turned out to be a more pleasant paradigm than RFW (3.92 and 2.67 points, respectively), and significant differences were found between the two. This result is particularly interesting, since a P300 speller is intended to establish a communication channel for people with severe motor impairments who cannot use speech or other alternative methods requiring muscle mobility [1]. Thus, these systems must be designed for daily use, meaning that the designer should not only consider performance measures, but also others such as the user's pleasure.

The study by Zhang et al. [4] presented a type of visual stimulus—those employed in our RFW paradigm—that offered promising accuracy performance on a P300-BCI visual. Nevertheless, the RFW paradigm may be less usable than the NP paradigm. Indeed, as mentioned above, in the NP paradigm, a pictogram or command can be used to facilitate communication through the system. Our findings for satisfaction support the view that the NP paradigm could have more advantages than the RFW paradigm.

5. Conclusions

The aim of the present work is to assess the use of two different types of stimuli in a visual ERP-based BCI under RCP for communication purposes: red faces surrounded by a white rectangle, which was reported to yield the best performance in previous work [4], and neutral pictures, which have also shown promising performance in a prior study [19]. In general terms, it seems that both types of stimuli tend to receive a positive evaluation for objective and subjective measurements; this means that both could be useful stimuli when implemented in a P300 speller to control such equipment as a home automation system or any other application that can improve the quality of life of patients with severe motor impairments. However, NP was found to be significantly more pleasant to use than RFW. Hence, in the absence of significant differences in measures related to performance, it seems appropriate that the choice of one paradigm over another should depend on subjective parameters, such as the pleasantness felt while using the system. After all, systems such as those for home automation are intended to be controlled by patients on a daily basis, so it is important to improve the overall user experience (since a system that is not pleasant to use may not be attractive for daily use, even if it gives adequate performance). Likewise, other characteristics such as the specific purpose of the system should be considered; for example, the NP paradigm allows for the use of different images such as pictograms as stimuli, and when these are related to the desired command (e.g., an image of a television if the user wants to use it) the system may be more intuitive and comfortable than one that uses unrelated images.

In short, the present work demonstrates the usefulness of the NP paradigm in terms of the higher pleasure of the user during system control compared to RFW, with no differences in performance. It also demonstrates the importance of evaluating not only performance but also the factors affecting the complete experience through a usability analysis, which is especially important to offer this technology to potential users in their daily lives.

Author Contributions: Conceptualization, R.R.-A. and V.L.-N.; methodology, R.R.-A. and V.L.-N.; formal analysis, Á.F.-R.; investigation, R.R.-A., C.D., J.M. (Jeanne Maigrot), J.M. (Juliette Meunier), H.T., V.L.-N. and J.-M.A.; resources, R.R.-A., V.L.-N. and J.-M.A.; data curation, Á.F.-R.; writing—original draft preparation, R.R.-A., Á.F.-R., C.D., J.M. (Jeanne Maigrot), J.M. (Juliette Meunier) and H.T.; writing—review and editing, V.L.-N. and J.-M.A.; visualization, Á.F.-R.; supervision, R.R.-A. and V.L.-N.; project administration, R.R.-A.; funding acquisition, R.R.-A. and J.-M.A. All authors have read and agreed to the published version of the manuscript.

Funding: This work was partially supported by the project PID2021-127261OB-I00 (SICODIS), funded by MCIN (Ministerio de Ciencia e Innovación)/AEI (Agencia Estatal de Investigación)/10.13039/501100011033/FEDER, UE (Fondo Europeo de Desarrollo Regional). The work was also partially supported by the University of Málaga (Universidad de Málaga).

Institutional Review Board Statement: The study was conducted in accordance with the Declaration of Helsinki and approved by the Ethics Committee of the University of Malaga (51-2019-H, 18 February 2020).

Informed Consent Statement: Informed consent was obtained from all subjects involved in the study.

Data Availability Statement: The data presented in this study are available on request from the corresponding author.

Acknowledgments: The authors would like to thank all participants for their cooperation.

Conflicts of Interest: The authors declare no conflict of interest. The funders had no role in the design of the study; in the collection, analyses, or interpretation of data; in the writing of the manuscript; or in the decision to publish the results.

References

1. Wolpaw, J.R.; Birbaumer, N.; McFarland, D.J.; Pfurtscheller, G.; Vaughan, T.M. Brain-computer interfaces for communication and control. *Clin. Neurophysiol.* **2002**, *113*, 767–791. [CrossRef]
2. Rezeika, A.; Benda, M.; Stawicki, P.; Gembler, F.; Saboor, A.; Volosyak, I. Brain-computer interface spellers: A review. *Brain Sci.* **2018**, *8*, 57. [CrossRef]
3. Nicolas-Alonso, L.F.; Gomez-Gil, J. Brain computer interfaces, a review. *Sensors* **2012**, *12*, 1211–1279. [CrossRef]
4. Zhang, X.; Jin, J.; Li, S.; Wang, X.; Cichocki, A. Evaluation of color modulation in visual P300-speller using new stimulus patterns. *Cogn. Neurodyn.* **2021**, *15*, 873–886. [CrossRef] [PubMed]

5. Nakanishi, M.; Wang, Y.; Chen, X.; Wang, Y.T.; Gao, X.; Jung, T.-P. Enhancing detection of SSVEPs for a high-speed brain speller using task-related component analysis. *IEEE Trans. Biomed. Eng.* **2018**, *65*, 104–112. [CrossRef] [PubMed]
6. Cao, L.; Xia, B.; Maysam, O.; Li, J.; Xie, H.; Birbaumer, N. A synchronous motor imagery based neural physiological paradigm for brain computer interface speller. *Front. Hum. Neurosci.* **2017**, *11*, 274. [CrossRef] [PubMed]
7. Polich, J. Updating P300: An integrative theory of P3a and P3b. *Clin. Neurophysiol.* **2007**, *118*, 2128–2148. [CrossRef]
8. Farwell, L.A.; Donchin, E. Talking off the top of your head: Toward a mental prosthesis utilizing event-related brain potentials. *Electroencephalogr. Clin. Neurophysiol.* **1988**, *70*, 510–523. [CrossRef] [PubMed]
9. Pasqualotto, E.; Simonetta, A.; Federici, S.; Olivetti Belardinelli, M. Usability evaluation of BCIs. *Assist. Technol. Res. Ser.* **2009**, *25*, 882. [CrossRef]
10. Guy, V.; Soriani, M.H.; Bruno, M.; Papadopoulo, T.; Desnuelle, C.; Clerc, M. Brain computer interface with the P300 speller: Usability for disabled people with amyotrophic lateral sclerosis. *Ann. Phys. Rehabil. Med.* **2018**, *61*, 5–11. [CrossRef]
11. *ISO 9241-11:2018*; Ergonomics of Human-System Interaction—Part 11: Usability: Definitions and Concepts. International Organization for Standardization: Geneva, Switzerland, 2018.
12. Bangor, A.; Kortum, P.T.; Miller, J.T. An empirical evaluation of the system usability scale. *Int. J. Hum. Comput. Interact.* **2008**, *24*, 574–594. [CrossRef]
13. Frøkjaer, E.; Hertzum, M.; Hornbaek, K. Measuring Usability: Are Effectiveness, Efficiency, and Satisfaction Really Correlated? In *CHI Letters, Proceedings of CHI 2000*; Association for Computing Machinery: New York, NY, USA, 2000. [CrossRef]
14. Ryan, D.B.; Townsend, G.; Gates, N.A.; Colwell, K.; Sellers, E.W. Evaluating brain-computer interface performance using color in the P300 checkerboard speller. *Clin. Neurophysiol.* **2017**, *128*, 2050–2057. [CrossRef] [PubMed]
15. Kellicut-Jones, M.R.; Sellers, E.W. P300 brain-computer interface: Comparing faces to size matched non-face stimuli. *Brain-Comput. Interfaces* **2018**, *5*, 30–39. [CrossRef]
16. Kaufmann, T.; Schulz, S.M.; Grünzinger, C.; Kübler, A. Flashing characters with famous faces improves ERP-based brain-computer interface performance. *J. Neural Eng.* **2011**, *8*, 056016. [CrossRef] [PubMed]
17. Li, Q.; Liu, S.; Li, J.; Bai, O. Use of a green familiar faces paradigm improves P300-speller brain-computer interface performance. *PLoS ONE* **2015**, *10*, e0130325. [CrossRef] [PubMed]
18. Li, S.; Jin, J.; Daly, I.; Zuo, C.; Wang, X.; Cichocki, A. Comparison of the ERP-Based BCI Performance Among Chromatic (RGB) Semitransparent Face Patterns. *Front. Neurosci.* **2020**, *14*, 54. [CrossRef]
19. Fernández-Rodríguez, Á.; Velasco- Álvarez, F.; Medina-Juliá, M.T.; Ron-Angevin, R. Evaluation of emotional and neutral pictures as flashing stimuli using a P300 brain-computer interface speller. *J. Neural Eng.* **2019**, *16*, 056024. [CrossRef]
20. Bühler, D.; Hemmert, F.; Hurtienne, J.; Petersen, C. Designing Universal and Intuitive Pictograms (UIPP)—A Detailed Process for More Suitable Visual Representations. *Int. J. Hum. Comput. Stud.* **2022**, *163*, 102816. [CrossRef]
21. Ron-Angevin, R.; Beasse, J.; Adrien, C.; Dupont, C.; Gall, M.L.; Meunier, J.; Lespinet-Najib, V.; Jean-Marc, A. Comparison of Two Paradigms Based on Stimulation with Images in a Spelling Brain-Computer Interface. In Proceedings of the BRAININFO 2022: The Seventh International Conference on Neuroscience and Cognitive Brain Information, Venice, Italy, 22–26 May 2022.
22. Schalk, G.; McFarland, D.J.; Hinterberger, T.; Birbaumer, N.; Wolpaw, J.R. BCI2000: A general-purpose brain-computer interface (BCI) system. *IEEE Trans. Biomed. Eng.* **2004**, *51*, 1034–1043. [CrossRef]
23. Velasco-Álvarez, F.; Sancha-Ros, S.; García-Garaluz, E.; Fernández-Rodríguez, Á.; Medina-Juliá, M.T.; Ron-Angevin, R. UMA-BCI Speller: An Easily Configurable P300 Speller Tool for End Users. *Comput. Methods Programs Biomed.* **2019**, *172*, 127–138. [CrossRef]
24. Lang, P.J.; Bradley, M.M.; Cuthbert, B.N. *International Affective Picture System (IAPS): Affective Ratings of Pictures and Instruction Manual*; Technical Report A-8; University of Florida: Gainesville, FL, USA, 2008.
25. Wolpaw, J.R.; Ramoser, H.; McFarland, D.J.; Pfurtscheller, G. EEG-based communication: Improved accuracy by response verification. *IEEE Trans. Rehabil. Eng.* **1998**, *6*, 326–333. [CrossRef] [PubMed]
26. Bangor, A.; Kortum, P.; Miller, J. Determining what individual SUS scores mean: Adding an adjective rating scale. *J. Usability Stud.* **2009**, *4*, 114–123.
27. Russell, J.A.; Weiss, A.; Mendelsohn, G.A. Affect Grid: A Single-Item Scale of Pleasure and Arousal. *J. Personal. Soc. Psychol.* **1989**, *57*, 493–502. [CrossRef]
28. IBM Corp. *IBM SPSS Statistics for Windows*; Version 24.0; IBM Corp.: Armonk, NY, USA, 2016.
29. Delorme, A.; Makeig, S. EEGLAB: An open source toolbox for analysis of single-trial EEG dynamics including independent component analysis. *J. Neurosci. Methods* **2004**, *134*, 9–21. [CrossRef] [PubMed]
30. Sellers, E.W.; Krusienski, D.J.; McFarland, D.J.; Vaughan, T.M.; Wolpaw, J.R. A P300 event-related potential brain-computer interface (BCI): The effects of matrix size and inter stimulus interval on performance. *Biol. Psychol.* **2006**, *73*, 242–252. [CrossRef] [PubMed]
31. Ron-Angevin, R.; Garcia, L.; Fernández-Rodríguez, Á.; Saracco, J.; André, J.M.; Lespinet-Najib, V. Impact of Speller Size on a Visual P300 Brain-Computer Interface (BCI) System under Two Conditions of Constraint for Eye Movement. *Comput. Intell. Neurosci.* **2019**, *2019*, 7876248. [CrossRef] [PubMed]
32. Li, M.; Yang, G.; Liu, Z.; Gong, M.; Xu, G.; Lin, F. The Effect of SOA on An Asynchronous ERP and VEP-Based BCI. *IEEE Access* **2021**, *9*, 9972–9981. [CrossRef]
33. Lu, Z.; Li, Q.; Gao, N.; Yang, J. The Self-Face Paradigm Improves the Performance of the P300-Speller System. *Front. Comput. Neurosci.* **2020**, *13*, 93. [CrossRef]

34. Eimer, M. The face-specific N170 component reflects late stages in the structural encoding of faces. *Neuroreport* **2000**, *11*, 2319–2324. [CrossRef]
35. Tian, Y.; Zhang, H.; Pang, Y.; Lin, J. Classification for Single-Trial N170 During Responding to Facial Picture with Emotion. *Front. Comput. Neurosci.* **2018**, *12*, 68. [CrossRef]

Disclaimer/Publisher's Note: The statements, opinions and data contained in all publications are solely those of the individual author(s) and contributor(s) and not of MDPI and/or the editor(s). MDPI and/or the editor(s) disclaim responsibility for any injury to people or property resulting from any ideas, methods, instructions or products referred to in the content.

sensors

MDPI

Article

Machine Learning Methods in Predicting Patients with Suspected Myocardial Infarction Based on Short-Time HRV Data

Dmytro Chumachenko [1,2], Mykola Butkevych [1], Daniel Lode [2], Marcus Frohme [2,*], Kurt J. G. Schmailzl [3] and Alina Nechyporenko [2,4]

1 Mathematical Modelling and Artificial Intelligence Department, National Aerospace University Kharkiv Aviation Institute, 61072 Kharkiv, Ukraine
2 Molecular Biotechnology and Functional Genomics Department, Technical University of Applied Sciences Wildau, 15745 Wildau, Germany
3 ccc. Center for Connected Health Care UG, 16818 Wustrau, Germany
4 Systems Engineering Department, Kharkiv National University of Radio Electronics, 61166 Kharkiv, Ukraine
* Correspondence: marcus.frohme@th-wildau.de

Abstract: Diagnosis of cardiovascular diseases is an urgent task because they are the main cause of death for 32% of the world's population. Particularly relevant are automated diagnostics using machine learning methods in the digitalization of healthcare and introduction of personalized medicine in healthcare institutions, including at the individual level when designing smart houses. Therefore, this study aims to analyze short 10-s electrocardiogram measurements taken from 12 leads. In addition, the task is to classify patients with suspected myocardial infarction using machine learning methods. We have developed four models based on the k-nearest neighbor classifier, radial basis function, decision tree, and random forest to do this. An analysis of time parameters showed that the most significant parameters for diagnosing myocardial infraction are SDNN, BPM, and IBI. An experimental investigation was conducted on the data of the open PTB-XL dataset for patients with suspected myocardial infarction. The results showed that, according to the parameters of the short ECG, it is possible to classify patients with a suspected myocardial infraction as sick and healthy with high accuracy. The optimized Random Forest model showed the best performance with an accuracy of 99.63%, and a root mean absolute error is less than 0.004. The proposed novel approach can be used for patients who do not have other indicators of heart attacks.

Keywords: myocardial infraction; heart rate variability; 10-second heart rate variability; diagnostics; machine learning; k-nearest neighbor classifier; radial basis function; decision tree; random forest

Citation: Chumachenko, D.; Butkevych, M.; Lode, D.; Frohme, M.; Schmailzl, K.J.G.; Nechyporenko, A. Machine Learning Methods in Predicting Patients with Suspected Myocardial Infarction Based on Short-Time HRV Data. *Sensors* **2022**, *22*, 7033. https://doi.org/10.3390/s22187033

Academic Editor: Daniele Cenni

Received: 24 August 2022
Accepted: 15 September 2022
Published: 17 September 2022

Publisher's Note: MDPI stays neutral with regard to jurisdictional claims in published maps and institutional affiliations.

Copyright: © 2022 by the authors. Licensee MDPI, Basel, Switzerland. This article is an open access article distributed under the terms and conditions of the Creative Commons Attribution (CC BY) license (https://creativecommons.org/licenses/by/4.0/).

1. Introduction

Every year, information technology is becoming increasingly established in all areas of activity. Rapidly gaining momentum in recent decades and progress against the background of the widespread introduction of computer information technologies have also embraced medicine. The global COVID-19 pandemic has dramatically accelerated the pace of digitalization, causing entire industries to be transformed [1]. Today, information systems in medicine are used more and more widely: from making a diagnosis to forecasting the resources necessary for the continuous operation of medical institutions.

In healthcare, there are two groups of innovations—evolutionary [2] and revolutionary [3]. Evolutionary information technologies (IT) solutions help improve the quality of existing services: automate examinations, book patients online, and conduct screenings. Revolutionary ones are associated with new models of medical services, such as telemedicine or the use of artificial intelligence in diagnostics. A big driver of digital transformations in medicine is also a large accumulation of data: case histories, clinical analyses, etc. [4].

In addition, the impetus in healthcare informatization is the development of the artificial intelligence industry. Tools, powered by artificial intelligence (AI), uncover meaningful relationships in raw data. They can be applied to all areas of medicine, including drug discovery, medical diagnosis, treatment decision-making, patient care, and financial transactions and decisions. Artificial intelligence can make it easier to identify patterns by helping researchers create dynamic patient cohorts for research and clinical trials [5]. Modern machine learning tools that use artificial neural networks to learn highly complex relationships or deep learning technologies often outperform human capabilities in performing medical tasks. AI-enabled systems are capable of solving complex problems that are common in modern clinical care.

An analysis of modern research shows a growing prospect of using data-driven medical solutions for smart homes, which will turn the living environment into an innovative clinical environment for the prevention and early diagnosis of common diseases [6]. The introduction of personalized medicine solutions into the living environment will significantly reduce the risks of diseases associated with aging, including cardiovascular diseases [7].

Cardiovascular disease (CVD) is the leading cause of death worldwide: no other disease causes as many deaths yearly as CVD [8]. An estimated proportion of CVD among all global death is 32% [9]. Over 75% of CVD deaths occur in low- and middle-income countries [10]. This is mainly because people in low- and middle-income countries with CVD have less access to effective health care.

The primary behavioral risk factors for cardiovascular disease and stroke are unhealthy diet, lack of physical activity, tobacco use, and diabetes [11]. Such risk factors can manifest as high blood pressure, high blood glucose, high blood lipids, and being overweight and obese. Most cardiovascular diseases can be effectively managed not only by preventive measures but also by early diagnosis [12].

The global COVID-19 pandemic has become another challenge for health systems around the world in combating CVD, as they are one of the main complications of this infection after respiratory manifestations [13]. People with cardiovascular diseases are believed to be more susceptible to infection because the new coronavirus uses the Angiotensin-converting enzyme 2 (ACE2) receptor to enter the cell. People with cardiovascular complications during COVID-19 seem to have high levels of ACE2 expression, and SARS-CoV-2 uses it to dock onto the body's cells to infect them [14].

The disastrous final route of coronary heart disease (CHD) in the world is myocardial infarction (MI) [15]. MI is damage to the heart muscle caused by an acute disruption of its blood supply due to blockage (thrombosis) of one of the heart's arteries with atherosclerotic plaque [16]. In this case, the affected cardiac muscle cells die by necrosis, and in the further course the necrotic district changes into a fibrous scar. Cell death begins within 20–40 min from the moment of cessation of blood flow in the coronary artery. The high prevalence and narrow time window demand new methods for diagnosing early-stage MI to prevent patient lethality. Therefore, developing models and methods for the early diagnosis of MI is an urgent task. This shall reduce the mortality rate from MI and open up access for countries and people who do not have special equipment and enough medical personnel to prevent MI. The models proposed in this study are based on statistical machine learning methods and do not require high computational power and special equipment. The use of the proposed models is possible on personal computers.

In order for a patient to be diagnosed with myocardial infarction, they must fulfill at least two of the following three criteria, according to the World Health Organization:

- Clinical history of chest discomfort consistent with ischemia, such as crushing chest pain;
- An elevation of cardiac markers in the blood (Troponin-I, CK-MB, Myoglobin);
- Characteristic changes on electrocardiographic tracings taken serially.

The significant electrocardiography (ECG) changes indicative of myocardial infarction are the elevation (in STEMI) or depression (NSTEMI) of the ST segment, as well as the inversion of the T wave (in NSTEMI) [17]. However, this requires the patient to visit the emergency department of a hospital or a doctor's office, which necessarily means some delay in diagnosis and treatment. A very easy and "at home" applicable ECG-based heart attack diagnosis would be desirable.

Given this, there is a crucial need to study new signs that have diagnostic value for the diagnosis of myocardial infarction. In this paper, we will study the impact of heart rate variability (HRV) time domain metrics, which are obtained from short 10-second samples of ECG. One more aim of the current research is to develop an effective machine learning model for MI diagnostics.

To achieve this aim, the following requirements need to be met:

- Current research on MI classification should be analyzed;
- Data should be analyzed;
- A machine learning models should be developed;
- Data should be prepared for the experimental study with developed machine learning models;
- Experimental evaluation of the developed models with open data on MI should be provided;
- Comparative analysis of obtained results with other methods and models should be provided.

The respective contribution of this study is two-fold—Firstly, the development of machine learning models based on a k-nearest neighbor classifier, radial basis function, decision tree, and random forest will allow for estimating accuracy of simple machine learning techniques for cardiovascular diseases classification; Secondly, the study of the diagnostic value of data, HRV, obtained from the original ECG signal, as an alternative to such characteristic changes in the ECG curve as ST-segment elevation or depression, T-wave inversion, confirming the diagnosis of MI. Moreover, research on HRV metrics is particularly interesting as they are derived from short 10-second ECG records.

The rest of this publication is structured as follows: Section 2 provides the current state of cardiovascular disease machine learning classification models and methods. Section 3, namely Materials and Methods, describes machine learning models based on k-nearest neighbors classifier, radial basis function, decision tree, and random forest methods. Section 4 provides analysis of the publicly available dataset PTB-XL and results of data preprocessing and preparation. Section 5 provides the results of experiments with developed models and the results of model optimization. Section 6 presents the conclusions and future work.

Given research is part of a complex intelligent information system for epidemiological diagnostics, developed within the project 2020.02/0404 "Development of intelligent technologies for assessing the epidemic situation to support decision-making within the population biosafety management" funded by National Research Foundation of Ukraine, the concept of which is discussed in [18].

2. Current State of Research

The most effective method for automated diagnosis of myocardial infarction is the analysis of electrocardiogram data. ECG is a method for analyzing the work of the heart based on the registration of electromagnetic field variations that occur in the heart muscle during the cardiac cycle [1]. The signal that reflects the nature of these variations is called an electro-cardio signal (ECS). Analysis of the ECS is a process of studying the ECG signal, aimed at detecting pathological abnormalities in its individual sections and determining the causes of these abnormalities.

The main problems that arise during the analysis of the ECG can be classified by reason of their occurrence into [19]:

- stochastic nature of the biological system under study;
- imperfection of the technical means of signal pickup. There are three main stages in the task of automated ECG analysis [20,21]:
- pre-processing stage, in which the signal is separated from interference;
- conversion stage, at which informative features of the signal are extracted;

- the stage of solving the problem, which generates the output signal according to the identified informative features.

The task of ECG classification is to identify informative signs and find their dependence on the corresponding heart disease or its absence. To date, the methods based on neural networks show the highest accuracy among the methods of automated CVD diagnostics. However, their disadvantage is the high computational complexity and the need for computing resources [22]. This is not feasible in the context of health care facilities in low- and middle-income countries, which account for most deaths. Therefore, machine learning methods not based on artificial neural networks are preferred in this study.

Authors of Ref. [23] have built classification models of MI using 192 lead body surface potential maps analysis. The most important features were used as input to a series of supervised classification models using Naive Bayes, Support Vector Machine, and Random Forest methods. The accuracy of the constructed models was 81.9% for Naive Bayes, 82.8% for Support Vector Machine, and 84.5% for Random Forest. However, using 192 leads for the detection is not practical for the detection of MI.

Polat et al. [24] have modified the k-nearest neighbors method and used it as a preprocessing approach before the classification. Artificial immune recognition system with a fuzzy resource allocation mechanism as a classifier, showed an accuracy of 87.0% for MI diagnosing. In Ref. [25], the decision tree and bagging based on decision tree models are proposed. The authors have used the database of 920 samples. The accuracy is 78.91% for decision tree and 81.41% for bagging. Ref. [26] proposes the modification of the decision tree method by nine voting equal frequency discretization gain ratio. Based on the data from 297 samples, authors obtained an accuracy of 84.1%.

Ref. [27] discussed two machine learning approaches to ECG classification. Models based on support vector machine and radial basis function network have shown accuracy 85.05% and 82.71%, respectively, using 5-fold cross-validation, and 85.05% and 82.24% using 10-fold cross-validation.

In Ref. [28], the ensemble method for heart diseases classification is proposed, which integrates k-means clustering with naïve Bayes. The best accuracy obtained for two clusters random row initial centroid selection is 84.5%. Chitra and Seenivasagam [29], to validate the developed cascaded neural network of CVD classification, have built the model based on a support vector machine and obtained an accuracy of 77.5% with it. Authors of [30] have proposed three machine learning models of heart disease detection. The accuracy obtained with gain ratio decision tree is 79.1%, the accuracy of Naïve Bayes method is 83.5%, and the accuracy of k-nearest neighbor method with $K = 19$ is 83.2%.

Authors of [31] used data from 143 cases, 79 of which were MI-related. The machine learning model based on k-nearest neighbors method showed an accuracy of 87.0% with $K = 4$. Authors of [32] have used the weighted vote-based ensemble technique to combine the results of Naive Bayes, decision tree based on information gain, decision tree based on Gini index, instance-based learner, and support vector machine algorithms. The accuracy of obtained ensemble model is 87.37%. Authors of [33] modified the proposed in the Ref. [26] method using nine voting equal frequency discretization with Gini index decision tree applied to the same dataset and obtained an accuracy of 85.3%.

In Ref. [34], ECG data taken for six seconds and ECG data taken the entire length of the data in two minutes are investigated. The developed model of modified K-nearest neighbors showed an accuracy of 71.2% with $K = 3$.

The comparative analysis of investigated researches is presented in Table 1.

Table 1. Comparison of accuracy of current researches.

Author, Source	Approach	Accuracy
Yuwono T., et al. [34]	Modified K-nearest neighbors with K = 3	71.2%
Chitra R., Seenivasagam V. [32]	Support vector machine	77.5%
Tu M.C., et al. [25]	Decision tree	78.91%
Shouman M., et al. [30]	Gain ratio decision tree	79.1%
Tu M.C., et al. [25]	Bagging based on decision tree	81.41%
Zheng H., et al. [23]	Naïve Bayes	81.9%
Ghumbre S., et al. [27]	Radial basis function network using 10-fold cross-validation	82.24%
Ghumbre S., et al. [27]	Radial basis function network using 5-fold cross-validation	82.71%
Zheng H., et al. [23]	Support vector machine	82.8%
Shouman M., et al. [30]	K-nearest neighbors with K = 19	83.2%
Shouman M., et al. [30]	Naïve Bayes	83.5%
Shouman M., et al. [26]	Equal frequency discretization gain ratio decision tree	84.1%
Zheng H., et al. [23]	Random forest	84.5%
Shouman M., et al. [28]	Ensemble: k-means with Naïve Bayes	84.5%
Ghumbre S., et al. [27]	Support vector machine	85.05%
Kirmani M.M., et al. [33]	Nine voting equal frequency discretization with Gini index decision tree	85.3%
Polat K., et al. [24]	Artificial immune recognition system	87.0%
Yuwono T., et al. [31]	K-nearest neighbor	87.0%
Bashir S., et al. [32]	Ensemble: Naive Bayes, decision tree based on information gain, decision tree based on Gini index, instance-based learner, support vector machine	87.37%

Ref. [35] analyzes recent research on classifying HRV data using machine learning models. However, most of them are focused on stress classification. In addition, this paper discusses the features of various durations of HRV records and metrics in the time and frequency domains in the context of diagnosing cardiovascular diseases. Thus, studying the effect of HRV indicators obtained on the basis of short 10-second ECG recordings on the accuracy of myocardial infarction classification based on machine learning models is of particular research interest.

The study of [36] investigated 10-second heart rate variability. The authors concluded that using the 10-second estimate would be of extreme benefit in assessing HRV as opposed to a need to record a 24-h ECG. However, no further investigations of its features have been provided.

Analysis of the current state of research on ECG processing and diagnosing MI using machine learning models also shows that data preprocessing is a crucial step to achieving a high degree of accuracy in the training model. Heterogeneous data also play a vital role in the accuracy of classifiers. Reviewed studies say that machine learning classifiers with preprocessed data show more accurate results than those without preprocessed data.

3. Materials and Methods

Within research, four models based on machine learning methods for the classification of patients with MI were developed. Machine learning models are based on k-nearest neighbors classifier, radial basis function, decision tree, and random forest.

3.1. K-Nearest Neighbor Classifier

The principle behind k-nearest neighbor method is to find a predetermined number of training samples closest in the distance to a new point and provide a value for the data [37]. Despite its simplicity, the k-nearest neighbor method has succeeded in many classification and regression problems, including the medical domain. Being a non-parametric method, it is often successful in classification situations where the decision limit is unclear.

Euclidean distance is a commonly used distance metric for continuous variables [38]. For discrete variables, such as text classification, you can use another metric, such as the overlap metric (or Hamming distance) [39]. In addition, k-nearest neighbors classifier is used with correlation coefficients such as Pearson and Spearman [40]. Often, classification accuracy can be greatly improved if the distance metric is learned using specialized algorithms, such as high-margin, nearest-neighbor, or neighborhood component analysis.

The disadvantage of the primary majority vote classification is that the class distribution is skewed. More frequent class examples tend to dominate the prediction of a new example since they tend to be spread among nearest neighbors due to their large number.

One way to overcome this problem is to weight the classification given the distance from the control point to each of its nearest neighbors. The class (or value in regression problems) of each of the k closest points is multiplied by a weight proportional to the reciprocal distance from that point to the control point. Another way to overcome skew is to abstract the data representation.

To classify the objects of the test sample, you must sequentially perform the following steps:

1. To calculate the distance to each of the objects in the training sample;
2. To select the object of the training sample, the distance to which is minimal;
3. The class of the classified object is the class that occurs among k nearest neighbors most often.

Euclidean distance in multidimensional feature space is calculated as follows:

$$d_{ab} = \sqrt{\sum_{i=1}^{n} (x_{ai} - x_{bi})^2}, \quad (1)$$

where a and b are points in n-dimensional space;

i is ordinal number of the feature;

x_{ai} and x_{bi} are coordinates of points a and b by the feature i.

The class with the most votes is assigned to the new element:

$$y_a(a, X, k) = argmax_{y \in Y} \sum_{i=1}^{k} \left(y_a^i = y\right), \quad (2)$$

where a is a new element (connection),

X is a training sample,

y is a class,

Y is a set of classes,

${y_a}^i$ is the class of i-th neighbor a,

k is the number of neighbors.

3.2. Radial Basis Function

In machine learning, a radial basis function is used in various kernel learning algorithms [41]. In particular, it is commonly used to classify support vector machines. The radial basis function kernel on two samples x and x', represented as feature vectors in some input space, is defined as:

$$K(x, x') = exp\left(\frac{- \parallel x - x' \parallel^2}{2\sigma^2}\right), \quad (3)$$

where $\parallel x - x' \parallel^2$ can be defined as the square of the Euclidean distance between two feature vectors,

σ is free parameter.

The equivalent definition includes the parameter $\gamma = \frac{1}{2\sigma^2}$:

$$K(x, x') = exp\left(-\gamma \parallel x - x' \parallel^2\right). \tag{4}$$

Since the value of the RBF kernel decreases with distance and ranges from zero (at the boundary) to one (when $x = x'$), it has a ready interpretation as a measure of similarity [42]. The feature space of the kernel has an infinite number of dimensions; for $\sigma = 1$, it grows:

$$\begin{aligned} exp\left(\tfrac{-1}{2}\parallel x - x' \parallel^2\right) = exp\left(\tfrac{2}{2}x^T x' - \tfrac{1}{2}\parallel x \parallel^2 - \tfrac{1}{2}\parallel x' \parallel^2\right) = \\ exp(x^T x') exp\left(\tfrac{-1}{2}\parallel x \parallel^2\right) exp\left(\tfrac{-1}{2}\parallel x' \parallel^2\right) = \\ \textstyle\sum_{j=0}^{\infty} \frac{(x^T x')}{j!} exp\left(\tfrac{-1}{2}\parallel x \parallel^2\right) exp\left(\tfrac{-1}{2}\parallel x' \parallel^2\right) = \\ \textstyle\sum_{j=0}^{\infty} \sum_{\sum n_i = j} \frac{(x^T x')}{j!} exp\left(\tfrac{-1}{2}\parallel x \parallel^2\right) exp\left(\tfrac{-1}{2}\parallel x' \parallel^2\right). \end{aligned} \tag{5}$$

3.3. Decision Tree

Decision Trees is a non-parametric supervised learning technique used for classification and regression [43]. The goal is to create a model that predicts the value of the target variable by learning simple decision rules derived from the characteristics of the data. The tree can be considered as a piecewise constant approximation.

Decision trees are trained on the data to approximate a sinusoid using a set of if-then-else decision rules. The deeper the tree, the more complex the decision rules and the better the model. The benefits of decision trees include:

- Easy to understand and interpret;
- Trees can be visualized;
- Requires little data preparation;
- Tree usage weights are the logarithmic number of data points used to train the tree. The disadvantages of decision trees include:
- Trained decision models can create highly complex trees that do not generalize well. To avoid this problem, mechanisms such as pruning, setting a minimum number of samples required in a leaf node, or setting a maximum tree depth is needed.
- Decision trees can be unstable because minor variations in the data can result in a completely different tree. The use of ensemble decision trees mitigates this problem.

The problem of learning an optimal decision tree is NP-complete in several aspects of optimum, even for simple concepts. Therefore, practical decision tree learning algorithms are based on heuristic algorithms such as the greedy algorithm, where locally optimal decisions are made at each node. Such algorithms cannot guarantee the return of a globally optimal decision tree. This can be mitigated by training multiple trees in an ensemble where features and samples are randomly sampled with replacement.

3.4. Random Forest

Random forest is a type of supervised machine learning algorithm based on ensemble learning [44]. Ensemble learning is a type of learning where you combine different types of algorithms or the same algorithm multiple times to form a more powerful prediction model. The random forest algorithm combines several algorithms of the same type, that is, several decision trees, resulting in a forest of trees, hence the name Random Forest. The random forest algorithm can be used for both regression and classification problems.

These two sources of randomness aim to reduce the variance of the forest estimate. Individual decision trees typically exhibit high variance and tend to overflow. The introduced randomness in forests yields decision trees with slightly isolated prediction errors. By taking the average of these predictions, some errors can be avoided. Random forests achieve reduced variance by combining diverse trees, sometimes at the cost of a slight increase in bias. In practice, the reduction in variance is often significant, giving an overall better model.

Benefits of using Random Forest include:

- The random forest algorithm is not biased because there are multiple trees, and each tree learns from a subset of the data. Basically, the random forest algorithm relies on the power of the "crowd"; therefore, the overall bias of the algorithm is reduced.
- The algorithm is stable. Even if a new data point is introduced into the data set, the overall algorithm is not significantly affected because the new data may affect one tree. However, it is challenging to affect all trees.
- The random forest algorithm works well if the sample contains both categorical and numerical features. The random forest algorithm also performs well when data are missing or have not been well scaled.

The main disadvantage of random forest is its complexity. The model takes out much more computational resources due to a large number of merged decision trees. Due to their complexity, they take much longer to train other similar algorithms.

4. Data Analysis and Preprocessing

For the experimental study, we used the open database PTB-XL [45], which was collected within the project PhysioNet [46]. PTB-XL is the to-date largest freely accessible clinical 12-lead ECG-waveform dataset comprising 21,837 records from 18,885 patients of 10 seconds length [47]. Two cardiologists annotate the ECG data as a multi-label dataset, where the diagnostic labels have been further grouped into superclasses and subclasses. The data set spans many diagnostic classes, including healthy individuals. The data also contain demographic metadata, additional diagnostic statements, and probabilities of diagnosis, which are manually annotated.

4.1. Input Data Description

The initial dataset consists of 15,014 records, 9528 normal, and 5486 myocardial infarction (MI) data, including ECG signals and metadata. The random patient examination ECG data is shown in Figure 1. One can single out a clear cyclically repeating pattern with some modifications, and a baseline drift, which must be eliminated using signal preprocessing techniques. The patient dataset also contains the corresponding metadata.

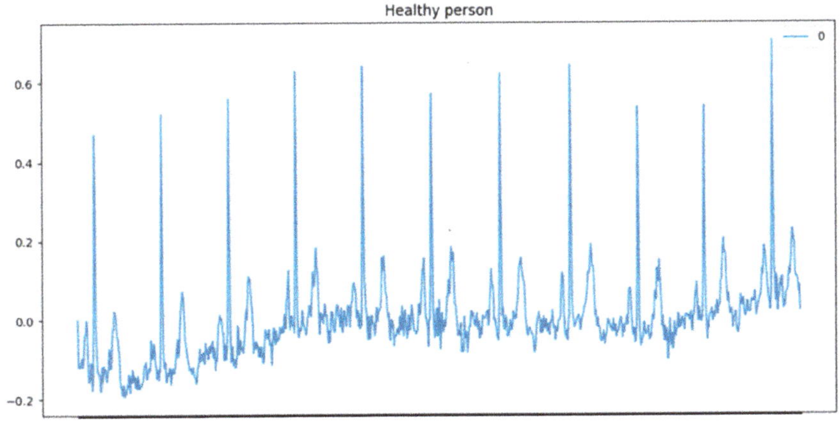

Figure 1. Example of ECG data.

The metadata can be divided into the following categories:

1. Identifiers: Each entry is identified by a unique ecg_id. The eligible patient is encoded via the patient ID; paths to the original recording (500 Hz) and the downsampled version of the recording (100 Hz);

2. General metadata: demographic and registration metadata such as age, gender, height, weight, nurse, site, device, and date of enrollment;
3. ECG operations: The main components are scp_codes (SCP-ECG operations as a dictionary with entries of the form statement: probability, where probability is set to 0 if unknown) and report (report string). Additional fields are heart_axis, infarction_stadium1, infarction_stadium2, validated_by, second_opinion, initial_autogenerated_report, and validated_by_human;
4. Signal metadata: signal quality such as noise (static_noise and burst_noise), baseline offset (baseline_drift), and other parameters such as electrodes_problems;
5. Figure 2 shows records of two random patients: normal and with MI.

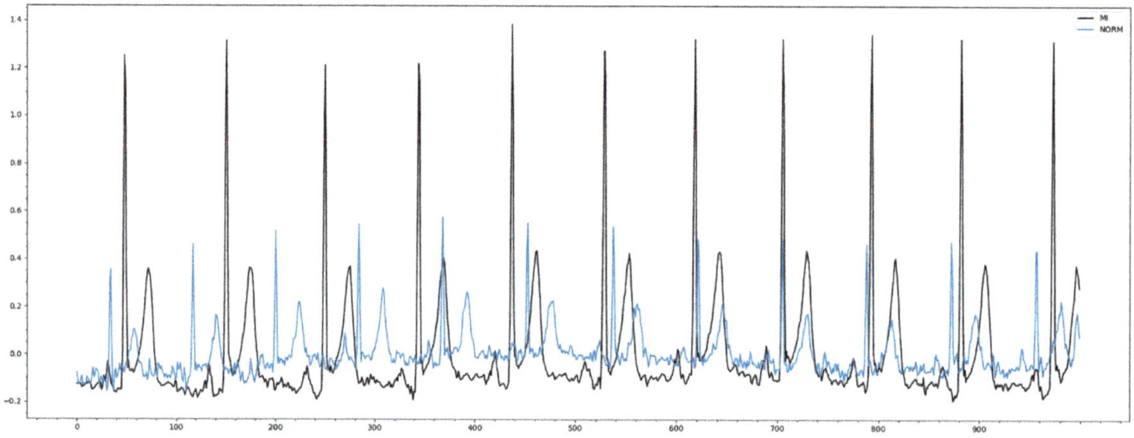

Figure 2. Example of ECG data (Norm and MI).

Figure 3 shows the basic data about the patient and information about the examination, biological data, and diagnosis.

4.2. ECG Data Analysis

Based on studies [48] regarding the diagnostic value of HRV indicators for the detection of cardiovascular diseases and, in particular, myocardial infarction, ECG signals were processed as follows: First, Heart Rate Variability (HRV) data were obtained from the original ECG signal, and then the time-domain metrics were calculated. HRV is the fluctuation in the time intervals between adjacent heartbeats that correspond to R peaks on the ECG signal. The time-domain metrics include inter-beat interval (IBI), heartbeats per minute (BPM), the standard deviation of the R peak to R peak intervals (SDNN), with the assumption that the data only come from the Normal to Normal sinus rhythm and the root mean square of successive differences (RMSSD). RMSSD is determined by first calculating each successive R peak to R peak. Afterward, each of these values is squared, and the results are averaged before finding the square root of the total. Equations (6) and (7) describe the SDNN and RMSSD calculations, respectively:

$$\mathrm{SDNN} = \sqrt{\frac{1}{N-1}\sum_{i=1}^{N}(X_i - \mu)^2} \tag{6}$$

$$\mathrm{RMSSD} = \sqrt{\frac{1}{N}\sum_{i=1}^{N-1}(X_i - X_{i+1})^2} \tag{7}$$

The extracted features for analysis are presented in Figure 4.

ecg_id	patient_id	age	sex	height	weight	nurse	site	device	recording_date	report	...	validated
1	15709.0	56.0	1	NaN	63.0	2.0	0.0	CS-12 E	1984-11-09 09:17:34	sinusrhythmus periphere niederspannung	...	True
2	13243.0	19.0	0	NaN	70.0	2.0	0.0	CS-12 E	1984-11-14 12:55:37	sinusbradykardie sonst normales ekg	...	True
3	20372.0	37.0	1	NaN	69.0	2.0	0.0	CS-12 E	1984-11-15 12:49:10	sinusrhythmus normales ekg	...	True
4	17014.0	24.0	0	NaN	82.0	2.0	0.0	CS-12 E	1984-11-15 13:44:57	sinusrhythmus normales ekg	...	True
5	17448.0	19.0	1	NaN	70.0	2.0	0.0	CS-12 E	1984-11-17 10:43:15	sinusrhythmus normales ekg	...	True
...	...	...	...	...	...	...	...	...	...	...	...	...
21833	17180.0	67.0	1	NaN	NaN	1.0	2.0	AT-60 3	2001-05-31 09:14:35	ventrikulÄre extrasystole(n) sinustachykardie ...	...	True
21834	20703.0	93.0	0	NaN	NaN	1.0	2.0	AT-60 3	2001-06-05 11:33:39	sinusrhythmus lagetyp normal qrs(t) abnorm ...	...	True
21835	19311.0	59.0	1	NaN	NaN	1.0	2.0	AT-60 3	2001-06-08 10:30:27	sinusrhythmus lagetyp normal t abnorm in anter...	...	True
21836	8873.0	64.0	1	NaN	NaN	1.0	2.0	AT-60 3	2001-06-09 18:21:49	supraventrikulÄre extrasystole(n) sinusrhythmu...	...	True
21837	11744.0	68.0	0	NaN	NaN	1.0	2.0	AT-60 3	2001-06-11 16:43:01	sinusrhythmus p-sinistrocardiale lagetyp norma...	...	True

Figure 3. Primary dataset review.

patient_id	ecg_id	bpm	ibi	sdnn	rmssd
15709.0	1.0	63.733	941.429	13.553	11.547
15709.0	1.0	63.425	946.000	13.565	21.602
15709.0	1.0	63.915	938.750	14.524	13.628
15709.0	1.0	63.966	938.000	14.000	17.638
15709.0	1.0	63.915	938.750	12.686	12.536
15709.0	1.0	64.133	935.556	16.405	18.028
15709.0	1.0	64.133	935.556	15.713	15.811
15709.0	1.0	63.927	938.571	13.553	10.801
15709.0	1.0	63.915	938.750	13.636	12.536
15709.0	1.0	64.133	935.556	15.713	15.811

Figure 4. Results of ECG signal processing.

4.3. Data Preprocessing

Both ECG signal data and metadata comprise noises, missing values, and other inconsistencies and require preprocessing. For these purposes, noise reduction techniques were applied to ECG signals. It allowed us to enhance ECG peaks, convolving synthetic QRS templates with the signal, and applying a notch filter, resulting in a strong signal-to-noise ratio.

The resulting disease classes were divided into negative and positive diagnosis values of 0 and 1, respectively. The MI class was classified as positive and assigned a value of 1, and the NORM class was classified as negative and assigned a value of 0.

The descriptive statistics shown in Figure 5 include statistics summarizing the major trend, variance, and distribution shape of the data set, excluding NaN values. The average age of patients is quite high with 52 years, and the 50% percentile is 54 years. It can be seen that half of the resulting sample are men and the other half are women. The majority of the 76% sample was re-validated by another physician. In addition, 10 ECG machines and 11 nurses participated in the process, which could affect the result.

	ecg_id	age	sex	weight	nurse \
count	11621.000000	11621.000000	11621.000000	11621.000000	11621.000000
mean	10699.334395	54.709572	0.506411	71.589187	2.180397
std	6320.722818	17.117032	0.499980	10.919290	3.089584
min	1.000000	2.000000	0.000000	5.000000	0.000000
25%	5195.000000	43.000000	0.000000	70.000000	0.000000
50%	10600.000000	56.000000	1.000000	71.589187	1.000000
75%	16124.000000	67.000000	1.000000	71.589187	2.180397
max	21837.000000	94.000000	1.000000	210.000000	11.000000

	site	validated_by	second_opinion	validated_by_human \
count	11621.000000	11621.000000	11621.000000	11621.000000
mean	1.481915	0.747013	0.028569	0.764564
std	4.242491	0.857500	0.166599	0.424289
min	0.000000	0.000000	0.000000	0.000000
25%	0.000000	0.000000	0.000000	1.000000
50%	1.000000	0.747013	0.000000	1.000000
75%	2.000000	1.000000	0.000000	1.000000
max	50.000000	11.000000	1.000000	1.000000

Figure 5. Data sample descriptive statistics.

When data values for a variable in observation are not stored, they are missing data or missing values. Missing data are common and can have a significant impact on the conclusions that can be drawn from the data. Figure 6 shows a list of missing values in the analyzed dataset.

	missing_count	missing_ratio
age	14	0.001205
weight	5998	0.516135
nurse	684	0.058859
site	9	0.000774
validated_by	5344	0.459857

Figure 6. Missed data in the obtained data sample.

Most cases of missing values are noted in the excess mortality columns, which will not be used for predictions or testing but only for analysis. They can be easily dropped. Other values seem to be missing because there were no specific observations or studies on certain days.

The following methods can be applied to fill in the missing values:

- Linear interpolation;
- Linear interpolation of neighboring values.

In our case, since all records are not related to each other, interpolation options are not suitable since, in this case, there is no task to preserve the original behavior.

Missing values can be filled in with:

- Parameters Outlier or Zero;
- Mean value;
- Median;
- Constant.

Filling missing values with a constant or zero is not sufficient and not a good option. Filling in missing values with the last value may give better results when using the mean or median. The median method was chosen because it has an advantage over the mean values in a situation where some values are anomalous and strongly bias the mean.

In order to calculate the correlation of the Pearson product with a moment, one first needs to determine the covariance of the two variables in question. Next, you need to calculate the standard deviation of each variable. The correlation coefficient is determined by dividing the covariance by the product of the standard deviations of the two variables. Based on the analysis of medical data, the following conclusions can be drawn:

- Age has an average negative correlation with diagnosis;
- Gender has a medium negative correlation with weight and a low positive correlation with diagnosis.

Both significantly influence the diagnosis parameters infraction stadium and heart axis.

Medical parameters have an average positive correlation with the diagnosis. The heart axis indicates the heart's position relative to the body and its inclination [49]. Abnormal values may indicate related diseases. In turn, infraction stadium is a disease in which the patient has a brain tumor–glioma [50].

5. Results

After preliminary analysis and data preparation, we obtained 13,480 records, and separate our data into training and validation data. We want to provide the model with as much training data as possible. However, we also want to ensure we have enough data to test the model. As the number of rows in the dataset increases, we can provide more data to the training set. Another critical parameter is data mixing.

In this study, the set was distributed in the ratio of 80% to 20% for training and validation data. A set of regression classifiers and machine learning models was defined for testing with this data set. Tests of statistical significance were carried out to check the validity of the result [51]. To do this, we evaluated the model 10 times and obtained the average values of accuracy and RMSE.

The results of the developed machine learning models are shown in Table 2.

Table 2. Results of simulation.

Machine Learning Model	Accuracy	Root Mean Square Error
K-nearest neighbors classifier	71.105%	0.289
Radial basis function	75.408%	0.245
Decision tree	89.867%	0.109
Random forest	97.774%	0.022

It can be concluded that, with a high probability, the Random Forest classifier gives the best results. The results shown are competitive with those presented in Table 1. To improve accuracy, it is necessary to select parameters to optimize the results obtained for classification by the Random Forest model.

The Random Forest classifier is trained using load aggregation, where each new tree is selected from a sample of load observations. Out of bag is the average error for each computed using tree predictions not contained in the corresponding load sample. This allows the built Random Forest model to fit and validate during training.

The main parameters to be adjusted when using these methods are n_estimators and max_features. The first is the number of trees in the forest. The more, the better, but also the more time it will take to calculate. In addition, the results will no longer improve significantly over a critical number of trees. The latter is the size of random feature subsets that should be considered when splitting a node. The more minor, the more significant the reduction in dispersion, but also the more significant the increase in bias. The empirically "correct" defaults are: max_features = None (always consider all features instead of a random subset) for regression problems and max_features = "sqrt" (using a random subset of size sqrt(n_features)) for classification problems (where n_features is the number of features in the data). Good results are often achieved by setting max_depth = None in combination with min_samples_split = 2 (i.e., when trees are fully developed). However, these values are usually not optimal and can result in models that consume a lot of RAM.

The best parameter values should be cross-validated. Cross-validation technology is the most common using the most common K-Fold CV method. Typically, the data are divided into training and validation. However, for the K-Fold technique, the training data are split into K different samples, which are called Fold. By iteratively selecting samples, the model accuracy result is evaluated. After that, the average performance is found, which is the validation metric.

In the case of RandomForest, the following can be controlled as parameters:

- The number of trees in the forest (n_estimators);
- The maximum depth of the tree (max_depth). If not, then the nodes are expanded until all leaves are clean or until all leaves contain less than min_samples_split samples;
- The minimum number of samples (min_samples) that must be in a leaf node. A split point at any depth will only be considered if it leaves at least min_samples_leaf training samples in each of the left and right branches. This can have the effect of smoothing the model.

In addition, having selected the optimal hyperparameters, we can determine the maximum number of max_features features used to find the best test data split ratio:

For "Sqrt" parameter:

$$max_{features} = \sqrt{n_{features}}. \tag{8}$$

For "Log2" parameter:

$$max_{features} = log n_{features}. \tag{9}$$

For "None" parameter:

$$max_{features} = n_{features}. \tag{10}$$

The results are presented in Figure 7.

Analyzing Figure 6, we can conclude that the best strategy for choosing max_features is "None" (all features are always considered instead of a random subset). However, this configuration requires more system resources. In addition, the critical number of trees (n_estimators) is around 300.

Figure 8 shows the matrix of correspondences between the provided and actual values of the constructed model.

The accuracy and absolute error of the data are checked against the validation data. The building and training of the model were carried out several times, and the accuracy was maintained at 99.629 %, which is 2% better than the first model tuning. Even when mixing data at the stages of data preparation, they do not affect the result in any way. The resulting stability can be explained by the properties of the tree structure of the algorithm.

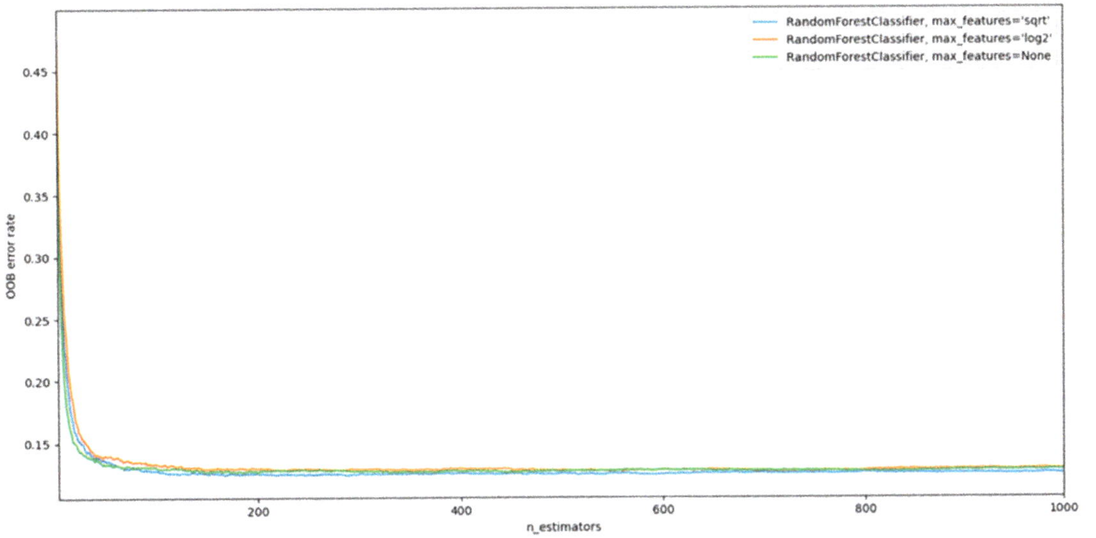

Figure 7. Results of Random Forest model with different parameters.

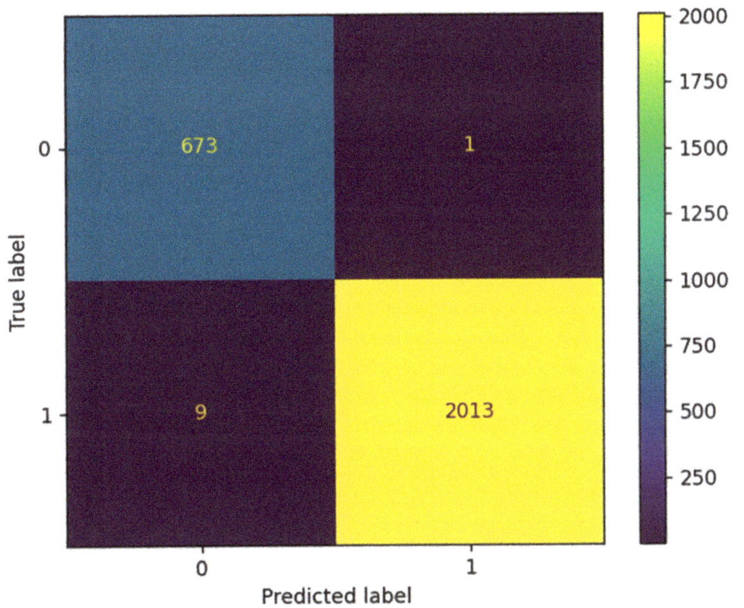

Figure 8. Correspondence matrix of the provided and actual values of the constructed model.

The accuracy of the optimized model is 99.629%, and root mean absolute error is 0.0037.

6. Conclusions

In existing studies in patients with myocardial infarction, the HRV time domain indicators are mainly isolated from long-term measurements from 5 min to 24 h. On short measurements, 10 s, the time parameters were not studied properly. However, this is an essential task because the protocol for examining patients with suspected MI says the patient should undergo an ECG for 10 s with 12 leads. In addition, this leads to the conclusion that HRV parameters have potential clinical value in cases where analysis of ECG data did not reveal changes in signal morphology. The evidence for correlations between changes in HRV parameters and ST-segment changes in the classic 12-lead ECG is still limited. Although the number of cases is still small and the results are based on ECG databases (and not on real-world clinical data), there are encouraging results on this and we are working on it ourselves.

An analysis of time parameters showed that the most significant parameters for diagnosing MI are sdnn, bpm, and ibi.

As part of this study, an experimental study was conducted on the data of the open PTB-XL dataset for patients with suspected MI. The results showed that, according to the parameters of the 10-second ECG, it is possible to classify patients with suspected MI as sick and healthy. Four machine learning methods were analyzed: k-nearest neighbors classifier, radial basis function, decision tree, and random forest. All methods showed high accuracy. However, the optimized Random Forest method showed an accuracy of 99.629%.

Unfortunately, we all know cases in which one or more pieces of the mosaic for the diagnosis of myocardial infarction ultimately proved to be false positives or negatives, and this is one reason for the need to be able to look at all the "big" pieces of the mosaic as far as possible: a typical clinical picture, typical ECG changes, and a troponin increase. For troponin, there are point-of-care tests, i.e., rapid tests made near the patient (they are not perfectly specific compared to the "normal" laboratory tests, but they allow very often a more in-depth assessment of the case). For ECG diagnostics, we believe that a possibility should be created to obtain the necessary information early and, if possible, already "at home" and by the patient himself or his relatives. Furthermore, for this, in our opinion, a classic 12-lead ECG is not suitable, but perhaps a wearable patch that allows HRV analysis. The analysis should be supported by AI. In addition, even if this information from HRV is also not 100% sensitive and specific, it seems plausible to combine it with a rapid troponin test. The affected patient who has acute chest pain could perform both at home, and an AI algorithm could derive a recommendation for action from the information now available (typical symptomatology yes/no, HRV suspicious of myocardial infarction yes/no, troponin elevated yes/no): Alert and transport to emergency department or visit primary care physician's office at the earliest possible date.

The proposed approach can be used for patients who do not have other indicators of heart attacks.

In the future, it is planned to conduct studies on each individual leading to determine the minimum number of leads required to obtain reliable results for the diagnosis of MI. This will allow the proposed methodology to be applied outside medical institutions and integrated into the smart home system. The proposed automated solution based on machine learning models is a practical addition to traditional diagnostic approaches and saves resources while supporting decision-making by doctors. This is especially important in the context of the global COVID-19 pandemic when healthcare resources are limited and patients do not always have access to their family doctors regularly.

Author Contributions: Conceptualization, D.C., K.J.G.S., and A.N.; methodology, D.C., M.F., and A.N.; software, M.B. and D.L.; validation, D.C., K.J.G.S., and A.N.; formal analysis, M.F.; investigation, D.C., M.F., K.J.G.S., and A.N.; resources, M.B. and D.L.; data curation, M.B. and D.L.; writing—original draft preparation, D.C., M.B., D.L., K.J.G.S., and A.N.; writing—review and editing, M.F.; visualization, M.B.; supervision, D.C. and A.N.; project administration, M.F.; funding acquisition, D.C. All authors have read and agreed to the published version of the manuscript.

Funding: This research received no external funding.

Institutional Review Board Statement: Not applicable.

Informed Consent Statement: Not applicable.

Data Availability Statement: The initial data used in this research are publicly available by the link https://physionet.org/content/ptb-xl/1.0.2/ (accessed on 21 August 2022).

Conflicts of Interest: The authors declare no conflict of interest.

References

1. Nagel, L. The influence of the COVID-19 pandemic on the digital transformation of work. *Int. J. Sociol. Soc. Policy* **2020**, *40*, 861–875. [CrossRef]
2. Sherer, S.A.; Meyerhoefer, C.D.; Shienberg, M.; Levick, D. Integrating commercial ambulatory electronic health records with hospital systems: An evolutionary process. *Int. J. Med. Inform.* **2015**, *84*, 683–693. [CrossRef] [PubMed]
3. Bonner, L. Prepare now for the digital health revolution. *Pharm. Today* **2021**, *27*, 24–29. [CrossRef]
4. Hoffman, J.; Mahmood, S.; Fogou, P.S.; George, N.; Raha, S.; Safi, S.; Schmailzl, K.J.; Brandalero, M.; Hubner, M. A Survey on Machine Learning Approaches to ECG Processing. In Proceedings of the 2020 Signal Processing: Algorithms, Architectures, Arrangements, and Applications (SPA), Poznan, Poland, 23–25 September 2020; pp. 36–41. [CrossRef]
5. Davenport, T.; Kalakota, R. The potential for artificial intelligence in healthcare. *Futur. Healthc. J.* **2019**, *6*, 94–98. [CrossRef] [PubMed]
6. Hu, R.; Linner, T.; Trummer, J.; Güttler, J.; Kabouteh, A.; Langosch, K.; Bock, T. Developing a Smart Home Solution Based on Personalized Intelligent Interior Units to Promote Activity and Customized Healthcare for Aging Society. *J. Popul. Ageing* **2020**, *13*, 257–280. [CrossRef]
7. Moses, J.C.; Adibi, S.; Angelova, M.; Islam, S.M.S. Smart Home Technology Solutions for Cardiovascular Diseases: A Systematic Review. *Appl. Syst. Innov.* **2022**, *5*, 51. [CrossRef]
8. Protulipac, J.M.; Sonicki, Z.; Reiner, Z. Cardiovascular disease (CVD) risk factors in older adults—Perception and reality. *Arch. Gerontol. Geriatr.* **2015**, *61*, 88–92. [CrossRef]
9. Smith, J.; Velez, M.P.; Dayan, N. Infertility, Infertility Treatment, and Cardiovascular Disease: An Overview. *Can. J. Cardiol.* **2021**, *37*, 1959–1968. [CrossRef]
10. Ogungbe, O.; Byiringiro, S.; Adeola-Afolayan, A.; Seal, S.M.; Himmelfarb, C.R.D.; Davidson, P.M.; Commodore-Mensah, Y. Medication Adherence Interventions for Cardiovascular Disease in Low- and Middle-Income Countries: A Systematic Review. *Patient Prefer. Adherence* **2021**, *15*, 885–897. [CrossRef]
11. Zhang, Y.-B.; Chen, C.; Pan, X.-F.; Guo, J.; Li, Y.; Franco, O.H.; Liu, G.; Pan, A. Associations of healthy lifestyle and socioeconomic status with mortality and incident cardiovascular disease: Two prospective cohort studies. *BMJ (Clin. Res. Ed.)* **2021**, *373*, n604. [CrossRef]
12. Capotosto, L.; Massoni, F.; De Sio, S.; Ricci, S.; Vitarelli, A. Early Diagnosis of Cardiovascular Diseases in Workers: Role of Standard and Advanced Echocardiography. *BioMed Res. Int.* **2018**, *2018*, 7354691. [CrossRef] [PubMed]
13. Pal, A.; Ahirwar, A.K.; Sakarde, A.; Asia, P.; Gopal, N.; Alam, S.; Kaim, K.; Ahirwar, P.; Sorte, S.R. COVID-19 and cardiovascular disease: A review of current knowledge. *Horm. Mol. Biol. Clin. Investig.* **2021**, *42*, 99–104. [CrossRef] [PubMed]
14. Talasaz, A.H.; Kakavand, H.; Van Tassell, B.; Aghakouchakzadeh, M.; Sadeghipour, P.; Dunn, S.; Geraiely, B. Cardiovascular Complications of COVID-19: Pharmacotherapy Perspective. *Cardiovasc. Drugs Ther.* **2021**, *35*, 249–259. [CrossRef] [PubMed]
15. Jayaraj, J.C.; Davatyan, K.; Subramanian, S.S.; Priya, J. Epidemiology of Myocardial Infraction. In *Myocardial Infraction*; IntechOpen: London, UK, 2018. [CrossRef]
16. Lu, L.; Liu, M.; Sun, R.R.; Zheng, Y.; Zhang, P. Myocardial Infarction: Symptoms and Treatments. *Cell Biophys.* **2015**, *72*, 865–867. [CrossRef]
17. Chartrain, A.G.; Kellner, C.P.; Mocco, J. Pre-hospital detection of acute ischemic stroke secondary to emergent large vessel occlusion: Lessons learned from electrocardiogram and acute myocardial infraction. *J. NeuroInterv. Surg.* **2018**, *10*, 549–553. [CrossRef]
18. Yakovlev, S.; Bazilevych, K.; Chumachenko, D.; Chumachenko, T.; Hulianytskyi, L.; Meniailov, I.; Tkachenko, A. The concept of developing a decision support system for the epidemic morbidity control. In Proceedings of the CEUR Workshop Proceedings 2020, the 3rd International Conference on Informatics & Data-Driven Medicine, Växjö, Sweden, 19–21 November 2020; Volume 2753, pp. 265–274.
19. Park, J.; An, J.; Kim, J.; Jung, S.; Gil, Y.; Jang, Y.; Lee, K.; Oh, I.-Y. Study on the use of standard 12-lead ECG data for rhythm-type ECG classification problems. *Comput. Methods Programs Biomed.* **2021**, *214*, 106521. [CrossRef]
20. Raeiatibanadkooki, M.; Quachani, S.R.; Khalilzade, M.; Bahaadinbeigy, K. Real Time Processing and Transferring ECG Signal by a Mobile Phone. *Acta Inform. Medica* **2014**, *22*, 389–392. [CrossRef]
21. Iqbal, M.N.; Bamhara, M.; Al Khambashi, M.; Alhassan, H.; Abd-Alhameed, R.; Eya, N.; Qahwaji, R.; Noras, J. Real-time signal processing of data from an ECG. In Proceedings of the Internet Technologies and Applications (ITA) 2017, Wrexham, UK, 12–15 September 2017; pp. 334–338. [CrossRef]
22. Quer, G.; Arnaout, R.; Henne, M.; Arnaout, R. Machine Learning and the Future of Cardiovascular Care: JACC State-of-the-Art Review. *J. Am. Coll. Cardiol.* **2021**, *77*, 300–313. [CrossRef]
23. Zheng, H.; Wzng, H.; Nugent, C.D.; Finlay, D.D. Supervised classification models to detect the presence of old myocardial infraction in body surface potential maps. In Proceedings of the Computers in Cardiology 2006, Valencia, Spain, 17–20 September 2006; pp. 265–268.

24. Polat, K.; Şahan, S.; Güneş, S. Automatic detection of heart disease using an artificial immune recognition system (AIRS) with fuzzy resource allocation mechanism and k-nn (nearest neighbour) based weighting preprocessing. *Expert Syst. Appl.* **2007**, *32*, 625–631. [CrossRef]
25. Tu, M.C.; Shin, D.; Shin, D. Effective Diagnosis of Heart Disease through Bagging Approach. In Proceedings of the 2009 2nd International Conference on Biomedical Engineering and Informatics, Tianjin, China, 17–19 October 2009; pp. 1–4. [CrossRef]
26. Shounam, M.; Turner, T.; Stocker, R. Using decision tree for diagnosing heart disease patients. In Proceedings of the 9th Australian Data Mining Conference 2011, Victoria, Australia, 1–2 December 2011; pp. 23–29.
27. Ghumbre, S.; Patil, C.; Ghatol, A. Heart disease diagnosis using support vector machine. In Proceedings of the International Conference on Computer Science and Information Technology 2011, Penang, Malaysia, 22–24 February 2011; pp. 84–88.
28. Shouman, M.; Turner, T.; Stocker, R. Integrating Naïve Bayes and K-means clustering with different initial centroid selection methods in the diagnosis of heart disease patients. In Proceedings of the Computer Science and Information Technologies 2012, Bangalore, India, 2–4 January 2012; pp. 125–137. [CrossRef]
29. Chitra, R.; Seenivasagam, V. Heart disease prediction system using supervised learning classifier. *Int. J. Softw. Eng. Soft Comput.* **2013**, *3*, 1–7.
30. Shouman, M.; Turner, T.; Stocker, R. Integrating clustering with different data mining techniques in the diagnosis of heart disease. *J. Comput. Sci. Eng.* **2013**, *20*, 1–10.
31. Yuwono, T.; Setiawan, N.A.; Nugroho, H.A.; Persada, A.G.; Prasojo, I.; Dewi, S.K.; Rahmadi, R. Decision Support System for Heart Disease Diagnosing Using K-NN Algorithm. In Proceedings of the International Conference on Electrical Engineering, Computer Science and Informatics 2015, Palembang, Indonesia, 19–21 August 2015; Volume 2, pp. 160–164. [CrossRef]
32. Bashir, S.; Qamar, U.; Khan, F.H. A Multicriteria Weighted Vote-Based Classifier Ensemble for Heart Disease Prediction. *Comput. Intell.* **2015**, *32*, 615–645. [CrossRef]
33. Kirmani, M.M.; Ansarullah, S.I. Prediction of heart disease using decision tree a data mining technique. *Int. J. Comput. Sci. Netw.* **2016**, *5*, 855–892.
34. Yuwono, T.; Franz, A.; Muhimmah, I. Design of Smart Electrocardiography (ECG) Using Modified K-Nearest Neighbor (MKNN). In Proceedings of the IEEE 2018 1st International Conference on Computer Applications & Information Security, Riyadh, Saudi Arabia, 4–6 April 2018; pp. 1–5. [CrossRef]
35. Ishaque, S.; Khan, N.; Krishnan, S. Trends in Heart-Rate Variability Signal Analysis. *Front. Digit. Health* **2021**, *3*, 639444. [CrossRef] [PubMed]
36. Hodgart, E.; Macfarlane, P.W. 10 second heart rate variability. In Proceedings of the Computers in Cardiology 2004, Chicago, IL, USA, 19–22 September 2004; pp. 217–220. [CrossRef]
37. Khamis, H.S.; Cheruiyot, K.W.; Kimani, S. Application of k-nearest neighbor classification in medical data mining. *Int. J. Inf. Commun. Technol. Res.* **2014**, *4*, 121–128.
38. Hu, L.-Y.; Huang, M.-W.; Ke, S.-W.; Tsai, C.-F. The distance function effect on k-nearest neighbor classification for medical datasets. *SpringerPlus* **2016**, *5*, 1304. [CrossRef]
39. Sahu, S.K.; Mishra, B.; Thakur, R.S.; Sahu, N. Normalized hamming k-nearest neighbor (NHK-nn) classifier for document classifi-cation and numerical result analysis. *Glob. J. Pure Appl. Math.* **2017**, *13*, 4837–4850.
40. Rovetta, A. Raiders of the Lost Correlation: A Guide on Using Pearson and Spearman Coefficients to Detect Hidden Correlations in Medical Sciences. *Cureus* **2020**, *12*, e11794. [CrossRef]
41. Alexandridis, A.; Chondrodima, E. A medical diagnostic tool based on radial basis function classifiers and evolutionary simulated annealing. *J. Biomed. Inform.* **2014**, *49*, 61–72. [CrossRef]
42. Shashua, A. Introduction to machine learning: Class notes 67577. *arXiv* **2009**, arXiv:0904.3664.
43. Dudkina, T.; Meniailov, I.; Bazilevych, K.; Krivtsov, S.; Tkachenko, A. Classification and prediction of diabetes disease using decision tree method. In Proceedings of the CEUR Workshop Proceedings 2021, Symposium on Information Technologies & Applied Sciences (IT&AS 2021), Bratislava, Slovakia, 5 March 2021; Volume 2824, pp. 163–172.
44. Sokoliuk, A.; Kondratenko, G.; Sidenko, I.; Kondratenko, Y.; Khomchenko, Y.; Atamanyuk, I. Machine Learning Algorithms for Binary Classification of Liver Disease. In Proceedings of the 2020 IEEE International Conference on Problems of Infocommunications. Science and Technology (PIC S&T), Kharkiv, Ukraine, 6–9 October 2020; IEEE: Piscataway, NJ, USA, 2020; pp. 417–421. [CrossRef]
45. Wagner, P.; Strodthoff, N.; Bousseljot, R.; Samek, W.; Schaeffter, T. *PTB-XL*, version 1.0.1. PTB-XL, A Large Publicly Available Electrocardiography Dataset. PhysioNet: Bristol, UK, 2020. [CrossRef]
46. Goldberger, A.; Amaral, L.; Glass, L.; Hausdorff, J.; Ivanov, P.C.; Mark, R.; Mietus, J.E.; Moody, G.B.; Peng, C.K.; Stanley, H.E. PhysioBank, PhysioToolkit, and PhysioNet: Components of a new research resource for complex physiologic signals. *Circulation* **2020**, *101*, e215–e220. [CrossRef] [PubMed]
47. Wagner, P.; Strodthoff, N.; Bousseljot, R.-D.; Kreiseler, D.; Lunze, F.I.; Samek, W.; Schaeffter, T. PTB-XL, a large publicly available electrocardiography dataset. *Sci. Data* **2020**, *7*, 154. [CrossRef] [PubMed]
48. Smith, S.J.M. EEG in the diagnosis, classification, and management of patients with epilepsy. *J. Neurol. Neurosurg. Psychiatry* **2005**, *76* (Suppl. 2), ii2–ii7. [CrossRef] [PubMed]
49. Malinova, V.; von Eckardstein, K.; Mielke, D.; Rohde, V. Diagnostic yield of fluorescence-assisted frame-based stereotactic biopsies of intracerebral lesions in comparison with frozen-section analysis. *J. Neuro-Oncology* **2020**, *149*, 315–323. [CrossRef] [PubMed]

50. Goodenberger, M.K.L.; Jenkins, R.B. Genetics of adult glioma. *Cancer Genet.* **2012**, *205*, 613–621. [CrossRef] [PubMed]
51. Zehra, T.; Anjum, S.; Mahmood, T.; Shams, M.; Sultan, B.A.; Ahmad, Z.; Alsubaie, N.; Ahmed, S. A Novel Deep Learning-Based Mitosis Recognition Approach and Dataset for Uterine Leiomyosarcoma Histopathology. *Cancers* **2022**, *14*, 3785. [CrossRef]

sensors

MDPI

Article

Affective Recommender System for Pet Social Network

Wai Khuen Cheng [1], Wai Chun Leong [1], Joi San Tan [1], Zeng-Wei Hong [2] and Yen-Lin Chen [3,*]

1 Faculty of Information and Communication Technology, Universiti Tunku Abdul Rahman, Kampar 31900, Perak, Malaysia
2 Department of Information Engineering and Computer Science, Feng Chia University, Taichung 40724, Taiwan
3 Department of Computer Science and Information Engineering, National Taipei University of Technology, Taipei 106344, Taiwan
* Correspondence: ylchen@mail.ntut.edu.tw

Abstract: In this new era, it is no longer impossible to create a smart home environment around the household. Moreover, users are not limited to humans but also include pets such as dogs. Dogs need long-term close companionship with their owners; however, owners may occasionally need to be away from home for extended periods of time and can only monitor their dogs' behaviors through home security cameras. Some dogs are sensitive and may develop separation anxiety, which can lead to disruptive behavior. Therefore, a novel smart home solution with an affective recommendation module is proposed by developing: (1) an application to predict the behavior of dogs and, (2) a communication platform using smartphones to connect with dog friends from different households. To predict the dogs' behaviors, the dog emotion recognition and dog barking recognition methods are performed. The ResNet model and the sequential model are implemented to recognize dog emotions and dog barks. The weighted average is proposed to combine the prediction value of dog emotion and dog bark to improve the prediction output. Subsequently, the prediction output is forwarded to a recommendation module to respond to the dogs' conditions. On the other hand, the Real-Time Messaging Protocol (RTMP) server is implemented as a platform to contact a dog's friends on a list to interact with each other. Various tests were carried out and the proposed weighted average led to an improvement in the prediction accuracy. Additionally, the proposed communication platform using basic smartphones has successfully established the connection between dog friends.

Keywords: affective recommendation; pet social network; emotion recognition model; dog barking recognition; deep learning

Citation: Cheng, W.K.; Leong, W.C.; Tan, J.S.; Hong, Z.-W.; Chen, Y.-L. Affective Recommender System for Pet Social Network. *Sensors* **2022**, *22*, 6759. https://doi.org/10.3390/s22186759

Academic Editor: Daniele Cenni

Received: 25 July 2022
Accepted: 2 September 2022
Published: 7 September 2022

Publisher's Note: MDPI stays neutral with regard to jurisdictional claims in published maps and institutional affiliations.

Copyright: © 2022 by the authors. Licensee MDPI, Basel, Switzerland. This article is an open access article distributed under the terms and conditions of the Creative Commons Attribution (CC BY) license (https://creativecommons.org/licenses/by/4.0/).

1. Introduction

With the emergence of the Internet of Things (IoT), the landing of smart homes in the new era is no longer impossible. Current smart home designs are smarter when integrated with recommender systems (RS) [1–8]. RS and the Internet of Things (RSIoT) are highly dependent on real-time resources, especially sensor data, not just interactions between users and items. The initial stages of acquiring data, especially from sensors, are critical as these data are preprocessed (removing noise or redundant features) and generated events by defining suitable rules. After that, the system is able to learn the pattern of the rules and provide recommendations that match users' preferences. Some smart systems [9–13] have been developed to promote efficient resource mapping through user habits. Habits are often formed when intentions are translated into actions and behaviors repeatedly [14]. Resource mapping efficiency can be achieved by gradually changing user habits through micro-moments and recommendations [15]. Most current systems are smarter than ones in the past because they leverage users' social networks and integrate this information with the system to provide preferred recommendations [11]. Furthermore, by considering the characteristics of users, a preferred system with an appropriate level of automation can be designed [16].

Users of smart homes are not limited to humans but animals. Most pet owners can only monitor their pets' behaviors through home security cameras, especially when they are not at home. However, some pets such as dogs require long-term close companionship with their owners. Dogs are highly social animals that easily form close attachments to their own species or other species [17]. There comes a time when a puppy or dog is separated from its owners and most of them learn to adjust to social isolation at home. However, some dogs later become sensitive to social isolation (left at home alone) and tend to develop separation anxiety, which can lead to excessive vocalization and disruptive behavior. Huasang et al. [18] proposed a multi-level hierarchical behavior monitoring system to detect separation anxiety symptoms in dogs. The purpose of the system is to automatically monitor the dogs and analyze their behaviors through a taxonomy consisting of three progressive levels. In the system, the Stacked Long Short-Term Memory (LSTM) is adopted to recognize postures through sensors. These postures are then interpreted by a Fuzzy Complex Event Processing (CEP) engine that detects the anxiety symptoms.

In this study, we developed a smart environment for domestic dogs that not only monitors dogs' behaviors but also integrates their social networks to relieve separation anxiety, especially for those that are left alone. People adopted dogs for stress reliever, companion, and protection purposes [19]. This phenomenon is more pronounced during the COVID-19 pandemic period. However, pandemic puppies turn into a big issue for many inexperienced owners. These puppies are deprived of socialization, which not only happens during the pandemic but gets worse once their inexperienced owners return to normal job routines as before the pandemic. After the lockdowns are lifted, dogs need a transition period to get used to being away from their owners. According to suggestions by dog behavior specialists and veterinarians [20–22], there are several ways to ease post-pandemic separation anxiety in dogs. Experts recommend dog owners to provide an environment in which the dogs can relax when nobody is home. Dog owners can also adopt some technology gadgets to monitor their dogs and make those gadgets as interactive toys to keep their brains and bodies moving when they are alone. All these approaches are useful in providing dogs with enrichment that can be enjoyed independently. Our proposed system aims to create a safe and comfortable place for dogs and implement the suggested approaches by the experts in solving dogs' separation anxiety issues. The proposed solution allows dogs from different households to communicate remotely by using a distributed system architecture with cloud computing adoption. The Real-Time Messaging Protocol (RTMP) servers are used by the social network platform to connect and communicate with their dog friends. In the system, on the other hand, the dogs' behaviors are predicted through emotion recognition and sound (barking) recognition, and this makes it possible to implement an efficient recommender system for dogs. A large number of images consisting of various dog expressions were collected and a dog expression classification model was trained using the Residual Neural Network (ResNet) [23]. Dogs' emotions are definable based on predicted expressions. Similarly, different audio files of dog barks were collected, and the sequential model was used to train a sound recognition classification model. Subsequently, the expression classifier was combined with the sound classifier using weighted average techniques to improve the behavior prediction results.

The main contributions of this paper are summarized as follows: (1) present a unique cloud-based smart environment dogs' social network architecture; (2) propose an affective recommender framework with dogs' emotion recognition and sound (barking) recognition; (3) proof of concept and verify the viability of the proposed dogs' social network architecture. The rest of the paper is organized as follows. In Section 2, related works on the RSIoT are presented. Section 3 illustrates the overall cloud-based dogs' social network architecture. The affective recommender framework and the dogs' emotion recognition model are discussed in Section 4. Section 5 presents the experiment results, and the conclusion is stated in the last section.

2. Related Work

With the advancement of technology and the pursuit of a better quality of life, smart home systems are rapidly gaining attention. The main purpose of most systems is to identify any proactive behavior of users in the current situation and recommend them a service that suits their habits [1]. The recommendations are constructed based on long-term studies of the repetitive patterns in users' daily lives [3]. In 2010, Parisa et al. [24] developed an unsupervised model to track and recognize activities in a smart environment. The Discontinuous Varied-Order Sequential Miner (DVSM) was proposed to determine activity patterns that might be discontinuous or in various order. The patterns were grouped together and represented using cluster centroids. Later, the boosted version of the hidden Markov model was used to represent the activities and recognize them in the environment.

Katharina [1] proposed a smart home system integrated with an unsupervised recommender system that predicted the relationships between users' actions through collected data. The system tries to predict the next action of the users and recommends some actions. Firstly, a formal model of the context which represents the multidimensional space was constructed. These contexts were the users' actions that related to each other which integrated with time elapsed and represented with tuples. These tuples were trained based on the basis of observed sensor events. An algorithm Dempster–Shafer theory which is similar to the Naïve Bayes was proposed to predict the next contexts based on the current action. A ranked list was provided as the output of the recommendation.

The Pervasive RS (PRS) was proposed by Naouar et al. [25] which represents the contexts in tuples. The data were collected through physical sensors including RFID, and later it was transformed into various contexts to build the user profile according to preferences. Preferences are actions that occur repeatedly and are relevant to each other. The Apriori algorithm was implemented to extract the relevant preferences that occurred from the database. A three-layer neural network based on back propagation was proposed to predict user preferences in a given context. Nirmalya and Chia proposed a model named Complex Activity Recognition Algorithm (CARALGO) which is based on probability theory [26]. The main idea is to decompose a complex activity into small atomic activities, and the context attributes are constructed so that each of these activities is associated with a specific weight depending on their relevance. The occurrence of the activities is decided by the threshold function. The number of ways to perform complex activities is derived through the binomial theorem.

Alexander et al. [27] discussed the new recommendation techniques that are relevant to real-world IoT scenarios including the IoT gateway. Smart homes with RS should be able to enhance the applicability of the equipment and optimize the usage of the resources. The SEQREQ was developed to recommend items by finding sequential patterns; it analyzes users with similar behaviors that share common sequences of actions. The idea is to find the common node sequences (which are similar to the actions) that are available in the workflow repository and list them in a look-up table. Then, similarity values are calculated between the actions and the common node sequences where values greater than zero will be recommended. It is important that the RS is able to recommend items based on the sequence of the activities.

The subjects of recognition are not limited to humans; they can also be animals such as horses [28]. The behaviors of both subjects were analyzed in order to recognize their actions and provide some recommendations. In terms of Animal Activity Recognition (AAR), it can be an owner that is monitoring their pet when they are not at home; it can also be the observation of wildlife in a natural environment. Basically, the processing pipeline of the AAR and the Human Activity Recognition (HAR) are quite similar to each other since they both capture the activity data through sensors, and the features from the activity data are extracted and further classified into a few groups [29]. The main difference between the AAR and the HAR is the input data and the output data they produce.

Cassim et al. [30] carried out a study to recognize the activity of dogs. It determines a set of activities that are connected to the behavioral patterns that identify dogs' behavior.

The dogs were required to wear a collar-worn accelerometer in order to collect their movements, such as body movements and response behaviors. Feature extraction was carried out using principal component analysis (PCA) and the k-nearest neighbor was implemented to classify the features. Yumi et al. [31] proposed research to study the AAR based on a first-person view from a dog. In this research, a GoPro camera was attached to the back of the dogs and recorded the activities that were carried out by them from their viewpoint. From the video recording, global and local features were extracted using various algorithms such as dense optical flow, local binary patterns, cuboid detector, and STIP detector. Global features were mainly captured from the dogs' motions, whereas local features were captured from motions other than the dogs. Visual words were integrated in order to increase the efficiency of the representation of the motion. Lastly, the support vector machine (SVM) is used to classify first-person animal activities through features.

Patricia, Javier, and Alejandro [32] developed a system that is able to track cats' location, posture, and field of view using a depth-based method. The Microsoft Kinect sensor, which is able to record both color and depth video, was set up to capture the motion of a cat. The depth value of a cat's pixel in each video frame was extracted and divided into different clusters using the k-mean algorithm. Different postures produced different depth values for every part (head, body, and tail) of the cat. A decision tree was constructed by considering different parameters to determine body postures and classify the clusters. Jacob et al. proposed a multitask learning (MTL) framework for embedded platforms to perform AAR [33]. This framework is able to solve multiple tasks simultaneously and explore connections among the tasks using the Relief algorithm. The dataset was collected from multiple sensors and features were extracted. To perform action (or task) classification, seven classification techniques including deep neural network (DNN) were implemented. DNN was able to provide promising results in this approach. Enrico et al. [34] studied horse gait activity recognition by capturing the data using the built-in accelerometer sensor in a smartwatch through a developed application. The smartwatch was placed on the saddle of a horse and the wrist of the rider. Each gait has distinctive characteristics, and its features were extracted using different algorithms such as neural networks, decision trees, k-neighbors, and support vector machines. The performances of the algorithms were compared and showed similar results.

Studies in pet emotion recognition and RS are still under exploration, and most of the existing works are mainly focused on dogs [35] and cats [36]. For instance, Quaranta et al. [36] noticed that different cats' vocalizations that they had recorded produced different patterns of sound waves. Each pattern of sound waves should represent a relevant cat condition. Similar research was presented by Varun et al. [37]. They presented a recommender framework with dog vocalization pattern recognition in their study. The authors gathered a number of vocalization patterns and taught the convolutional neural networks to recognize dog emotions. Bhupesh et al. [38] noticed that animals express different types of expressions on their faces in different scenarios. The authors managed to run several experiments to assess their hypothesis on sheep and rats. They observed the animals' noses, ears, whiskers, and eyes react differently when receiving different levels of stimulation. In addition, Cátia Caeiro et al. [39] also inspected dogs' facial expressions under different scenarios. They discovered dogs showed a higher level of facial expression in conditions such as "fear" and "happy", but not "frustrated".

In recent years, deep learning has been widely used for various recognition applications as it is able to provide promising outputs with sufficient training through large amounts of data [40–42]. Through the training process, it is able to capture the relationship between the data itself [43]. Mohammed et al. [44] proposed a novel approach that was implemented through the deep belief network (DBN) to train the activities and recognition. The actions were collected using accelerometers and gyroscope sensors. Based on the sensor data, multiple features were extracted, and the kernel principal component analysis (KPCA) was used to reduce the data dimension before training. Jacob et al. [45] studied the AAR by focusing on unsupervised representation learning. It aimed to recognize activities from the raw motion data (unlabeled) that was collected online using an accelerometer. Various features were extracted from the collected data using algorithms and further classified into different activities. Algorithms such as PCA, sparse autoencoders (SAE), and convolutional deep belief network (CDBN) were implemented to extract features, while the support vector machine (SVM) was used to perform the activity classification. The performances of these algorithms were compared and evaluated using F1 measures. Rosalie Voorend [29] implemented a variational autoencoder (VAE) to perform feature extraction and a sequential classifier to classify the activity. The autoencoder was proposed to deal with unsupervised representation learning and it has not been extensively explored in the AAR. However, the output that the autoencoder produced is not satisfying enough when compared to the statistical approach. This is probably because the loss function in the VAE is not optimized. Coherence within the input data which causes the representations to be unable to be extracted properly is needed as well. Enkeleda et al. [46] proposed deep convolutional neural networks (ConvNets) to recognize the activity of livestock animals without feature extraction. The proposed network has four layers and each layer consists of different operations. Different hyperparameters were adjusted and their performances were compared.

3. Dogs' Social Network Architecture

Figure 1 illustrates the overall implementation of a pet social network on a cloud computing platform. First, an Android app was developed with social networks and sensing capabilities (e.g., cameras and microphones). The social networking app serves as the interface layer to allow owners to register their dogs and connect other users' profiles to their pets' networks. The mobile app can detect dogs' movements and capture their images and sounds via live streaming which is connected to its own Real-Time Messaging Protocol (RTMP) server when the dog is near the device. The captured frames (images and audio clips) will then be uploaded to the Ubuntu VM instance hosted in the Google Cloud Platform. Those images and audio clips are uploaded through POST requests to the Node.js RESTful API. After receiving the files, Node.js saves the image and audio files into the "/images" and "/audios" directories, respectively. The affective recommender engine will be triggered by a python script (dogEmotionClassifier.py) in order to grab those relevant image and audio files. Dogs' facial expressions and barking analysis are performed at this stage, and the predicted results will be returned to Node.js. The RESTful API stores the predicted result in the MySQL database and further obtains a recommended action from the database records according to the respective input.

For instance, if the predicted result is "sick" for the dog's condition, the MySQL database should return the owner's email; additionally, an alert message will be delivered to the owner. On the other hand, if the predicted result shows "boring", the interface of the Android device will be switched on and connect to one of the dog's friends in its network. When an active account (dogs that are near their respective devices through sensing) is chosen, dogs are able to meet each other, and the barking records from both sides will be shared when they are captured. Furthermore, Google data studio is used to compile and visualize dogs' conditions. Dog owners can even access an interactive dashboard and monitor their pets remotely through the system.

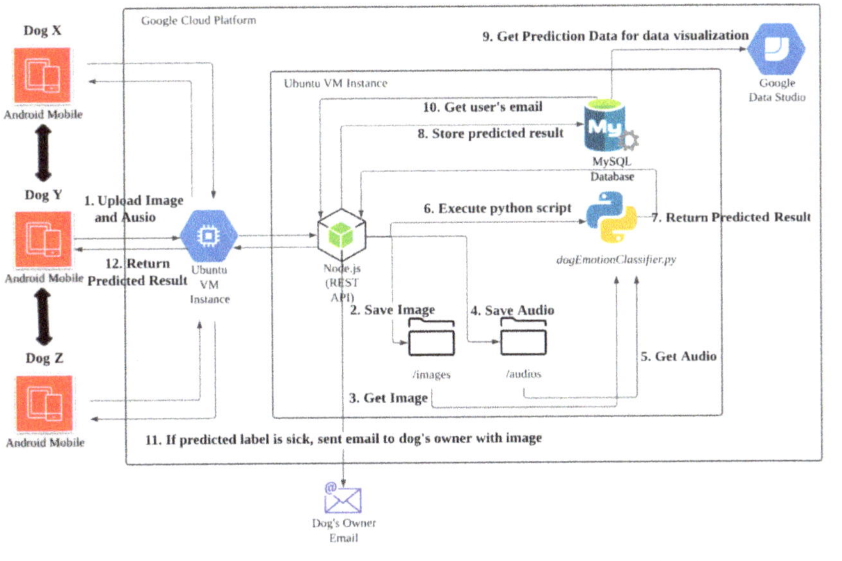

Figure 1. The proposed cloud-based smart environment dogs' social network architecture.

4. Affective Recommender Framework

The proposed affective recommender engine aims to provide an alert or early notification services to inexperienced dog owners through the dog's facial expression and barking analysis. There are several alternatives or auxiliary elements for assessing dogs' expression and behavior, such as ear and tail positions, mouth conditions, and body postures [47]. However, the facial expression of animals is still the richest channel that is used for expressing emotions [48]. Recognizing these visual signal expressions as emotional communication is important because emotions describe the internal state that is influenced by the central nervous system in response to an event [49]. Most experienced dog owners can equally identify the explicit dog's facial expression; thus, these human experts help in verifying the recognition performances easily later [50]. In addition to facial expressions, acoustic parameters such as dog barks showed promising performance in recognition tasks. Dog barking analysis can achieve more than human-level performance when classifying the context of a dog's bark [51]. The motivation for the proposed affective recommender engine is to combine both dog facial expressions and barking analysis for better dog emotion recognition. The recommender engine consists of the following modules, as shown in Figure 2.

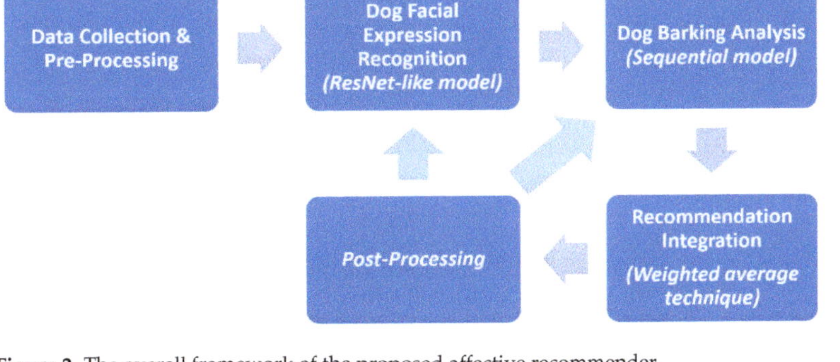

Figure 2. The overall framework of the proposed affective recommender.

4.1. Data Collection and Pre-Processing

Before training, images of dogs with various expressions were collected and divided into three categories: happy, angry, and sick. The collection of the images was performed according to the description in [49] as shown in Table 1. A Python script with an automated bot was written to download images of dogs from Google Images and save them in local storage. Images that were not related to the categories were removed, and the images were resized to a specific resolution of 224 × 224, as shown in Figure 3. To start building the recognition model, images were split into training, validation, and test data. Since the dataset was small, data augmentation was performed to replace the original batch of images with a randomly transformed batch.

Table 1. Dog facial expressions (happy, angry, and sick) characteristics [49].

Facial Expression	Eyes	Ears	Mouth/Teeth
Happy	Wide open, merry looking, raised eyebrows	Perked-up and forward, or relaxed	Mouth relaxed and slightly open, teeth covered, excited panting, possible lip-licking
Angry	Narrow or staring challengingly	Forward or back, close to head	Lips open, drawn back to expose teeth bared in a snarl, possible jaw snapping
Sick	Eyelids semi-closed with tearing, raised eyebrows, simulating large eyes, sad gaze	Distance between ears tends to widen	Contracted, giving the appearance of wrinkles on the face

Figure 3. Sample images of dogs with various expressions.

4.2. Dogs' Facial Expression Recognition

The idea of the deep learning algorithm Residual Neural Network (ResNet) [23] was adopted to train the image recognition engine due to its robust performance in image recognition. As described in the paper [23], the residual learning was integrated into every few stacked layers, which is known as the building block shown in the equation below:

$$y = \mathcal{F}(x, \{W_i\}) + x \quad (1)$$

where x and y are the input and output vectors of the layers considered, and $\mathcal{F}(x, \{W_i\})$ is the multiple convolutional layers in the residual block of the ResNet. To demonstrate the feasibility of the proposed framework, a ResNet-like model which consists of twelve layers (as shown in Figure 4b) was implemented. The ResNet-like model consists of four residual blocks, each of which consists of two convolutional layers and batch normalization, as shown in Figure 4a. In each convolutional layer, the filters are 32 and 64, respectively. There are two convolutional layers included after the two residual blocks of the filter size 32. To construct the model, the Adam optimizer [52] that performs fast optimization efficiently was chosen. In addition, the sparse categorical cross entropy was selected as the loss function where a single integer was labeled for each category rather than a whole vector. The expression "happy" is labeled as 0, "angry" is labeled as 1, and "sick" is labeled as 2. Global average pooling and a dense layer were implemented at the end of the model.

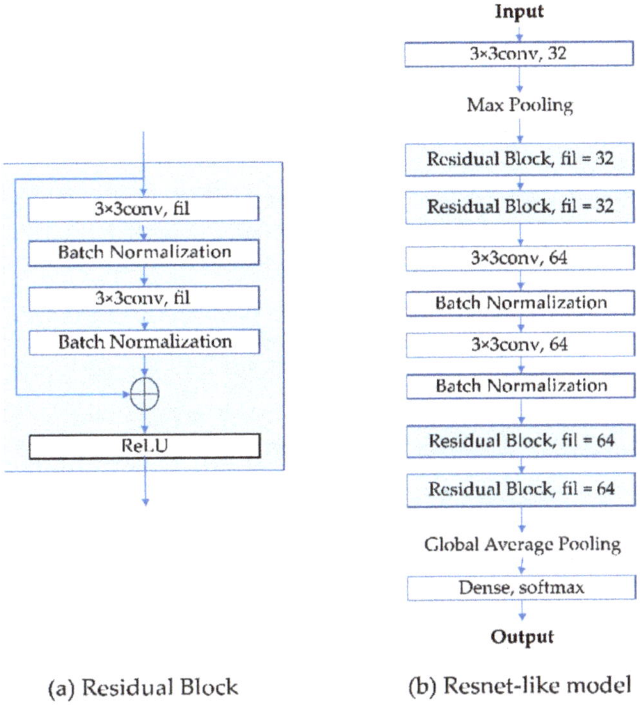

Figure 4. The ResNet-like model consists of twelve convolutional layers. (**a**) Residual block of the ResNet-like model with two convolutional and batch normalization layers, and (**b**) the whole structure of the ResNet-like model.

As shown in Figure 5, the code in the first block shows the function that generates a ResNet-like network. The second block indicates a function of the ImageDataGenerator that performs the data augmentation over the original batch images. The output from the data augmentation is selected during the training stage with the convolutional neural network (CNN) model. To determine the hyperparameters of the ResNet-like model, successive experiments were conducted. The details of the experiments will be discussed in Section 5. From the trained model, the emotions of dogs in input images are able to be identified based on the predicted values.

```python
def identity_block(n_f,x):
    shortcut = x
    x = Conv2D(n_f,(3,3),strides=(1,1),padding='same',kernel_initializer='he_normal',activation='relu')(x)
    x = BatchNormalization()(x)
    x = Conv2D(n_f,(3,3),strides=(1,1),padding='same',kernel_initializer='he_normal')(x)
    x = BatchNormalization()(x)
    x = Add()([shortcut,x])
    x = Activation('relu')(x)
    return x

def conv_block(n_f,x):
    x = Conv2D(n_f,(3,3),strides=(2,2),padding='same',kernel_initializer='he_normal',activation='relu')(x)
    x = BatchNormalization()(x)
    x = Conv2D(n_f,(3,3),strides=(2,2),padding='same',kernel_initializer='he_normal',activation='relu')(x)
    x = BatchNormalization()(x)
    return x
```

```python
datagen = ImageDataGenerator(width_shift_range=0.2,
                             height_shift_range=0.2,
                             horizontal_flip=True,
                             zoom_range=0.2, ######### modified
                             rotation_range=45)
```

```python
inputs = Input(shape=img_data[0].shape)
```

Figure 5. Building the residual block of the ResNet-like model.

4.3. Dog Barking Analysis

After performing dogs' facial expression recognition, a deep learning-based Sequential model was proposed to analyze dog barks. This study focuses on three types of dog barks: "bow-wow," "growling," and "howling." Each bark corresponds to an expression in the previous dog expression recognition, in which "bow wow" is happy, "growling" is angry, and "howling" is sick. A Python script was also written to download all the required dog barking video files from Google AudioSet and convert them to audio file format (WAV). Later, a software called Audacity was used to study the audio spectrum containing the desired barks, in which the patterns were identified and labeled, as shown in Figure 6. For "bow-wow" class labels, there were two audio spectrums with a gap between the barks. For "growling," the audio spectrum bounced up and down due to the vibrating sound that a dog makes. For the "howling" class label, the audio spectrum remained constant when the dog howled.

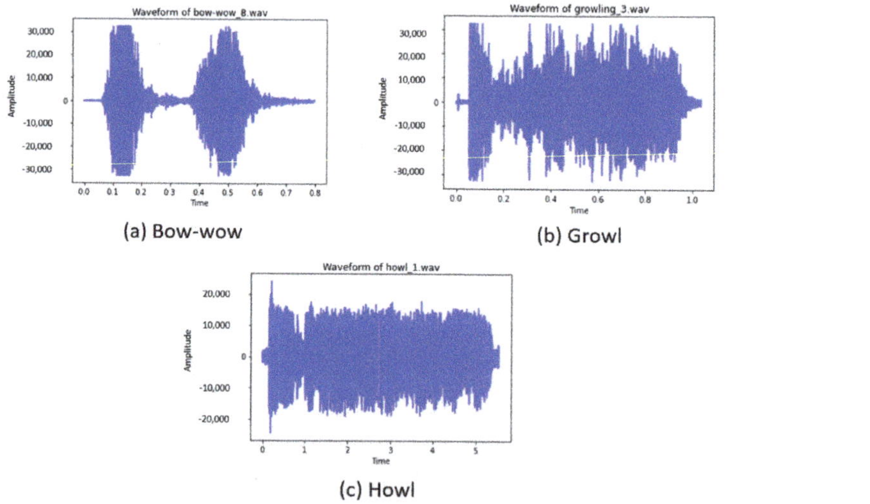

Figure 6. Audio Spectrums of dog barks. (**a**) Dog barks with a bow-wow sound, (**b**) dog barks with a growl sound, and (**c**) dog barks with a howl sound.

According to the identified patterns, the training dataset was prepared in the pre-processing stage: (1) audio features were extracted from audio files in all directories, and (2) class labels were inserted for each relevant dataset. Once the dataset was completed, a sequential model with four layers was constructed to classify the dog barks. The best epochs for the classification model will be discussed in Section 5. From the trained model, the expressions of the dog from the audio can be identified based on the predicted values.

4.4. Recommendation Integration and Post-Processing

A hybrid solution that integrated dogs' facial expressions and barking analysis was presented earlier. Subsequently, a weighted average technique was adopted to combine the outputs from two predictions. A weighted average function as shown in Figure 7 was chosen. In general, the proposed recommender system involves three sub-stages in predicting dogs' behavior: the first sub-stage performs dog image recognition; the second sub-stage operates dog bark recognition; the third sub-stage integrates both recognition outputs with a weighted average technique, as shown in Figure 8.

```
def average(a, b):
    combinedList = []
    combinedList.append(a.tolist())
    combinedList.append(b.tolist())

    dogPredArr = np.array(combinedList)
    predAvg = np.average(dogPredArr, axis=0, weights=[7,3])

    return predAvg
```

Figure 7. Functions are used to perform weighted average calculations.

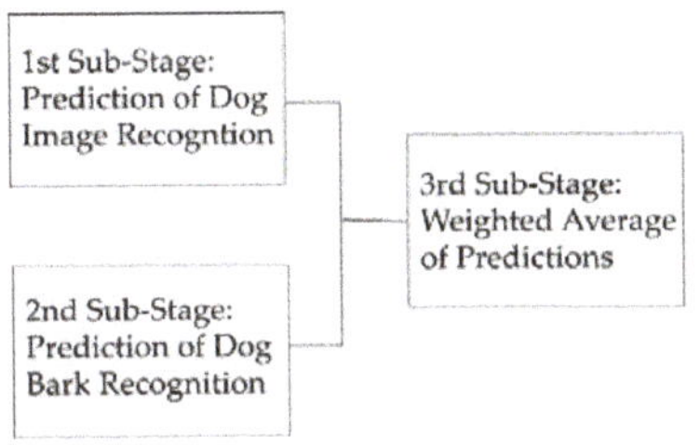

Figure 8. Three sub-stages to predict dogs' behavior.

The prediction outputs from the two trained models were combined to improve the result. Each input produces its predicted value for each category ("bow-wow," "growling," and "howling") from both models. A weighted average was implemented to calculate the weight of the predictions. The calculation is shown in the equation below:

$$Average\ weight = \frac{(7 * x) + (3 * y)}{10} \quad (2)$$

where x is the predicted value for a specific category of dogs' facial expression recognition, y is the predicted value for a specific category of dog barking recognition, and x is corresponding to y. The dogs' facial expression recognition model is weighted higher than the dog barking recognition model because it has higher accuracy. By comparing the average weights of the categories, the one with the highest values will be the predicted dog emotion or behavior.

As illustrated in Figure 2, the prediction outputs from the recommendation integration will provide feedback to respective recognition models in post-processing. The feedback includes user satisfaction and respective confidence values for further recommendation engine improvement and fine-tuning. The performance of the proposed affective recommender framework will be shown in Section 5.

4.5. Building Dogs' Social Network

As mentioned earlier, a social network for dogs is proposed to relieve separation anxiety, especially for those dogs that are left alone. A distributed system architecture is proposed to enable dogs to communicate with each other remotely, as shown in Figure 9. As described in Section 3, the developed mobile app in this study not only predicts dogs' behavior but also connects with other users' remote RTMP servers for interaction. Rather than installing complicated equipment, the proposed application allowed any household with dogs to create a smart home environment for their pets by setting up a mobile phone. Owners create a dog account in the application by providing the required information such as username, password, email, RTMP IP, and port, as shown in Figure 10a. When the account is completed, dogs can have their own friends, just like humans, and their owners can add them to the friend list, as shown in Figure 10b.

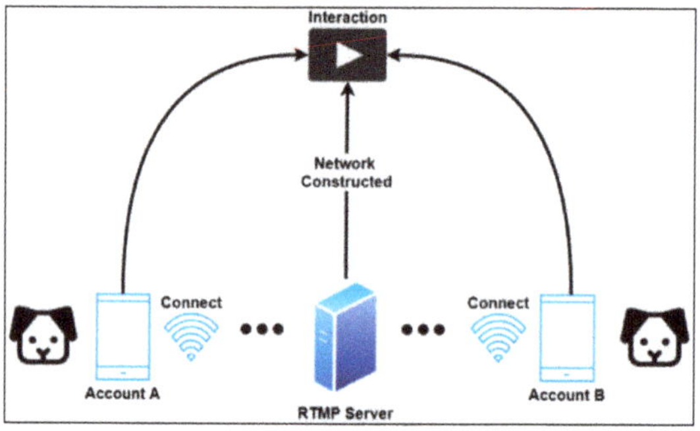

Figure 9. The distributed system architecture of proposed dogs' social network.

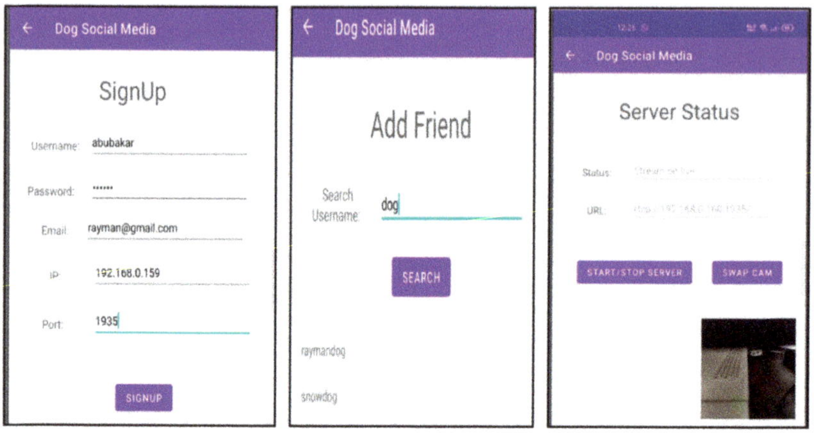

(a) Login (b) Add friend (c) Start server

Figure 10. User interface of the developed dogs' social network application: (**a**) login page of the application to set up live streaming, (**b**) add friend into the database, and (**c**) start the RTMP server.

In order to make a call, there are two important actions: (1) the RTMP server for streaming needs to be activated, as shown in Figure 10c and, (2) the system must check whether the selected friend's RTMP service is available as well. If it is available, the connection starts to be established and the system prepares the video and audio for live streaming on both sides. This is an automated process if the system detects the dog is "boring" and needs a friend. Figure 11a shows the user interface of the developed mobile app allowing a manual call. It enables the dog owner to manually make a call, just in case there is a need. If the connection to the friend's RTMP server is successful, the real-time video will be displayed and the audio function will be turned on, as shown in Figure 11b. In the platform setting, the mobile app captures the video and audio from the other side and uploads those data to the cloud for the dog's behavior training. As shown in Figure 11c, a dog with an unhealthy condition is detected; thus, an alert and notification email are sent to the owner to warn him about the dog's emotional condition.

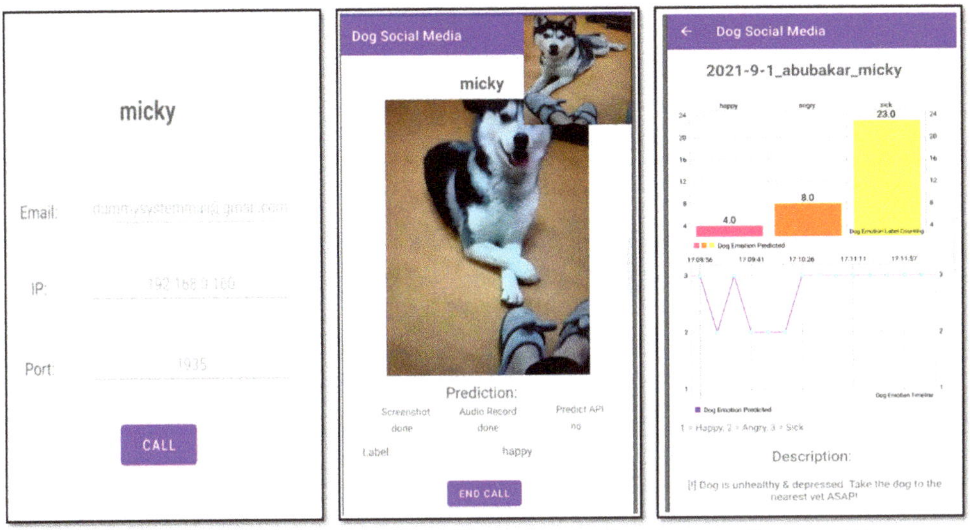

(a) Details of a friend (b) User Interface of a call (c) Dog conditions

Figure 11. User interface to make a friend call and show the dog's conditions: (**a**) contact a friend in the name list, (**b**) connect to a dog friend and live streaming, and (**c**) analysis report of the dog emotion.

5. Testing and Discussion

Various experiments have been carried out to train the deep learning models, as mentioned in Section 4, for the proposed affective recommendation engine.

5.1. Dog's Emotion Recognition

The ResNet-like was implemented to recognize dogs' facial expressions (as described in Section 4.2), and various tests were performed to determine its hyperparameters. Initially, hyperparameters of 200 epochs, batch size of 16, and 0.0005 learning rate were set for training with various dog images as described in Section 4.2. As shown in Figure 12, two sets of images were involved: (1) the dataset of images with a size of 636 for training, 80 for validation and 80 for testing. (2) The dataset of images with a size of 384 for training, 48 for validation and 48 for testing. Based on training prediction results (as shown in Table 2), the accuracies of using fewer images for validation and testing were 70.83% and 66.67%, whereas the accuracies of using more data for validation and testing were 73.75% and 72.50%. The testing was performed using the testing dataset and the accuracy rate of the dataset with fewer images reached 33.33%, which is much lower than the training prediction result, which indicates that overfitting has occurred. The result improved to 53.75% when the dataset with more images was tested. This shows that building the model using the dataset with more images has improved the recognition performance.

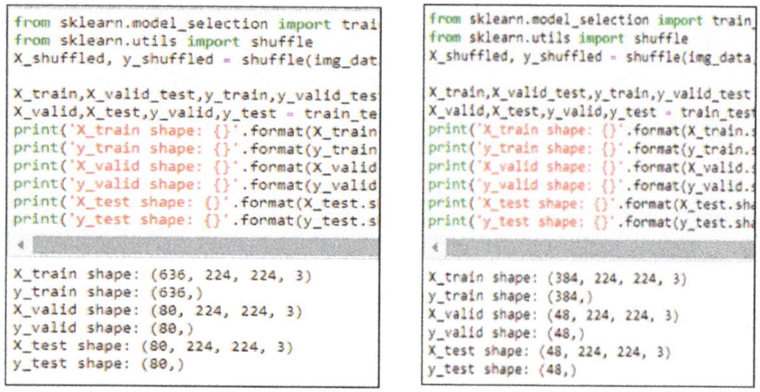

(**a**) (**b**)

Figure 12. Different numbers of images for training, validation and testing in ResNet-like. (**a**) Dataset with more images and, (**b**) Dataset with fewer images.

Table 2. The comparison of performance between smaller and larger sample sizes of data using ResNet-like batch 16. Larger sample size of data has better performance (as highlighted) compared to less data sample size.

	Evaluation Metric	Less Data Sample Size	More Data Sample Size
Training	Accuracy	70.83%	**73.75%**
	Loss	0.8192	**0.8289**
Validation	Accuracy	66.67%	**72.50%**
	Loss	0.8482	**0.6038**
Test	Accuracy	33.33%	**53.75%**
	Loss	0.8482	**0.6038**

Next, the test is continued by tuning the hyperparameters using the dataset with more images, as shown in Table 3. From the table, different learning rates with different numbers of epochs in two common batch sizes (16 and 32) were examined. First, learning rates ranging from 0.0001 to 0.1 with 50 epochs were tested with the batch sizes to determine the appropriate rate. During the training prediction, training loss, validation loss, training accuracy, and validation accuracy were obtained for both batch sizes. Graphs are also plotted, as shown in Figures 13 and 14. In the figures, the loss and accuracy for learning rates of 0.01 and 0.1 are not ideal when compared to the learning rates of 0.001 and 0.0001, where the loss is higher, and the accuracy is lower. When comparing the performance of all learning rates, the learning rate of 0.0001 shows continuous and steady improvement for both batch sizes. For example, in the validation loss, learning rates of 0.001, 0.01, and 0.1 fluctuate more than the learning rate of 0.0001, as shown in Figures 13 and 14. In other words, the learning rate of around 0.0001 is appropriate for the training of this model, where learning rates of between 0.0001 and 0.0005 are set for both batch sizes with the observation by increasing the number of epochs gradually for the next tuning step.

Table 3. Testing of the ResNet-like model with different hyperparameters using dataset with more images. The numbers in bold number are the hyperparameters discovered to build the ResNet-like model in this system.

Learning Rate	Batch Size of 16		Batch Size of 32	
0.1	50 epochs	**200 epochs**	50 epochs	200 epochs
0.01				
0.001				
0.0001				

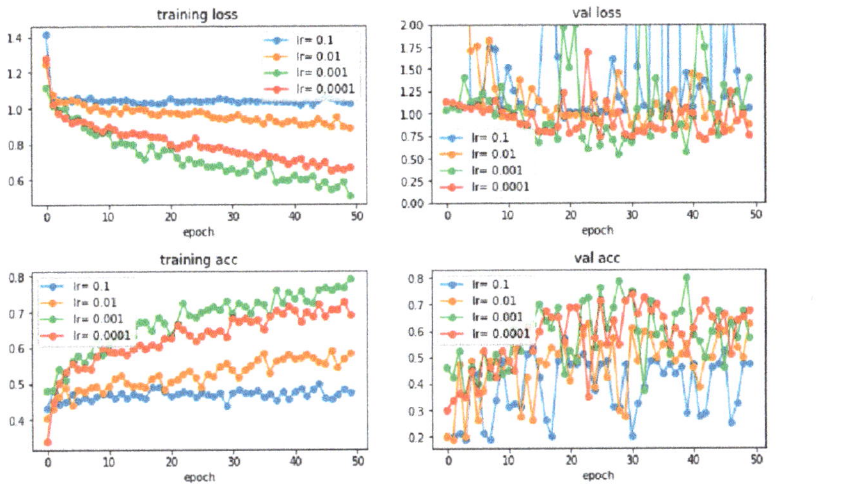

Figure 13. The training loss, validation loss, training accuracy, and validation accuracy of learning rates range from 0.0001 to 0.1 with 50 epochs and a batch size of 16.

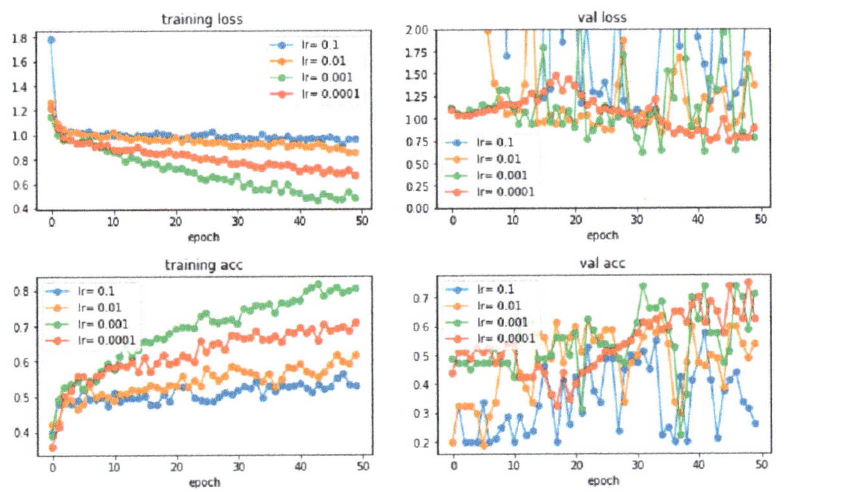

Figure 14. The training loss, validation loss, training accuracy, and validation accuracy of learning rates range from 0.0001 to 0.1 with 50 epochs and a batch size of 32.

Figures 15 and 16 are the training prediction results with batch sizes of 16 and 32. From the figures, the loss and accuracy fluctuate and become consistent starting around

75 epochs. The training and validation losses are consistent when the number of epochs with the batch size of 16 increases (as shown in Figure 15), whereas the training and validation loss function values deviate from each other with the batch size of 32 as shown in Figure 16. Later, the testing was conducted using the test dataset, and the accuracy and loss of batch size 16 reached 53.75% and 0.6038 while the accuracy and loss of batch size 32 reached 43.75% and 0.6629. As shown in Table 4, the result reveals that the model trained with batch size 16 is better than the batch size of 32 as it achieves better accuracy and lower loss. In summary, a learning rate of between 0.0001 and 0.0005, 200 epochs, and a batch size of 16 are the hyperparameters discovered to build the ResNet-like model in this system. The model is compared to VGG16 [53] as well when using the same settings of hyperparameters to evaluate the performance. The tests were also carried out in batch 16 and batch 32 for VGG16 and compared with ResNet-like in Table 4. As noticed in the table, the overall performance of ResNet-like is better than VGG16 since all the accuracies for VGG16 are less than 50% and the loss values are larger than 1.

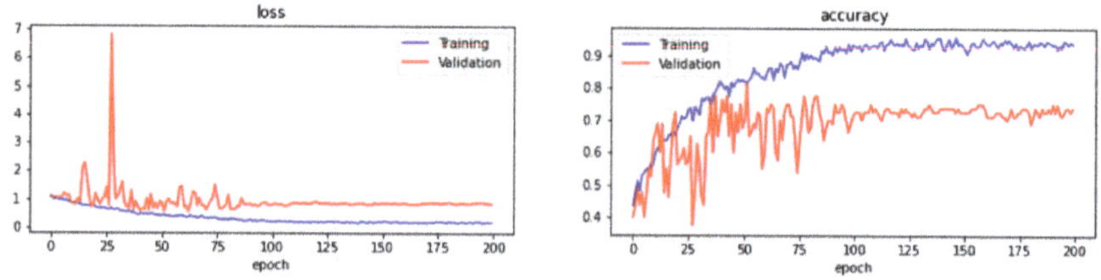

Figure 15. The training loss, validation loss, training accuracy, and validation accuracy of learning rates range from 0.0005 with maximum epochs of 200 and batch size of 16.

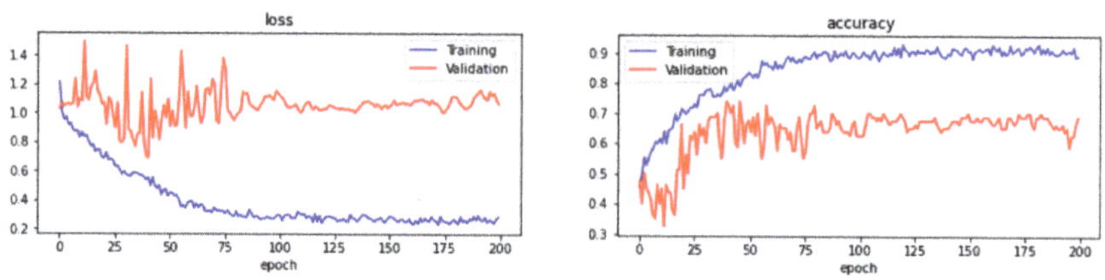

Figure 16. The training loss, validation loss, training accuracy, and validation accuracy of learning rates range from 0.0005 with maximum epochs of 200 and batch size of 32.

Table 4. The comparison of performances between hyperparameter batches 16 and 32. ResNet-like trained with batch size 16 is better than the batch size of 32 and it is also better than VGG16 as highlighted.

Hyper-Parameter		Evaluation Metric	ResNet-like	VGG16
Batch 16	Training	Accuracy	**73.75%**	47.50%
		Loss	**0.8289**	1.0408
	Validation	Accuracy	**72.50%**	47.50%
		Loss	**0.6038**	1.0408
	Test	Accuracy	**53.75%**	35.27%
		Loss	**0.6038**	1.0408
Batch 32	Training	Accuracy	**68.75%**	47.50%
		Loss	**1.0672**	1.0408
	Validation	Accuracy	**72.50%**	47.50%
		Loss	**0.6629**	1.0408
	Test	Accuracy	**43.75%**	38.16%
		Loss	**0.6629**	1.0408

With the constructed model, the test proceeded on the sample of dog images to predict dog emotions, as shown in Figure 17. The images of a dog named Luna were collected and tested on the model. Luna's emotions were predicted correctly in all images.

Figure 17. Prediction results of the emotions of a dog named Luna through the constructed ResNet-like model. Luna's emotion is predicted correctly in all images.

5.2. Dog Barking Emotion Recognition and Weighted Average for Dogs' Behavior Prediction

A sequential model was implemented to recognize dog barks (as described in Section 4.3) and simple tests were performed to determine its hyperparameters. Initially, 100 epochs and a batch size of 32 were set for training, and the validation of the training became consistent after starting a second epoch based on observation. Then, the model was tested with the test dataset, and the classification accuracy showed 75%. As shown in Figure 18, the model is able to predict the types of dog barks based on the provided audio test files.

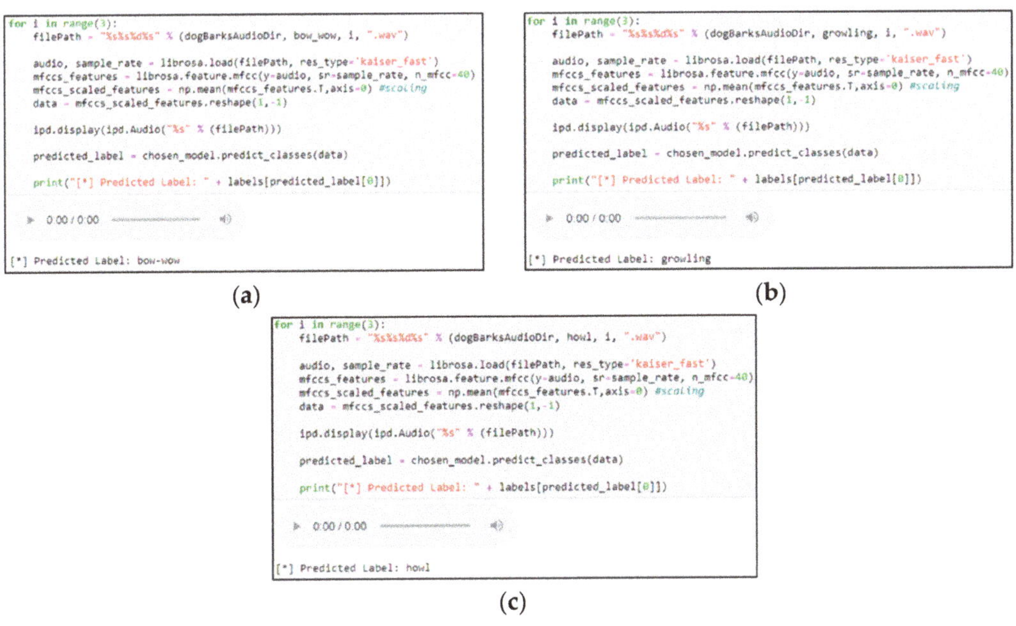

Figure 18. Prediction results of three types of dog barks on the test dataset using a trained model. (**a**) Prediction of bow-wow, (**b**) Prediction of growling and, (**c**) Prediction of Howl.

As explained in Section 4, the predicted outputs of the two trained models were combined through a weighted average using Equation (2) to enhance the prediction of dogs' behavior. The predicted output showed the dogs' behaviors which have been categorized as happy, angry, and sick. A total of 70 sample data files for each class label were prepared for testing, which are 210 dog images and 210 dog barking audio files in total. Figure 19 shows the accuracy of the predicted output for dog emotions, dog barking, and weighted average. The weighted average had the highest accuracy with 201 samples (95.70%) correctly predicting the dogs' behavior, while 187 images (89%) and 192 barks (91.40%) correctly predicted the dogs' behavior. Figure 20 shows three samples of the test data that correctly predict the dogs' behavior through the weighted average. In summary, the combination of dog emotions and dog barking improves the prediction accuracy of dogs' behavior in the three categories of happy, angry, and sick.

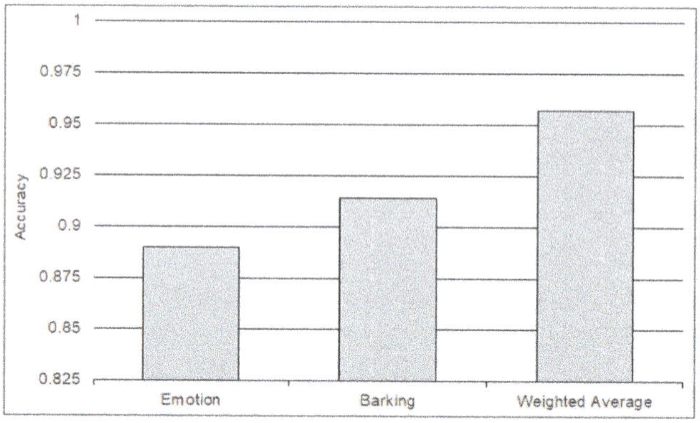

Notes: Sample Size: 210 data (Each label 70 data)

Evaluation Metric	Emotion	Barking	Weighted Average
No. of Correct Prediction	187	192	201
Accuracy	89.05%	91.43%	95.71%

Figure 19. Prediction accuracy obtained from dogs' emotions, dog barking, and weighted average. The weighted average which combines the prediction value of dogs' emotion and dog barking shows the highest accuracy when compared to emotion and barking.

(a) (b) (c)

Figure 20. Samples of test data (angry, happy, and sick) using the weighted average technique. The outputs shows (**a**) Dog is angry, (**b**) Dog is happy and, (**c**) Dog is sick.

6. Conclusions

Dogs are good companions for humans; they have a close relationship with their owners. However, dogs may face separation anxiety when they are apart from their owners for a long period of time and even develop disruptive behavior. Therefore, a novel cloud-based smart environment dog social network is proposed to solve this problem for dogs that live around the household. A mobile app for smartphones was developed to predict the dogs' behavior, and smartphones are used as communication devices to connect with different dog friends from different households. The ResNet-like model is used for dog emotion recognition in predicting dogs' behavior. A series of experiments were carried out to determine the hyperparameters of the ResNet-like model which found a learning rate of between 0.0001 and 0.0005, 200 epochs, and a batch size of 16. The proposed model was able to achieve 53.75% accuracy a 60.38% loss. The sequential model is used for dog barking recognition to predict the dog's behavior as well. The model was tested with the test dataset and the classification accuracy was shown to be 75%. Later, the weighted average technique (a combination of the prediction values of dog emotion recognition and dog barking recognition) was chosen to improve the prediction output, and it achieved an accuracy of 95.70%. On the other hand, the RTMP server is implemented as a platform to connect dog friends in a list using smartphones. Once RTMP is established, dogs can interact with each other, and it will trigger notification messages to owners once a sick dog is detected. In future work, dog pose recognition could be included to further improve the classification accuracy of the proposed affective recommender system. Due to the limitations of current data acquisition, multimodal training datasets should be applied for subsequent experiments to improve the recognition output. Furthermore, we may concentrate on the validity of the proposed system for various types of dogs and environments. The feasibility of the proposed solution could be one of the research directions.

Author Contributions: W.K.C. and W.C.L. investigated the ideas, review the systems and methods, and wrote the manuscript. J.S.T. provided the survey studies and methods. W.K.C. conceived the presented ideas and wrote the manuscript with support from Y.-L.C., Z.-W.H. provided suggestions on the experiment setup and provided the analytical results. W.K.C. and Y.-L.C. both provided suggestions on the research ideas, analytical results, wrote the manuscript, and provided funding support. All authors have read and agreed to the published version of the manuscript.

Funding: This work was funded by the Ministry of Science and Technology in Taiwan, under grant numbers MOST-109-2628-E-027-004-MY3, MOST-111-2218-E-027-003, and MOST-110-2622-8-027-006.

Institutional Review Board Statement: Not applicable.

Informed Consent Statement: Not applicable.

Data Availability Statement: There are no data applicable in this study.

Acknowledgments: The authors would like to thank Chun-Ting Chen for her English editing.

Conflicts of Interest: The authors declare no conflict of interest.

References

1. Rasch, K. An Unsupervised Recommender System for Smart Homes. *J. Ambient Intell. Smart Environ.* **2014**, *6*, 21–37. [CrossRef]
2. Ojagh, S.; Malek, M.R.; Saeedi, S.; Liang, S. A Location-Based Orientation-Aware Recommender System Using IoT Smart Devices and Social Networks. *Future Gener. Comput. Syst.* **2020**, *108*, 97–118. [CrossRef]
3. Mishra, P.; Gudla, S.K.; ShanBhag, A.D.; Bose, J. Alternate Action Recommender System Using Recurrent Patterns of Smart Home Users. In Proceedings of the 2020 IEEE 17th Annual Consumer Communications & Networking Conference (CCNC), Las Vegas, NV, USA, 10–13 January 2020; IEEE: Piscataway, NJ, USA, 2020; pp. 1–6.
4. Gladence, L.M.; Anu, V.M.; Rathna, R.; Brumancia, E. Recommender System for Home Automation Using IoT and Artificial Intelligence. *J. Ambient Intell. Humaniz. Comput.* **2020**, 1–9. [CrossRef]
5. Altulyan, M.; Yao, L.; Wang, X.; Huang, C.; Kanhere, S.S.; Sheng, Q.Z. A Survey on Recommender Systems for Internet of Things: Techniques, Applications and Future Directions. *Comput. J.* **2021**, *65*, 2098–2132. [CrossRef]
6. Liu, H.; Zheng, C.; Li, D.; Shen, X.; Lin, K.; Wang, J.; Zhang, Z.; Zhang, Z.; Xiong, N.N. EDMF: Efficient Deep Matrix Factorization with Review Feature Learning for Industrial Recommender System. *IEEE Trans. Ind. Inform.* **2021**, *18*, 4361–4371. [CrossRef]
7. Liu, H.; Zheng, C.; Li, D.; Zhang, Z.; Lin, K.; Shen, X.; Xiong, N.N.; Wang, J. Multi-Perspective Social Recommendation Method with Graph Representation Learning. *Neurocomputing* **2022**, *468*, 469–481. [CrossRef]

8. Li, D.; Liu, H.; Zhang, Z.; Lin, K.; Fang, S.; Li, Z.; Xiong, N.N. CARM: Confidence-Aware Recommender Model via Review Representation Learning and Historical Rating Behavior in the Online Platforms. *Neurocomputing* **2021**, *455*, 283–296. [CrossRef]
9. Rodríguez Fernández, M.; Cortés García, A.; González Alonso, I.; Zalama Casanova, E. Using the Big Data Generated by the Smart Home to Improve Energy Efficiency Management. *Energy Effic.* **2016**, *9*, 249–260. [CrossRef]
10. Hossain, M.S.; Rahman, M.A.; Muhammad, G. Cyber–Physical Cloud-Oriented Multi-Sensory Smart Home Framework for Elderly People: An Energy Efficiency Perspective. *J. Parallel Distrib. Comput.* **2017**, *103*, 11–21. [CrossRef]
11. Lye, G.X.; Cheng, W.K.; Tan, T.B.; Hung, C.W.; Chen, Y.-L. Creating Personalized Recommendations in a Smart Community by Performing User Trajectory Analysis through Social Internet of Things Deployment. *Sensors* **2020**, *20*, 2098. [CrossRef] [PubMed]
12. Wang, R.; Liu, Y.; Zhang, P.; Li, X.; Kang, X. Edge and Cloud Collaborative Entity Recommendation Method towards the IoT Search. *Sensors* **2020**, *20*, 1918. [CrossRef] [PubMed]
13. Cheng, W.K.; Ileladewa, A.A.; Tan, T.B. A Personalized Recommendation Framework for Social Internet of Things (SIoT). In Proceedings of the 2019 International Conference on Green and Human Information Technology (ICGHIT), Kuala Lumpur, Malaysia, 15–17 January 2019; IEEE: Piscataway, NJ, USA, 2019; pp. 24–29.
14. Gardner, B. A Review and Analysis of the Use of 'Habit'in Understanding, Predicting and Influencing Health-Related Behaviour. *Health Psychol. Rev.* **2015**, *9*, 277–295. [CrossRef] [PubMed]
15. Alsalemi, A.; Sardianos, C.; Bensaali, F.; Varlamis, I.; Amira, A.; Dimitrakopoulos, G. The Role of Micro-Moments: A Survey of Habitual Behavior Change and Recommender Systems for Energy Saving. *IEEE Syst. J.* **2019**, *13*, 3376–3387. [CrossRef]
16. Yang, H.; Lee, W.; Lee, H. IoT Smart Home Adoption: The Importance of Proper Level Automation. *J. Sens.* **2018**, *2018*, 6464036. [CrossRef]
17. McCrave, E.A. Diagnostic Criteria for Separation Anxiety in the Dog. *Vet. Clin. N. Am. Small Anim. Pract.* **1991**, *21*, 247–255. [CrossRef]
18. Wang, H.; Atif, O.; Tian, J.; Lee, J.; Park, D.; Chung, Y. Multi-Level Hierarchical Complex Behavior Monitoring System for Dog Psychological Separation Anxiety Symptoms. *Sensors* **2022**, *22*, 1556. [CrossRef]
19. Pet Ownership in Asia. Available online: https://insight.rakuten.com/pet-ownership-in-asia/ (accessed on 17 August 2022).
20. How to Manage Anti-Social Behavior in Your Pandemic Dog. Available online: https://www.nextavenue.org/separation-anxiety-in-dog/ (accessed on 27 August 2022).
21. 6 Ways To Ease Post-Pandemic Separation Anxiety in Pets | Mars, Incorporated. Available online: https://www.mars.com/news-and-stories/articles/6-ways-ease-post-pandemic-separation-anxiety-pets (accessed on 27 August 2022).
22. Shannon, L. *Dog Gone: How to Handle Your Pet's Post—Covid Separation Anxiety*; The Guardian: London, UK, 2020.
23. He, K.; Zhang, X.; Ren, S.; Sun, J. Deep Residual Learning for Image Recognition. In Proceedings of the IEEE Conference on Computer Vision and Pattern Recognition, Las Vegas, NV, USA, 27–30 June 2016; pp. 770–778.
24. Rashidi, P.; Cook, D.J.; Holder, L.B.; Schmitter-Edgecombe, M. Discovering Activities to Recognize and Track in a Smart Environment. *IEEE Trans. Knowl. Data Eng.* **2010**, *23*, 527–539. [CrossRef]
25. Belghini, N.; Gouttaya, N.; Bouab, W.; Sayouti, A. Pervasive Recommender System for Smart Home Environment. *Int. J. Appl. Inf. Syst.* **2016**, *10*, 1–7. [CrossRef]
26. Thakur, N.; Han, C.Y. A Context-Driven Complex Activity Framework for Smart Home. In Proceedings of the 2018 IEEE 9th Annual Information Technology, Electronics and Mobile Communication Conference (IEMCON), Vancouver, BC, Canada, 1–3 November 2018; IEEE: Piscataway, NJ, USA, 2018; pp. 801–806.
27. Felfernig, A.; Polat-Erdeniz, S.; Uran, C.; Reiterer, S.; Atas, M.; Tran, T.N.T.; Azzoni, P.; Kiraly, C.; Dolui, K. An Overview of Recommender Systems in the Internet of Things. *J. Intell. Inf. Syst.* **2019**, *52*, 285–309. [CrossRef]
28. Corujo, L.A.; Kieson, E.; Schloesser, T.; Gloor, P.A. Emotion Recognition in Horses with Convolutional Neural Networks. *Future Internet* **2021**, *13*, 250. [CrossRef]
29. Voorend, R.W.A. *Deep Unsupervised Representation Learning For Animal Activity Recognition*; University of Twente: Enschede, The Netherlands, 2021.
30. Ladha, C.; Hammerla, N.; Hughes, E.; Olivier, P.; Ploetz, T. Dog's Life: Wearable Activity Recognition for Dogs. In Proceedings of the 2013 ACM International Joint Conference on Pervasive and Ubiquitous Computing, Zurich, Switzerland, 8–12 September 2013; pp. 415–418.
31. Iwashita, Y.; Takamine, A.; Kurazume, R.; Ryoo, M.S. First-Person Animal Activity Recognition from Egocentric Videos. In Proceedings of the 2014 22nd International Conference on Pattern Recognition, Stockholm, Sweden, 24–28 August 2014; IEEE: Piscataway, NJ, USA, 2014; pp. 4310–4315.
32. Pons, P.; Jaen, J.; Catala, A. Developing a Depth-Based Tracking System for Interactive Playful Environments with Animals. In Proceedings of the 12th International Conference on Advances in Computer Entertainment Technology, Iskandar, Malaysia, 16–19 November 2015; pp. 1–8.
33. Kamminga, J.W.; Bisby, H.C.; Le, D.V.; Meratnia, N.; Havinga, P.J. Generic Online Animal Activity Recognition on Collar Tags. In Proceedings of the 2017 ACM International Joint Conference on Pervasive and Ubiquitous Computing and Proceedings of the 2017 ACM International Symposium on Wearable Computers, Maui, HI, USA, 11–15 September 2017; pp. 597–606.
34. Casella, E.; Khamesi, A.R.; Silvestri, S. Smartwatch Application for Horse Gaits Activity Recognition. In Proceedings of the 2019 IEEE International Conference on Smart Computing (SMARTCOMP), Washington, DC, USA, 12–15 June 2019; IEEE: Piscataway, NJ, USA, 2019; pp. 409–416.

35. Siniscalchi, M.; Quaranta, A.; Rogers, L.J. Hemispheric Specialization in Dogs for Processing Different Acoustic Stimuli. *PLoS ONE* **2008**, *3*, e3349. [CrossRef] [PubMed]
36. Quaranta, A.; d'Ingeo, S.; Amoruso, R.; Siniscalchi, M. Emotion Recognition in Cats. *Animals* **2020**, *10*, 1107. [CrossRef] [PubMed]
37. Totakura, V.; Janmanchi, M.K.; Rajesh, D.; Hussan, M.T. Prediction of Animal Vocal Emotions Using Convolutional Neural Network. *Int. J. Sci. Technol. Res.* **2020**, *9*, 6007–6011.
38. Singh, B.K.; Dua, T.; Sharma, D.P.; Changare, A.A. Animal Emotion Detection and Application. In *Data Driven Approach towards Disruptive Technologies*; Springer: Berlin/Heidelberg, Germany, 2021; pp. 449–460.
39. Caeiro, C.; Guo, K.; Mills, D. Dogs and Humans Respond to Emotionally Competent Stimuli by Producing Different Facial Actions. *Sci. Rep.* **2017**, *7*, 15525. [CrossRef]
40. Liu, T.; Yang, B.; Liu, H.; Ju, J.; Tang, J.; Subramanian, S.; Zhang, Z. GMDL: Toward Precise Head Pose Estimation via Gaussian Mixed Distribution Learning for Students' Attention Understanding. *Infrared Phys. Technol.* **2022**, *122*, 104099. [CrossRef]
41. Liu, H.; Fang, S.; Zhang, Z.; Li, D.; Lin, K.; Wang, J. MFDNet: Collaborative Poses Perception and Matrix Fisher Distribution for Head Pose Estimation. *IEEE Trans. Multimed.* **2021**, *24*, 2449–2460. [CrossRef]
42. Liu, H.; Liu, T.; Zhang, Z.; Sangaiah, A.K.; Yang, B.; Li, Y. ARHPE: Asymmetric Relation-Aware Representation Learning for Head Pose Estimation in Industrial Human–Computer Interaction. *IEEE Trans. Ind. Inform.* **2022**, *18*, 7107–7117. [CrossRef]
43. Zhang, S.; Yao, L.; Sun, A.; Tay, Y. Deep Learning Based Recommender System: A Survey and New Perspectives. *ACM Comput. Surv. CSUR* **2019**, *52*, 1–38. [CrossRef]
44. Hassan, M.M.; Uddin, M.Z.; Mohamed, A.; Almogren, A. A Robust Human Activity Recognition System Using Smartphone Sensors and Deep Learning. *Future Gener. Comput. Syst.* **2018**, *81*, 307–313. [CrossRef]
45. Kamminga, J.W.; Le, D.V.; Havinga, P.J.M. Towards Deep Unsupervised Representation Learning from Accelerometer Time Series for Animal Activity Recognition. In Proceedings of the 6th Workshop on Mining and Learning from Time Series, MiLeTS, San Diego, CA, USA, 24 August 2020.
46. Bocaj, E.; Uzunidis, D.; Kasnesis, P.; Patrikakis, C.Z. On the Benefits of Deep Convolutional Neural Networks on Animal Activity Recognition. In Proceedings of the 2020 International Conference on Smart Systems and Technologies (SST), Osijek, Croatia, 14–16 October 2020; IEEE: Piscataway, NJ, USA, 2020; pp. 83–88.
47. Ferres, K.; Schloesser, T.; Gloor, P.A. Predicting Dog Emotions Based on Posture Analysis Using DeepLabCut. *Future Internet* **2022**, *14*, 97. [CrossRef]
48. Neethirajan, S. Happy Cow or Thinking Pig? Wur Wolf—Facial Coding Platform for Measuring Emotions in Farm Animals. *AI* **2021**, *2*, 342–354. [CrossRef]
49. Mota-Rojas, D.; Marcet-Rius, M.; Ogi, A.; Hernández-Ávalos, I.; Mariti, C.; Martínez-Burnes, J.; Mora-Medina, P.; Casas, A.; Domínguez, A.; Reyes, B. Current Advances in Assessment of Dog's Emotions, Facial Expressions, and Their Use for Clinical Recognition of Pain. *Animals* **2021**, *11*, 3334. [CrossRef] [PubMed]
50. Blumrosen, G.; Hawellek, D.; Pesaran, B. Towards Automated Recognition of Facial Expressions in Animal Models. In Proceedings of the IEEE International Conference on Computer Vision Workshops (ICCVW), Venice, Italy, 22–29 October 2017; pp. 2810–2819.
51. Hantke, S.; Cummins, N.; Schuller, B. What Is My Dog Trying to Tell Me? The Automatic Recognition of the Context and Perceived Emotion of Dog Barks. In Proceedings of the 2018 IEEE International Conference on Acoustics, Speech and Signal Processing (ICASSP), Calgary, AB, Canada, 15–20 April 2018; IEEE: Piscataway, NJ, USA, 2018; pp. 5134–5138.
52. Kingma, D.P.; Ba, J. Adam: A Method for Stochastic Optimization. *arXiv* **2014**, arXiv:14126980.
53. Simonyan, K.; Zisserman, A. Very Deep Convolutional Networks for Large-Scale Image Recognition. *arXiv* **2014**, arXiv:14091556.

sensors

MDPI

Article

A Design Framework of Exploration, Segmentation, Navigation, and Instruction (ESNI) for the Lifecycle of Intelligent Mobile Agents as a Method for Mapping an Unknown Built Environment

Junchi Chu [1,2], Xueyun Tang [1,3] and Xiwei Shen [1,*]

1 School of Architecture, University of Nevada, Las Vegas, NV 89154, USA
2 Department of Computer Science, Brown University, Providence, RI 02912, USA
3 Interior Architecture Department, Rhode Island School of Design, Providence, RI 02903, USA
* Correspondence: xiwei.shen@unlv.edu

Abstract: Recent work on intelligent agents is a popular topic among the artificial intelligence community and robotic system design. The complexity of designing a framework as a guide for intelligent agents in an unknown built environment suggests a pressing need for the development of autonomous agents. However, most of the existing intelligent mobile agent design focus on the achievement of agent's specific practicality and ignore the systematic integration. Furthermore, there are only few studies focus on how the agent can utilize the information collected in unknown build environment to produce a learning pipeline for fundamental task prototype. The hierarchical framework is a combination of different individual modules that support a type of functionality by applying algorithms and each module is sequentially connected as a prerequisite for the next module. The proposed framework proved the effectiveness of ESNI system integration in the experiment section by evaluating the results in the testing environment. By a series of comparative simulations, the agent can quickly build the knowledge representation of the unknown environment, plan the actions accordingly, and perform some basic tasks sequentially. In addition, we discussed some common failures and limitations of the proposed framework.

Keywords: artificial intelligence; autonomous agent; unknown built environment; hierarchical framework; path finding; robotic system design

Citation: Chu, J.; Tang, X.; Shen, X. A Design Framework of Exploration, Segmentation, Navigation, and Instruction (ESNI) for the Lifecycle of Intelligent Mobile Agents as a Method for Mapping an Unknown Built Environment. *Sensors* **2022**, *22*, 6615. https://doi.org/10.3390/s22176615

Academic Editor: Daniele Cenni

Received: 3 August 2022
Accepted: 27 August 2022
Published: 1 September 2022

Publisher's Note: MDPI stays neutral with regard to jurisdictional claims in published maps and institutional affiliations.

Copyright: © 2022 by the authors. Licensee MDPI, Basel, Switzerland. This article is an open access article distributed under the terms and conditions of the Creative Commons Attribution (CC BY) license (https://creativecommons.org/licenses/by/4.0/).

1. Introduction

Intelligent agent development theory is a widely discussed topic in the AI community. Its impact on several industries as a core driver for cutting-edge technologies is considered significant. Russel and Norvig [1] claim an agent is anything that can be viewed as perceiving its environment through sensors and sensors acting upon that environment through actuators. The agent has intelligent behaviors, mathematically an agent function, to specify the choice of actions by perceiving the surrounding environment. Franklin [2] offers a taxonomy of autonomous agents through various properties: environment, sensing capabilities, actions, drives, and action selection architecture. Hence, this research examines how an agent's awareness and perception of its current environment influence the analysis of its knowledge system. This research considers the agent has an objective function as a goal to achieve, the measurement of how well an agent can execute human commands to maximize the expected value of the object function. A system integration reveals the value of designing a sequence of modules to help intelligent agents to complete goals.

In comparison to the current literature on intelligent agent design [3–8], the proposed ESNI model indicates alternative benefits and suggests its potential: first, the agent can learn knowledge without supervision and is highly autonomous to acquire information without significant effort from users. Second, the agent proves its autonomous states by utilizing its explored information to build a novel knowledge system.

The main contribution of this research is to design a novel framework by using ESNI system integration for an embodied agent for use in a residential plan in a real-world situation (See Figure 1). This agent can learn the surroundings without supervision, and autonomously build its knowledge based on the algorithm integrated into the system. In this innovative framework, this agent can understand human commands and perform simple actions given its navigation trajectory and can mobilize freely with a grid environment. In general, this research discusses a novel approach for autonomous service robots to explore and navigate unknown environments by experimenting within a typical residential grid-oriented plan environment.

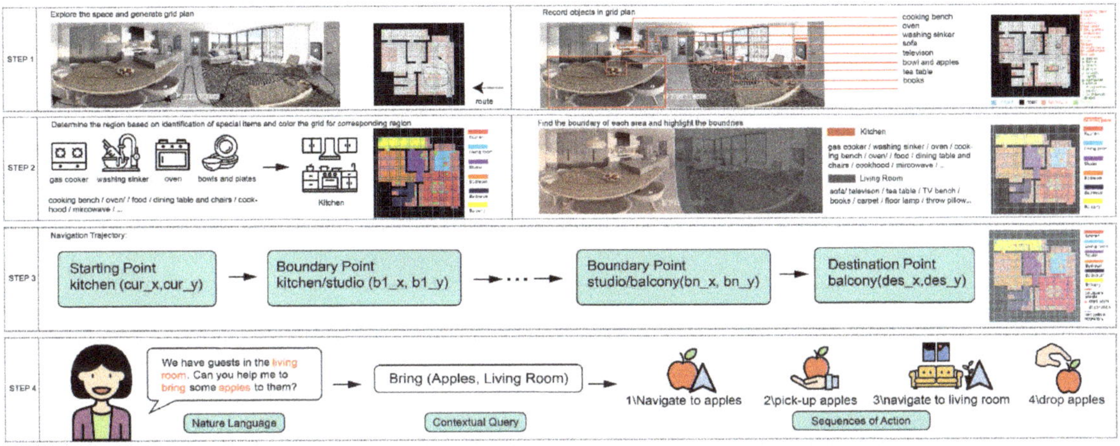

Figure 1. Overview of the ESNI design, step 1: maps floor plans to grid world, step 2: exploration and section segmentation to obtain all information, step 3: use boundary points to generate navigation trajectories, step 4: from NLP to contextual queries to process commands.

Several applications of intelligent mobile agent in robotics have proved the importance of agent's functionality and utilization, whereas the adaptiveness of an agent in different environment has not been extensively studied. The current research focuses on realistic industrial AI application in known or pre-defined environment. For example, the dialog intelligent agent has a specific function of communicating. A surgical robot deals with surgical procedures by using robotics systems. Unfortunately, neither of these raise a research question about how the intelligent agent investigates the unfamiliar settings, acquires useful information from the unknown build environment, learns the knowledge representations by processing and analyzing the data collected. The research question for the paper is how the intelligent agent builds a knowledge representation system to utilize for basic task.

2. Literature Review

2.1. Intelligent Agents in Unknown Environments

This section focuses on the literature covering the theoretical study of intelligent mobile agents' property in unknown environments during the past 10 years and presents an overview of the current state-of-art methods for intelligent mobile agents utilized in four different modules.

According to the International Organization for Standardization (ISO 8373) [9], intelligent robots require "a degree of autonomy", and involve the "ability to perform intended tasks based on current state and sensing, without human intervention". If there is no human activity involved, the machine can be considered intelligent when it makes decisions based on the interpretation of its recognition from the unknown environment. Furthermore, with a powerful computational resource and the ability to perform deep analytics, intelligent agent should extend human's cognition while dealing with the complex uncertainty and equivocality, instead of simply replacing human's contribution, Jarrahi [10] enhanced that.

The literature review is summarized respectively for 4 ESNI stages. The most recent methods in each stage are compared and discussed in Section 2.2. Given the combination of the existing methods, a knowledge gap that the lack of connections between how the agent utilizes information in unknown build environment and the achievement of basic task is discovered and thus it is necessary to propose a novel theoretical framework in Section 2.3.

2.1.1. Exploration

Intelligent agent exploration tasks involve a level of uncertainty that is typically constrained by ways to acquire information in an unknown environment. Most of these exploration methods have been classified into two categories: unsupervised exploration [11–16] and supervised exploration [17–22]. For the former, Mirco [11] casts the exploration problems to an entropy maximization equation, induced by reinforcement learning algorithms. They develop "alphaMEPOL" as a policy gradient algorithm to optimize the results and empirically discussed the performance of algorithms in continuous environments. Niroui [12] proposed a novel "partially observable Markov decision process" POMDP approach that divided exploration into subtasks with a well-defined reward function in a time-sensitive unknown environment for rescuing. Fickinger [13] provided 2 policies (Explore and Control) adversarial agents and outperformed the current state-of-art unsupervised exploration methods in exploration and zero-shot transfer tasks.

Raul Mur-Artal [14] worked on the development of exploration theory based on SLAM, he upgraded the original version to ORB-SLAM, a feature-based SLAM, allows wide baseline loop closing and re-localization, and includes fully automatic initialization. For supervised exploration methods, Kaushik [17] presented a model-free policy-based approach called Exploration from Demonstration (EfD); where human supervision is necessary to provide guidance for search space exploration. Sudo [18] proposed a whiteboard model as an intermediate information storage that allows the autonomous agents to study the location information at each node in a distributed structure.

2.1.2. Segmentation

Grid map segmentation is a method to convert undefined places into meaningful semantic sub-areas, and traditional methods mainly focused on divisions of floor plans into individual grid units to ease computational efforts. Bormann [23] made a survey of a collection of the current segmentation methods: morphological segmentation, distance transform-Based segmentation, Voronoi graph-based segmentation, and feature-based segmentation. However, all those methods encounter a common problem of instability with smaller area and lower compactness. Markus [24] proposed an approach of using occupancy grid mapping to derive instance based semantic maps by utilizing an improved method where occupancy grid mapping adapts to dynamic environments by applying an offline and an online technique to learn the parameters from agent's observations. Saarinen [25] combined the OGM method with an independent Markov Chain, a two-states transition to have representations that encoded occupancy of the cells in a dynamic environment. Fermin-Leon [26] proposed an alternative approach called contour-based topological segmentation by finding the exact convex decomposition of the contour instead of depending on the disarticulation of grid cells.

2.1.3. Navigation

At a minimum, the autonomous mobile agent's basic functionality should incorporate a high-level navigation technology. Two major approaches are dealt with: traditional navigation methods and A* based navigation methods. In the former, the agents are expected to acquire information of surrounding landmarks and update their position as a recognition of self-localization to ensure the correctness of navigation. To successfully navigate to any destination assigned, a path-finding algorithm is essential, thus Jiaoyang [27] propose a new framework rolling-horizon collision resolution (RHCR) by applying multi-agent path finding (MAPF) solver to avoid collision between agents. The paper extended the research into simulated warehouse instance and successfully outperformed the existing approaches. Quoc Huy [28] discussed safe path finding by using a combination of three techniques: particle filter, Bézier curves, and support vector machine. Frantisek [29] takes advantage of the modified A* algorithm, an algorithm that is composed of a heuristic estimation function and a cost function, for the purpose of searching for the shortest path while avoiding obstacles. Zhang [30] promoted an A* Dijkstra integration method to avoid collision and deadlocks, in which they select the optimal output from the result of these two algorithms. Gang [31] proposed a geometric A* algorithm applied to automated guided vehicle (AGV) by setting up the filter functions to eliminate invalid nodes that cause irregular paths, with a focus on solving issues on many nodes, long distance and large turning angles.

2.1.4. Instruction

A key bottleneck in deployments of robots is for users to provide commands for intelligent agents, since training an agent to process open-ended natural language is challenging. Jayant Krishnamurthy [32] introduced logical semantics with perception (LSP), an experimental model for grounded language acquisition that maps two physical objects in two applications: scene understanding and geographical question answering. Hatori [33] designed an integration of neural network architectures including SSD (image chopping), CNN (image classification), LSTM (language processing), and MLP (prediction), which takes images and languages as input to resolve ambiguity in spoken instructions; and he claims this model can deal with physical objects with unconstrained natural language instructions. Thao [34] found deficiencies in the above two models and developed a method that requires the verb to specify a task, and proposed an intermediate transition deemed as contextual query, information represented with a restricted format that maps specific commands to agents. Eric [35] proposed a second intermediate transition based on the usage of the contextual query to simplify the command; a refined structure called lifted linear temporal logic. The twohop transition reduced the complexity of mapping and proved to be efficient for agents to receive instructions from humans by mapping from the functional CQ to specific tasks.

2.2. Method Comparison

The limitations discussed above indicate a research gap in understanding different perspectives for intelligent agents. In the overall analysis of methods, the current literature on exploration indicates over-exploration, wandering issues, and high dependency on human supervision. Over-exploration means these methods are restricted in spatially limited areas; and hence the overview map of the environment can be construed as partially correct. Wandering issues create additional computational time and appear to reduce the efficiency of the agent's exploration process. Human supervision is reliable for correcting labels but losing the meaning of automatic exploration in an unknown environment. Furthermore, it is a paradox that human supervision is accountable in an unknown environment due to lack of prior data.

In terms of segmentation, most approaches suffer from low accuracy due to the inability to capture the variance from dynamic environments. In dynamic environments, the features surrounding the agent may indicate homogeneous changes and cause unpredictability. Another point is the importance of the known robot pose in the OGM method. The agent's map construction would be incorrect in the OGM method. The classic OGM method must convert and categorize all coordinates as either filled or unfilled. However, identifying a semi-transparent object, such as the glass is problematic and requires a non-binary representation. There is no OGM-related method that clusters a group of points or defines a group of points as a boundary to define a spatial region and hence, the categorization of areas in the unknown environment is unable to be identified. Researchers have considered navigation methods. However, some challenges appear: simulation environments are not able to represent the real-world applications; and determining an optimal path during mobility is challenged by physical obstacles while minimizing the traversal cost. The classic A* algorithm is widely used in the intelligent agent's navigation system, but some drawbacks emerge: the heuristic value determines the performance of A* algorithm. A* is not able to operate efficiently within dynamic environments and with the operating assumption that every action has a fixed cost, this would not be verifiable in a real world. Additionally, the direct application of the A* algorithm to find paths within the structured environment is computationally high.

To deliver effective commands to the agents, the current state-of-arts reveals these drawbacks: disconnected relationships in terms of logic, poor performance due to unseen vocabularies, and limited availability of task types. Natural languages have rules to express thoughts, but not necessarily in a well-formed structure. Using machines to capture the pattern of languages is challenging, and it is even harder with the issue of ambiguity. However, to behave in an "intelligent" way means an intelligent agent must have the confirmation of actions map to the expression, so the significance of finding an accurate transition between spoken languages to specified tasks brings the topic worth discussion.

2.3. Theoretical Framework

Given the apparent lack of research to integrate the four methods indicated earlier to design an intelligent agent, this research provides an opportunity to experiment and formulate a new framework (See Figure 2). Instead of considering the methods as discrete individual modules, this work examines the potential derived by the inter-connections and tangents among the respective methods. This research seeks to contribute to the emergent body of knowledge on experimental work and applied research in AI, especially robotics as autonomous entities as dynamic agents within the studies of the built environment. The proposed novel framework expands the discourse on robot design as active agents to gather data in the built environment, has the potential for systematizing robot design, influences the process of autonomous agents, and becomes a systematic and standardized robot design. The anticipation of another research will focus on the trade-off of local optimization for each module and aims to achieve global optimization. Another possible direction is to extend other perspectives to improve the efficiency of functionality or the diversity of functionality types. Second, for application wise, the framework maps from a working environment to an unknown home-type environment, thus having a significant impact on the promotion of barrier-free communication agents to provide end-to-end service to users. It reduces the complexity and redundancy of communication between machines and humans and is committed to improving human-computer interaction experience. The proposed models and algorithms will contribute to the development and progress in intelligent agent theoretical study and guide the implementation for promising real-world application.

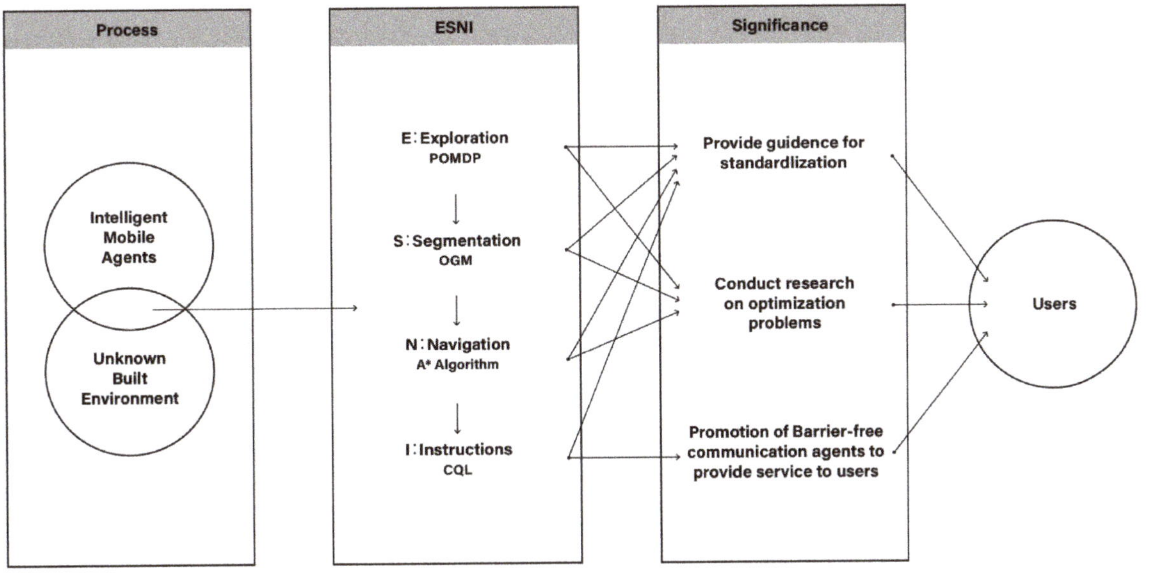

Figure 2. Theoretical framework.

2.4. *Key Terms in Proposed Method*

2.4.1. POMDP

Partially observable Markov decision process (POMDP) [36] is a stochastic process that describes the discrete states for embodied agents in an environment. Formally, Markov decision process (MDP) is represented by a tuple, (S, A, T, R, γ), where S is all possible states, action domain is the set of all possible actions, transition function T = T(s, a, s′) = Probability(s′ | s, a) is a set of conditional probabilities between states, R is a reward function, and γ is the discount factor. In MDP, all states are known to the agent to make an optimal solution, but in POMDP, the agent only has an observation O to relate the states. The tuple representing POMDP has additional two elements, O and Z: the observations space O is the set of all possible observations and Z is the set of conditional observation probabilities [37].

2.4.2. Contextual Query Language

The contextual query language is a formal language that retrieves information in a well-defined structure [38]. The query, in general, must be human-readable and writable while maintaining the featured information of complex original languages. The meaning of using contextual query in this environment is to map from a border generalization of human instructions to a selection of a well-defined task. More specifically, the verb and noun words are extracted as keywords parameters in the model.

2.4.3. Occupancy Grid Mapping

The occupancy grid mapping refers to a representation that the continuous space in the environment was partitioned into different cells. Each cell has a Bayesian probability to indicate whether it is 1 (occupied) or 0 (empty). The representation is simply the situation of mobile robots and has become the dominant paradigm for embodied agent environments [39]. Occupancy grid mapping (OGM) method is used in module: segmentation to convert the simulation into grid map.

2.4.4. A* Algorithm

A* Algorithm is a graph traversal and pathfinding computer algorithm that is widely used in games development. The advantage of A* algorithm is that it can search the shortest path from source to destination with hindrances. A* algorithm selects the path with a minimum cost function f(n) = g(n) + h(n), in which g(n) is the cost from the source node to the current node, and h(n) is a heuristic function that estimates from the current point to the destination point. The selection of heuristic functions depends on the problem, but in general, Manhattan distance or Euclidean distance are the two most popular heuristic functions to use [40].

3. Research Method

3.1. Jargon

Symbols: Before elaborating on methodologies to address the tasks in this project, important symbols and terminology definitions are given as follows. The collections of all moveable objects in the grid world are denoted as $M = \{m_1, m_2, m_3 \ldots, m_n\}$. The targeted grid's coordinates are denoted as ($gird_x$, $grid_y$) and the target object mj's coordinates are denoted as (mj_x, mj_y). The agent's current location is denoted as (cur_x, cur_y).

3.1.1. Sections

The collection of sections is denoted as: $S = \{s_1, s_2, \ldots, s_n\}$. For each grid cell c, a section s_j, $j \leq x$ is an accumulated grid cell cluster, $c \in S$, that can represent the functionality of a house, for example, kitchen, living room, bathroom, etc.

3.1.2. Boundary Points

The collections of points lay between two different sections. For any point that is identified as boundary points b_n, there exists a neighbor point that belongs to a section that is different from bn's section. Mathematically, it supposes a boundary point (bn_x, bn_y), then $\forall$ $(bn_x, bn_y) \in s_p$, $\exists (bn_x + a, bn_y + b) \in s_q$, such that $a, b \in \{-1, 0, 1\}$, $a + b \in \{-1, 1\}$ and $p \neq q$.

3.1.3. Section-Wise Path

A section path is from one section to another section. The purpose of section pathfinding is to find a path at high levels to help the agent's navigation trajectory. For example, to navigate a point from n kitchen to the bathroom, the section path is Kitchen $\rightarrow$ Studio $\rightarrow$ Bedroom $\rightarrow$ Bathroom. The section path can be generated by the BFS algorithm discussed in the section segmentation paragraph.

3.1.4. Walking Values in Exploration

Walking value is a term to describe the frequency of the agent visit on a particular grid. A dictionary that stores the information of each grid to associate a walking value is created with initiation of all walking values to 0. The program increases the value by 1 if and only if the agent visits that position.

3.1.5. Bin Values

The highest bin value determines the belonging of a grid cell. Each grid cell has six bins since there are six room types in the simulation. The formula to calculate bin value is provided in Section 3.4.

3.1.6. Ground Truth Knowledge Dictionary

Knowledge dictionary is a data structure that stores key value pairs to indicate the belongings of each object. For example, [TV] = {Bedroom, Living Room}, [Chair] = {Kitchen, Living Room, Studio}, [Bed] = {Bedroom}.

3.1.7. Maximum Allowed Steps (MAS)

This terminology is used in the experiment part to describe how many steps the agent is allowed to explore the unknown environment. In the program, the agent's movement is suspended when the maximum allowed steps is reached.

3.1.8. Percentage of Occupied Cells (POC)

This terminology is to describe how many space the agent has occupied or covered in the experiment. The percentage of occupied cells range from 0% to 100%.

3.2. Agent Model and Set Up

As Figure 3, this research assumes that the agent has no prior information of a floor plan. The computer program converted the architectural floor plans into the digital 40 × 40 grid world by using the OGM method. Moreover, it defined a white area as a walkable space for the agent. Any space other than white color is non-walkable. The initial position of the agent is located at the right corner of the building, with no prior knowledge of the environment. When initializing the environment, objects are instantiated in the zones by sampling the attribute distributions in the knowledge dictionary, which captures an ontology of locations and objects in the building. Objects are classified as key fixture objects (bedroom), non-key fixture objects (sofa, table, and TV bench, etc.), and movable items (such as apple, bowls, chairs, etc.).

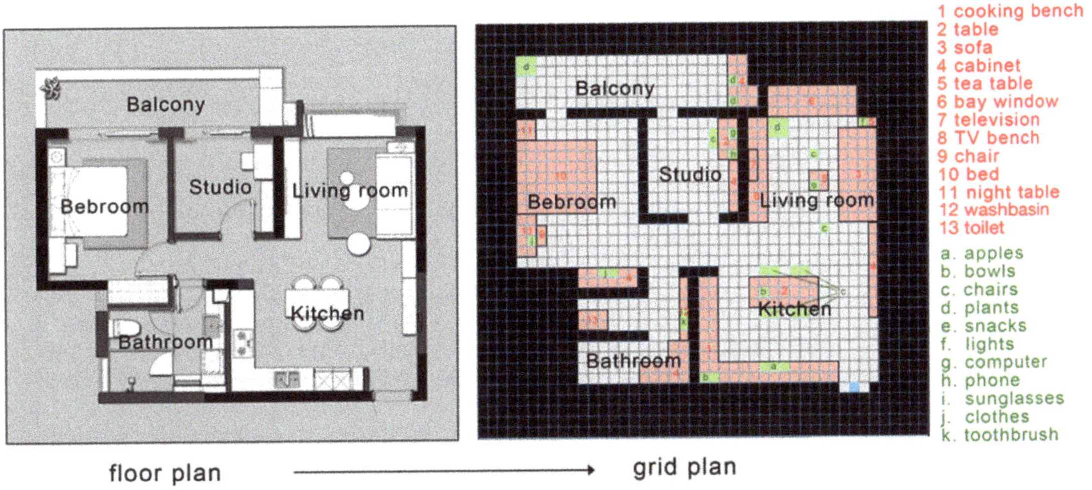

Figure 3. A partition from house plan to grid cells.

3.3. Exploration

The goal in this step is for the agent to cover maximum unvisited spaces, identify all the boundary points, and build the unweighted connectivity graph to generate a pathfinding algorithm. The algorithm (See Algorithm 1) can be described in two steps: initially, a dictionary WALKING_VALUES can map from all positional and walkable grids to a numerical value. Then the program marks a grid that has been visited by increasing the value by 1 if the agent has stepped over or passed that grid. In this simulated environment, the agent has four legal options for the next move, UP, LEFT, DOWN, RIGHT. The potential choice of the next move will be removed if the agent will collide with an object or obstacles. Based on four possible move options, the agent chooses the move which has the minimum walking value in the dictionary. Intuitively, a move directs the agent to a grid that has a higher walking value than others, meaning that the grid is more frequently visited compared to others. The agent will choose the next move of the grid with the lowest value among the four directions.

In the two dimensional grid world, the POMDP tuple has (S, A, T, R, γ, O, Z), where S describes states as all possible coordinates in the unknown environment. "A" has four legal directions: UP, DOWN, LEFT, RIGHT. "T" is the probability transition between states, so whatever action are chosen could cause the walking values for each grid. "R" is the reward function, a positive reward can lower walking value, and no reward for staying. γ is the discount factor with a default value of 0.9. The observation "O" is partial, thus the agent has no information for any state that is not adjacent. Z stands for the conditional observation probabilities depending on "O".

Algorithm 1: Agent Exploration Algorithm

Agent's current positions ($\mathbf{cur_x}$, $\mathbf{cur_y}$) Initiate a dictionary: walking_values
While Maximum Allowed Steps has not been reached do
    Add Adjacent S: states of $\mathbf{cur_x}$, $\mathbf{cur_y}$ to walking_values
    Choose Moves <——A: ["UP","DOWN","LEFT","RIGHT"]
    Remove move if collides with fixture objects & Update T, O, Z
  Loop each move in Adjacent MOVES do
    Move = argmin (walking_values [move]) & Update R
    Walking_values [($\mathbf{cur_x}$, $\mathbf{cur_y}$)] + = 1

3.4. *Section Segmentation*

The agent uses a section segmentation model that guesses the area correspondingly to distinguish the different functionalities of the areas (such as balcony, studio, kitchen, etc.) for every single grid cell. After agents explore all spaces in the grid world, all item information must be recorded and the next procedure for the agent is to predict areas for each grid. The agent uses an area prediction algorithm to determine which area a grid belongs to.

The environment is divided into six basic types according to the functions of the general home space: living room, bedroom, bathroom, studio, kitchen, and balcony. All elements in the simulation are classified into three categories, key objects, non-key objects, and moveable items (See Table 1). A key fixture object refers to a signature object that determines the functionality of a section, for example, the bathroom has a toilet as a key object and a bed must be in the bedroom. The simulation has three key fixture objects, the toilet, bed, and gas cooker that can be regarded as the key fixture objects for deciding the room types. However, for the other three types, lack of key fixture objects requires us to use other objects to measure the space types. Non-key fixture objects refer to an object that increases the probability of a grid cell to be assigned as a section, for example, TV as a common non-key fixture object can be found in living room and bedroom. Moveable item refers to any item that can appear in any room during daily usage, for example, it would not be surprising to find a teacup in the living room or bedroom.

Table 1. Groups of space types and with association of key objects.

	Section	Key Fixture Object	Non-Key Fixture Object	Moveable Item
Section with Key Fixture Object	Bedroom	Bed	Closet, Wardrobe, Night Table, Bed Lamp	Pillows, Toothbrush, Bowls, Plates, Condiment Bottles, Floor Lamp, Chair, Books, Laptop, Plants, Vase, Apple, Bananas, Pears, Oranges, Eggs, Snacks
	Bathroom	Toilet	Washing Sinker, Bathtube, Shower Head	
	Kitchen	Gas Cooker	Oven, Mircowave, Kitchen Washing Sinker, Cookhood, Refrigerator	
Section without Key Fixture Object	Living Room	N/A	Sofa, Television, Tea Table, TV Bench	
	Studio	N/A	Bookcase, Desktop PC	
	Balcony	N/A	Big Window, Curtain, Wind Chime	

Applying the above classification principles, the agent assigns each grid a room type based on the probability of what functionality this room should be, by calculating the surrounding objects. There are a few restrictions to be noticed. First, each object will contribute a non-negative bin value v to grid cell c. v is determined by the reciprocal of Euclidean distance times the factor N. A higher Euclidean distance results in a smaller bin value contribution. Second, for any grid cell c, object o does not contribute bin values if and only if a path between c and o exists wall objects. Third, any moveable item or non-fixture object has a contribution of 0 bin value.

The overview of the area prediction algorithms is described as follows (See Algorithm 2). For each grid, 6 sections are associated with 6 possible assignments, thus corresponding to 6 bins. For every single grid cell, the agent calculates the Euclidean distance between each object and the grid cell and adds up the weighted contribution value to the corresponding bin for each item associated with the bin. The contribution value is defined as the reciprocal of Euclidean distance by times with a factor N, to the associated bin to account for how possibly this grid belongs to the section. The default N = 50 for key objects, N = 1 for non-key objects, and N = 0 for moveable items. The bin that has the highest value determines the ownership of the grid cell. Mathematically, the bin calculation formula can be represented as the following:

$$\text{Bin\_Value} = \sum_{j=1}^{j=n} \frac{1}{N\left((grid_x - mj_x)^2 + \left(grid_y - mj_y\right)^2\right)} \tag{1}$$

$$\text{N} = \begin{cases} 50 & if\ Key\ Fixture\ Objects \\ 1 & if\ Non-Key\ Fixture\ Objects \\ 0 & if\ Moveable\ Item \end{cases} \tag{2}$$

Algorithm 2: Agent Section Segmentation Algorithm

```
While existing a grid without section segmentation do
        Initiate a histogram: Hist_Area that has 6 bins: balcony, living room, bedroom,
bathroom, kitchen, studio, and bedroom
      For each object in grid world do
            Calculate Bin_Value
            Loop for each bin in Hist_Area do
                  If bin in Ground Truth Knowledge[o] then
                        Bin += Bin_Value
      Area = argmax (Hist_Area)
```

3.5. Navigation Trajectory

In the grid world, exploration helps to build unweighted connectivity (Figure 4) graphs to instruct the agent to detect a path from its current location to the destination point in the section-wise levels. A deterministic algorithm (Breadth-First Search) [41] can determine the shortest path from the current section to the destination section in a high level, for example, Kitchen→ Studio→ Bedroom→ Bathroom. The agent uses the section segmentation function to locate its current section and destination section. The navigation strategy for the agent is to move to the boundary points from the current section to the next section's boundary points in the section-wise path until reaching the destination.

For example, consider a simulation of virtual environment in a bedroom, a task requires the agent to generate a two-hop navigation trajectory from bedroom to balcony. In the high levels (section-wise), the navigation task can be split into four sub-tas: from bedroom, to kitchen, studio, and balcony. In the low level, which means for each navigation sub-task within a particular section, there could be many objects appearing in the section as obstacles. To find the shortest path to avoid obstacles, A* algorithm can navigate in the low-level. The A* algorithm selects the path that minimizes f(n) = g(n) + h(n), where n represents the next points, g(n) is the cost of the path from the boundary point to n, and h(n) is a heuristic function that estimates the cost of the cheapest path from n to the next boundary point. Figure 5 demonstrates how the agent navigates from the kitchen to the balcony, the blue arrows are the navigation path and red bars are boundary points between sections.

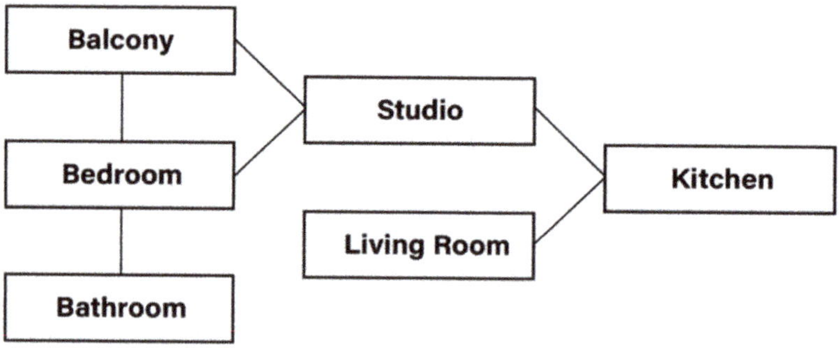

Figure 4. Section connectivity unweighted graph.

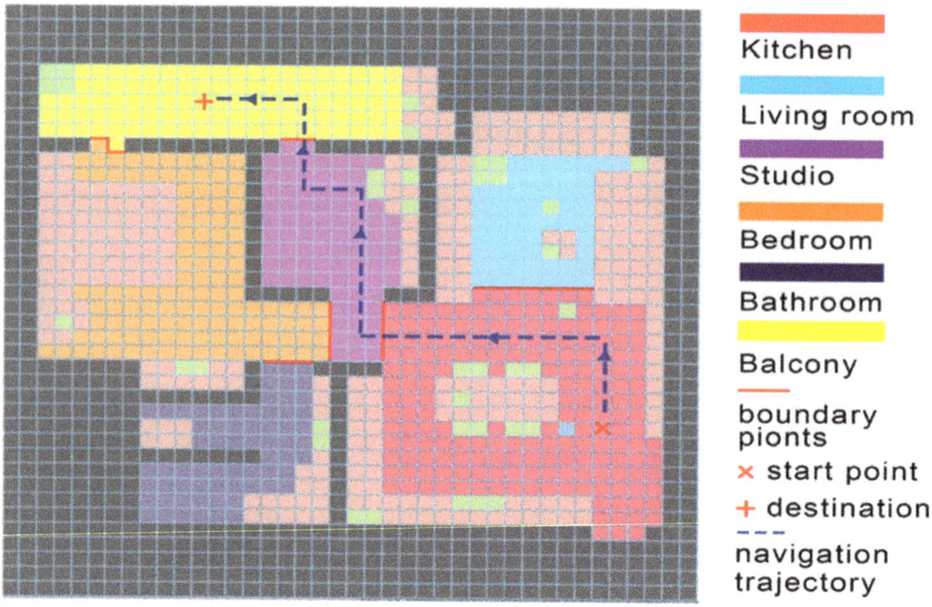

Figure 5. A navigation trajectory from kitchen to balcony.

3.6. From NLP to Contextual Query

CQ extracts the keywords from the original language and discards unrelated information to ensure the efficiency of information retrieval. CQ can serve as an intermediate transition from natural language to machine-readable code. For example, the agent cannot have a clear map of what task it needs to complete when a human says: "I haven't eaten since last night, I am hungry. Can you give me an apple?" The agent executes the following CQ: bring the apple to the living room. A template CQ filled with parameters will map a specific task, thus the agent understands what type of task is being requested. Table 2 lists examples from NLP→ CQ→ Tasks. CQ can be parameterized to sub-tasks, so the component of a single CQ has two parts, the template and the parameter. For the sake of simplicity, four fundamental CQ templates are defined as follows: Bring (parameter 1) to (parameter 2), Navigate (parameter 1), Find (parameter 1), and Swap (parameter 1). The last CQ template has no parameters and can be used as a default status to maintain the static for the agent. Each class of contextual query corresponds with a task parameterized by a goal. NLTK [42] as a pre-build package can tokenize and tag the queries, then analyze the sentence composition to extract the goal parameters. The goal is to implement custom transformation mapping from each contextual query class to goal-parameterized tasks.

Table 2. Examples of NLP -> CQ -> Tasks.

Examples of Mapping between NLP→CQ
• Task 1: I want a banana. I am at bedroom Bring [Banana, Bedroom] • Task 2: Can you come to my bedroom to serve? Navigate [Bedroom] • Task 3: Hey, where is my computer? I can't find it. Find [Computer] • Task 4: Hey, I want to take a shower. Can you swap my cloth and toothbrush? Swap [Cloth, toothbrush]
CQ→Sequences of Action
• Bring [Banana, Bedroom] = Find [Banana]→Navigate [Banana]→Pickup [Banana]→Navigate [Bedroom]→Drop [Banana) • Navigate [Bedroom) = Navigate [Bedroom] • Find [Computer] = Navigate [Computer] • Swap [Cloth, toothbrush] = Find [Cloth]→Navigate [Cloth]→Pickup [Cloth]→Find [Toothbrush]→Navigate [Toothbrush]→Pickup [Toothbrush]→Drop [Cloth]→Drop [Toothbrush]

4. Experiment

The experiment demonstrates the trade-off between steps taken by the agent and the sections he can explore in the grid world. The grid world simulation has limitations in terms of dimensionality and size. Theoretically, partial exploration of the environment always results in inaccurate segmentation. The maximum allowed step (MAS) is passed as an input for each iteration. The agent stops the exploration and performs the section segmentation once the maximum allowed steps are reached.

4.1. Exploration Experiment

For the exploration experiment, the agent starts at MAS of 50, and the computer program increases MAS by 50 each time, until 2000, to find an optimal MAS point where the percentage of occupied cells (POC) converges. The average results of 20 iterations are taken for each epoch of MAS.

4.2. Segmentation Experiment

For the section segmentation experiment, the same measurement metric was used as the exploration experiment, to get the results of section segmentation in different values of POC. Information in the previous stage is utilized for this stage.

4.3. Navigation Experiments

The computer program generates different results of navigation trajectories, given the information of segmentation stage. In this simulation, testing in 2000 random pairs of 2 points and comparing the correct label path with the generated navigation trajectories by the intelligent mobile agent are carried out.

4.4. Instruction Experiment

Testing on different inputs under various contexts reveals that the language does not have any pattern, but the NLTK model still extracts the accurate information as parameters to put in the function prototype. 2000 short human commands are tested in this experiment, and the result is measured based on how accurately the agent can capture the keywords and identify the correct task prototype in the dialogue, by comparing with the correct label of the parameters.

5. Results

5.1. Results for Exploration

When the maximum allowed steps are relatively low, the agent hovers around in those areas near the starting point. If MAS was increased, the agent should ultimately step in other sections that he has not previously covered. A percentage of covered space with respect to the maximum allowed steps can demonstrate the result of exploration. Figure 6 shows the percentage that the agent has covered or explored, given the maximum allowed steps. To achieve approximately 80% explored space in the grid cells, the agent has a MAS of 750. The curve converges to 100%, which means that the agent has fully explored all the grids in this environment when the maximum allowed steps are around 1600 steps.

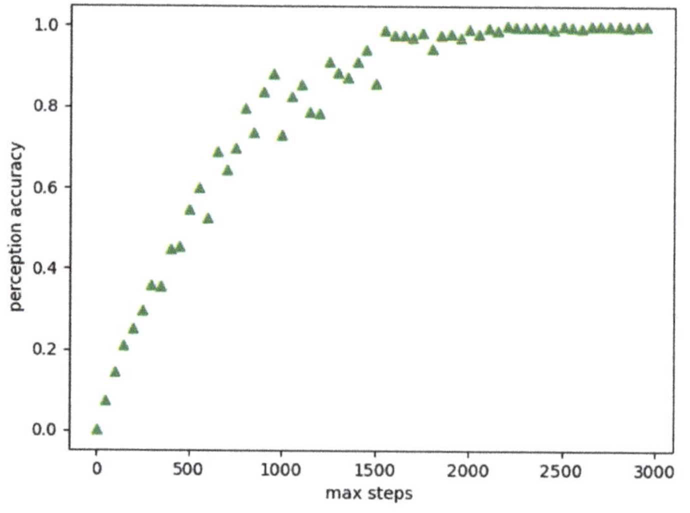

Figure 6. Relationship between MAS and POC.

5.2. Results for Section Segmentation

The number of sections the agent can predict accurately is determined by the number of cells that the agent can occupy. Theoretically, all sections are segmented well if the agent can fully explore the grid cells and observe all objects in the environment. Figure 7 demonstrates the segmentation results at different POC, based on the average results of 20 experiments. The grid cells that the agent segments with the ground truth label and the error of segmentation as the ratio of incorrect labels of segmentation/total grid cells in the simulation were compared. Figure 8 shows the relationship between POC and error of segmentation. The error of segmentation is 69% when POC is 10%, which means the agent recognizes most of the grid cells as part of kitchen. When POC is 30%, the agent can reach to most of grid cells in the living room and can recognize the living room cells correctly. The agent can recognize 3–4 sections correctly when POC is 50%, and only mark one section wrong (mostly the bathroom was incorrectly labelled since it is the latest places to be visited). The agent's error rate reduces to 3% if POC is close to 90%.

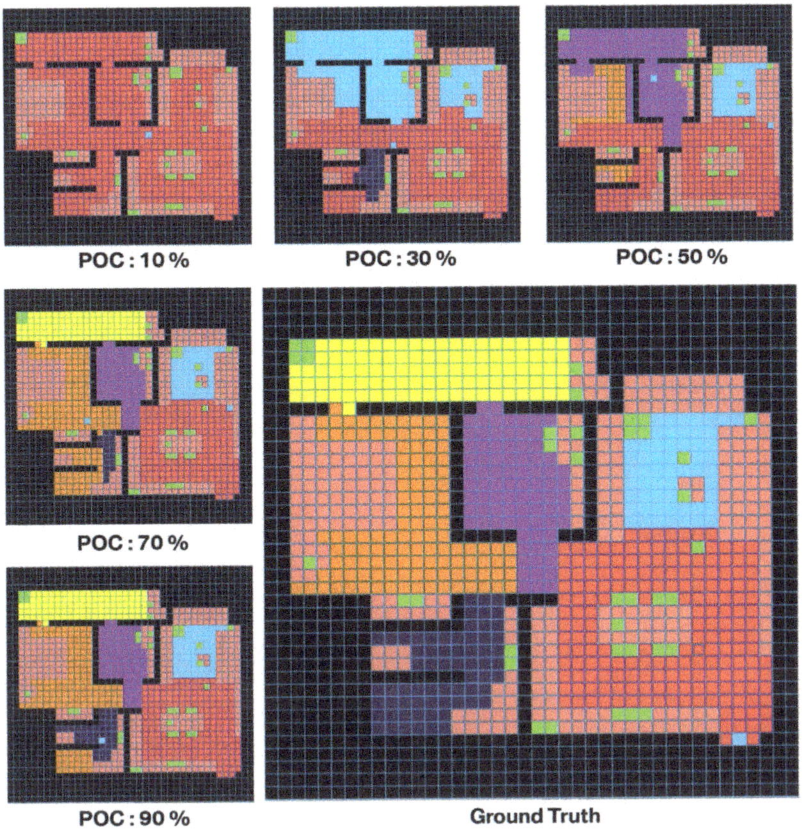

Figure 7. The result of section segmentation for different POC in the computer program visualization. The color blocks representation can be found at Figure 5.

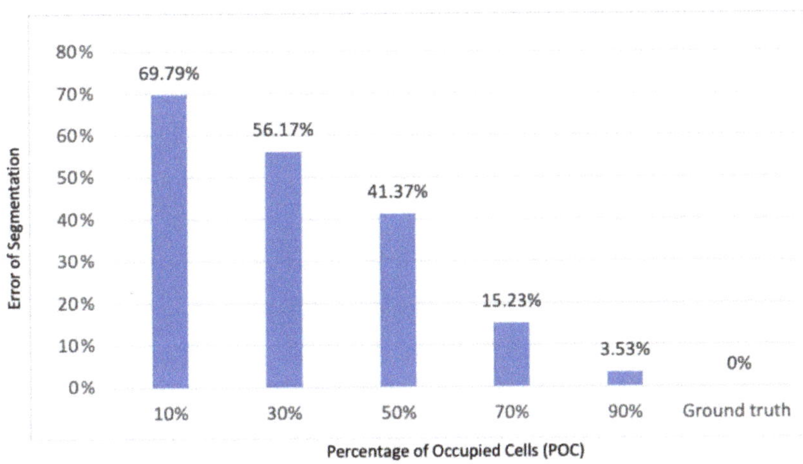

Figure 8. Twenty trials experiments' section recognized on different MAS.

5.3. Results for Generating Navigation Trajectories

Navigation trajectories vary in the result of segmentation, thus also have dependency on the MAS. In the simulation, with a high segmentation error rate, the section connectivity unweighted graph missed component of the grid world. Based on 2000 trails of random pair point's navigation, the system bears a certain degree of segmentation error rate, that is, up to 12% the agent maintains the correctness of path finding. For a 50% error rate, the correctness significantly drops to 25%, which indicates that the missing component in section connectivity weighted graph is imperative for the success of generating a correct navigation trajectory (See Figure 9).

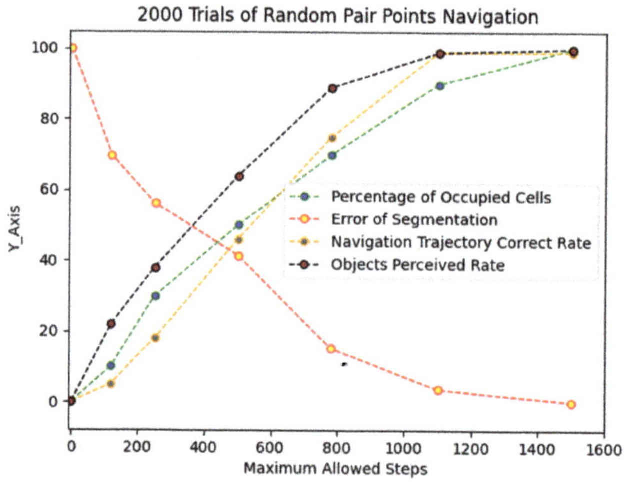

Figure 9. 2000 trails of random pair points experiment.

5.4. Results for Instruction Experiment

In this stage, testing is focused on the correctness of task prototype and capture of corresponding parameters. Even though the command inputs are short sentence, different perplexities of language affect the performance of the model. The comparison from the results of the instruction model with respect to the ground truth labelling reveals common failures that the agent cannot solve because of the ambiguity of the language. If the contextual query is missing parameters or mapping to wrong task prototype, the sequence of action cannot be generated correctly.

6. Discussion and Conclusions

A problem in traditional intelligent agent design is the lack of a lifecycle of how the agent can be autonomous by collecting information from the unknown environment. We introduced a framework of intelligent agent design and provided several algorithms during the pipeline for the agent to perform tasks. The framework has novel points for discovering hidden information during the section segmentation and generating a navigation trajectory. The autonomous agent processes human languages and utilizes the trajectory to perform several actions. We use different MAS to know the trade-off between explorations and exploitation. Finally, we discussed the limitations of the proposed approach and the extended research in the relative areas.

The experiment shows that ESNI system integration successfully achieve the goal of being autonomous by sequentially executing each stage one by one. The navigation trajectories accuracy almost approaches to 100% with approximately 1300 steps of exploration. Compared to the previous frameworks, the ESNI can be used as a design framework that meshed resources together, thus the feasibility and autonomy can be ensured.

One of the limitations of this work is that the intelligent agent can provide the service under the scope of the home environment since all elements are discussed based on smart home implementation. Thus, this framework cannot be applied or need to be revised largely for other places such as metro stations, schools, or nursing homes. Moreover, our project relies on the assumption that the agent can move freely in the grid world, not considering the physics of the agent, but the agent can collide with obstacles that are never seen before in the memory. Another limitation is that the tasks type is limited due to the form of CQ being in primitive structures and the agents cannot process a more complicated dialogue with ambiguity.

The future research of the ESNI framework need to be addressed in the following three perspectives: First, the optimization problem is a big topic that needs to be researched. For each stage of the ESNI, the local optimization does not guarantee a global optimization. Second, how should we evaluate the efficiency of the framework in terms of aiming to reach a global optimization? The evaluation metrics (runtime, entropy, energy conversion rate, etc.) must be carefully selected or established from the perspective of energy saving. Last but not the least, the current format of the intelligent agent can successfully perform basic task, but still being far away from a truly autonomous agent that have a more advanced reasoning system and decision-making mechanism.

Author Contributions: Writing—original draft, J.C.; Writing—review & editing, X.T. and X.S. All authors have read and agreed to the published version of the manuscript.

Funding: This article were supported by the UNLV University Libraries Open Article Fund.

Conflicts of Interest: The authors declare no conflict of interest.

References

1. Russell, S.; Norvig, P. AI a modern approach. *Learning* **2005**, *2*, 4.
2. Franklin, S.; Graesser, A. Is It an agent, or just a program? A taxonomy for autonomous agents. In *International Workshop on Agent Theories, Architectures, and Languages*; Springer: Berlin/Heidelberg, Germany, 1997; pp. 21–35.
3. Jie, Y.; Pei, J.Y.; Jun, L.; Yun, G.; Wei, X. Smart home system based on iot technologies. In Proceedings of the 2013 International Conference on Computational and Information Sciences, Shiyang, China, 21–23 June 2013; pp. 1789–1791.
4. Sepasgozar, S.; Karimi, R.; Farahzadi, L.; Moezzi, F.; Shirowzhan, S.; Ebrahimzadeh, S.M.; Hui, F.; Aye, L. A Systematic Content Review of Artificial Intelligence and the Internet of Things Applications in Smart Home. *Appl. Sci.* **2020**, *10*, 3074. [CrossRef]
5. Jivani, F.D.; Malvankar, M.; Shankarmani, R. A Voice Controlled Smart Home Solution With a Centralized Management Framework Implemented Using AI and NLP. In Proceedings of the 2018 International Conference on Current Trends towards Converging Technologies (ICCTCT), Coimbatore, India, 1–3 March 2018; pp. 1–5. [CrossRef]
6. Montoya, J.H.; Winther, K.T.; Flores, R.A.; Bligaard, T.; Hummelshøj, J.S.; Aykol, M. Autonomous in-telligent agents for accelerated materials discovery. *Chem. Sci.* **2020**, *11*, 8517–8532. [CrossRef] [PubMed]
7. Jeon, H.; Oh, H.R.; Hwang, I.; Kim, J. An intelligent dialogue agent for the IoT home. In Proceedings of the Work-Shops at the Thirtieth AAAI Conference on Artificial Intelligence, Phoenix, AZ, USA, 12–13 February 2016.
8. Binos, T.; Bruno, V.; Adamopoulos, A. Intelligent agent based framework to augment warehouse man-agement systems for dynamic demand environments. *Australas. J. Inf. Syst.* **2021**, *25*. [CrossRef]
9. Panesar, S.; Cagle, Y.; Chander, D.; Morey, J.; Fernandez-Miranda, J.; Kliot, M. Artificial intelligence and the future of surgical robotics. *Ann. Surg.* **2019**, *270*, 223–226. [CrossRef] [PubMed]
10. Jarrahi, M.H. Artificial intelligence and the future of work: Human-AI symbiosis in organizational de-cision making. *Bus. Horiz.* **2018**, *61*, 577–586. [CrossRef]
11. Mutti, M.; Mancassola, M.; Restelli, M. Unsupervised reinforcement learning in multiple envi-ronments. In Proceedings of the AAAI Conference on Artificial Intelligence, Virtual, 22 February–1 March 2022; Volume 36, pp. 7850–7858.
12. Niroui, F.; Sprenger, B.; Nejat, G. Robot exploration in unknown cluttered environments when dealing with uncertainty. In Proceedings of the 2017 IEEE International Symposium on Robotics and Intelligent Sensors (IRIS), Ottawa, ON, Canada, 5–7 October 2017; pp. 224–229.
13. Fickinger, A.; Jaques, N.; Parajuli, S.; Chang, M.; Rhinehart, N.; Berseth, G.; Levine, S. Explore and Control with Adversarial Surprise. *arXiv* **2021**, arXiv:2107.07394.
14. Mur-Artal, R.; Montiel, J.M.M.; Tardos, J.D. ORB-SLAM: A versatile and accurate monocular SLAM system. *IEEE Trans. Robot.* **2015**, *31*, 1147–1163. [CrossRef]
15. Steinmann, R.; Seydoux, L.; Beaucé, E.; Campillo, M. Hierarchical exploration of continuous seismo-grams with unsupervised learning. *J. Geophys. Res. Solid Earth* **2022**, *127*, e2021JB022455. [CrossRef] [PubMed]

16. Péré, A.; Forestier, S.; Sigaud, O.; Oudeyer, P.Y. Unsupervised learning of goal spaces for intrinsically motivated goal exploration. *arXiv* **2018**, arXiv:1803.00781.
17. Subramanian, K.; Isbell, C.L., Jr.; Thomaz, A.L. Exploration from demonstration for interactive reinforcement learning. In Proceedings of the 2016 International Conference on Autonomous Agents & Multiagent Systems, Singapore, 9–13 May 2016; pp. 447–456.
18. Sudo, Y.; Baba, D.; Nakamura, J.; Ooshita, F.; Kakugawa, H.; Masuzawa, T. An agent explora-tion in unknown undirected graphs with whiteboards. In Proceedings of the Third International Workshop on Reliability, Availability, and Security, Zurich, Switzerland, 25–28 July 2010; pp. 1–6.
19. Zhu, D.; Li, T.; Ho, D.; Wang, C.; Meng, M.Q.H. Deep reinforcement learning supervised au-tonomous exploration in office environments. In Proceedings of the 2018 IEEE International Conference on Robotics and Automation (ICRA), Brisbane, QLD, Australia, 21–25 May 2018; pp. 7548–7555.
20. Otsu, K.; Tepsuporn, S.; Thakker, R.; Vaquero, T.S.; Edlund, J.A.; Walsh, W.; Agha-Mohammadi, A.A. Supervised autonomy for communication-degraded subterranean exploration by a robot team. In Proceedings of the 2020 IEEE Aerospace Conference, Big Sky, MT, USA, 7–14 March 2020; pp. 1–9.
21. Bai, S.; Chen, F.; Englot, B. Toward autonomous mapping and exploration for mobile robots through deep supervised learning. In Proceedings of the 2017 IEEE/RSJ International Conference on Intelligent Robots and Sys-tems (IROS), Vancouver, BC, Canada, 24–28 September 2017; pp. 2379–2384.
22. Pathak, D.; Agrawal, P.; Efros, A.A.; Darrell, T. Curiosity-driven exploration by self-supervised prediction. In Proceedings of the International Conference on Machine Learning, Sydney, Australia, 6–11 August 2017; pp. 2778–2787.
23. Bormann, R.; Jordan, F.; Li, W.; Hampp, J.; Hägele, M. Room segmentation: Survey, implemen-tation, and analysis. In Proceedings of the 2016 IEEE International Conference on Robotics and Automation (ICRA), Stockholm, Sweden, 16–21 May 2016; pp. 1019–1026.
24. Hiller, M.; Qiu, C.; Particke, F.; Hofmann, C.; Thielecke, J. Learning topometric semantic maps from occupancy grids. In Proceedings of the 2019 IEEE/RSJ International Conference on Intelligent Robots and Systems (IROS), Macau, China, 3–8 November 2019; pp. 4190–4197.
25. Saarinen, J.; Andreasson, H.; Lilienthal, A.J. Independent markov chain occupancy grid maps for representation of dynamic environment. In Proceedings of the 2012 IEEE/RSJ International Conference on Intelligent Robots and Systems, Vilamoura-Algarve, Portugal, 7–12 October 2012; pp. 3489–3495.
26. Fermin-Leon, L.; Neira, J.; Castellanos, J.A. Incremental contour-based topological segmenta-tion for robot exploration. In Proceedings of the 2017 IEEE International Conference on Robotics and Automation (ICRA), Singapore, 29 May–3 June 2017; pp. 2554–2561.
27. Li, J.; Tinka, A.; Kiesel, S.; Durham, J.W.; Kumar, T.S.; Koenig, S. Lifelong multi-agent path finding in large-scale warehouses. In Proceedings of the AAAI Conference on Artificial Intelligence, Virtual, 2–9 February 2021; Volume 35, pp. 11272–11281.
28. Do, Q.H.; Han, L.; Nejad, H.T.N.; Mita, S. Safe path planning among multi obstacles. In Proceedings of the 2011 IEEE intelligent vehicles symposium (IV), Baden, Germany, 5–9 June 2011; pp. 332–338.
29. Duchoň, F.; Babinec, A.; Kajan, M.; Beňo, P.; Florek, M.; Fico, T.; Jurišica, L. Path planning with modi-fied a star algorithm for a mobile robot. *Procedia Eng.* **2014**, *96*, 59–69. [CrossRef]
30. Zhang, Z.; Zhao, Z. A multiple mobile robots path planning algorithm based on A-star and Dijkstra algorithm. *Int. J. Smart Home* **2014**, *8*, 75–86. [CrossRef]
31. Tang, G.; Tang, C.; Claramunt, C.; Hu, X.; Zhou, P. Geometric A-star algorithm: An improved A-star algorithm for AGV path planning in a port environment. *IEEE Access* **2021**, *9*, 59196–59210. [CrossRef]
32. Krishnamurthy, J.; Kollar, T. Jointly learning to parse and perceive: Connecting natural language to the physical world. *Trans. Assoc. Comput. Linguist.* **2013**, *1*, 193–206. [CrossRef]
33. Hatori, J.; Kikuchi, Y.; Kobayashi, S.; Takahashi, K.; Tsuboi, Y.; Unno, Y.; Tan, J. Interactively picking real-world objects with unconstrained spoken language instructions. In Proceedings of the 2018 IEEE International Conference on Robotics and Automation (ICRA), Brisbane, QLD, Australia, 21–25 May 2018; pp. 3774–3781.
34. Nguyen, T.; Gopalan, N.; Patel, R.; Corsaro, M.; Pavlick, E.; Tellex, S. Robot object retrieval with con-textual natural language queries. *arXiv* **2020**, arXiv:2006.13253.
35. Hsiung, E.; Mehta, H.; Chu, J.; Liu, X.; Patel, R.; Tellex, S.; Konidaris, G. Generalizing to New Domains by Mapping Natural Language to Lifted LTL. *arXiv* **2021**, arXiv:2110.05603.
36. Spaan, M.T. Partially observable Markov decision processes. In *Reinforcement Learning*; Springer: Berlin/Heidelberg, Germany, 2012; pp. 387–414.
37. Garg, N.P.; Hsu, D.; Lee, W.S. Despot-Alpha: Online Pomdp Planning with Large State and Observation Spaces. Robotics: Science and Systems. 2019. Available online: https://adacomp.comp.nus.edu.sg/pub_post/despot-%CE%B1-online-pomdp-planning-with-large-state-and-observation-spaces/ (accessed on 2 August 2022).
38. Dibia, V. Neuralqa: A usable library for question answering (contextual query expansion+ bert) on large datasets. *arXiv* **2020**, arXiv:2007.15211.
39. Collins, T.; Collins, J.; Ryan, D. Occupancy grid mapping: An empirical evaluation. In Proceedings of the 2007 Mediterranean Conference on Control Automation, Athens, Greece, 27–29 June 2007; pp. 1–6.

40. Motwani, P.; Sharma, R. Comparative study of pothole dimension using machine learning, manhattan and euclidean algorithm. *Int. J. Innov. Sci. Res. Technol.* **2020**, *5*, 165–170.
41. Bundy, A.; Wallen, L. Breadth-first search. In *Catalogue of Artificial Intelligence Tools*; Springer: Berlin/Heidelberg, Germany, 1984; p. 13.
42. Loper, E.; Bird, S. Nltk: The natural language toolkit. *arXiv*, **2002**, arXiv:cs/0205028.

sensors

MDPI

Article

Deep Learning-Based Subtask Segmentation of Timed Up-and-Go Test Using RGB-D Cameras

Yoonjeong Choi, Yoosung Bae, Baekdong Cha and Jeha Ryu *

School of Integrated Technology, Gwangju Institute of Science and Technology (GIST), Gwangju 61005, Korea
* Correspondence: ryu@gist.ac.kr

Abstract: The timed up-and-go (TUG) test is an efficient way to evaluate an individual's basic functional mobility, such as standing up, walking, turning around, and sitting back. The total completion time of the TUG test is a metric indicating an individual's overall mobility. Moreover, the fine-grained consumption time of the individual subtasks in the TUG test may provide important clinical information, such as elapsed time and speed of each TUG subtask, which may not only assist professionals in clinical interventions but also distinguish the functional recovery of patients. To perform more accurate, efficient, robust, and objective tests, this paper proposes a novel deep learning-based subtask segmentation of the TUG test using a dilated temporal convolutional network with a single RGB-D camera. Evaluation with three different subject groups (healthy young, healthy adult, stroke patients) showed that the proposed method demonstrated better generality and achieved a significantly higher and more robust performance (healthy young = 95.458%, healthy adult = 94.525%, stroke = 93.578%) than the existing rule-based and artificial neural network-based subtask segmentation methods. Additionally, the results indicated that the input from the pelvis alone achieved the best accuracy among many other single inputs or combinations of inputs, which allows a real-time inference (approximately 15 Hz) in edge devices, such as smartphones.

Keywords: timed up-and-go test; TUG subtask segmentation; deep learning; temporal convolutional network

Citation: Choi, Y.; Bae, Y.; Cha, B.; Ryu, J. Deep Learning-Based Subtask Segmentation of Timed Up-and-Go Test Using RGB-D Cameras. *Sensors* **2022**, *22*, 6323. https://doi.org/10.3390/s22176323

Academic Editor: Daniele Cenni

Received: 3 July 2022
Accepted: 18 August 2022
Published: 23 August 2022

Publisher's Note: MDPI stays neutral with regard to jurisdictional claims in published maps and institutional affiliations.

Copyright: © 2022 by the authors. Licensee MDPI, Basel, Switzerland. This article is an open access article distributed under the terms and conditions of the Creative Commons Attribution (CC BY) license (https://creativecommons.org/licenses/by/4.0/).

1. Introduction

The timed-up-and-go (TUG) test is an effective tool for assessing an individual's functional mobility that includes the most basic and essential activities in daily life, such as standing up, walking, turning, and sitting back [1,2]. Because the TUG test is relatively easy, it has been widely tested for individuals with balance and gait impairments, such as Parkinson's disease (PD) [3], total knee arthroplasty (TKA) [4], stroke [5], multiple sclerosis (MS) [6], lumbar degenerative disc disease [7], lower limb amputations [8], chronic obstructive pulmonary disease (COPD) [9], and cognitive decline [10], to assess their risk of falling [2,3].

In a typical TUG test, an individual starts sitting on the chair, rises from the chair, walks a set distance of 3 m or 5 m, turns around, returns to the chair, and finally sits down. An observer records the total time taken for the whole TUG test using a stopwatch as a metric to explain the overall mobility [11–13]. As the TUG test contains various subtasks like sit-to-stand, walking, turning, and stand-to-sit, the robust and accurate elapsed time of each TUG subtask can provide important clinical information, such as elapsed time and speed of each TUG subtask, gait speed, cadence (number of steps per minute), and the stance (percentage of the gait cycle) of subjects [2,3,11–15]. These can be used not only to assist clinicians in clinical interventions but also to distinguish patients' functional recovery [3,16]. More specifically, Salarian et al. [3] introduced the instrumented TUG (iTUG), which used portable inertial sensors to automatically detect and segment subtasks into two groups: early stages of PD versus age-matched control group. By comparing the two groups, they found significant differences between them in three subtasks of the iTUG, even though the total time had no significant difference in distinguishing the performance of the two groups. Furthermore, they showed that early PD subjects had unusual features, such as slow turning, arm swing, cadence, and trunk rotation, during straight walking [3]. Ansai et al. [16] also found that in the case of falls with mild cognitive impairment (MCI) and in non-fallers with Alzheimer's disease (AD), there was a significant difference in turn-to-sit subtask performance, even though no other difference was found in the total consumption time between the two groups. Therefore, patients who could not be classified according to total time could be classified through subtask segmentation.

A typical TUG test is performed under a clinician's supervision, which may require both physical and mental human efforts. Moreover, because the manual measurement using a stopwatch is a subjective observation, it could be inaccurate and inconvenient [1]. Furthermore, visiting a hospital to perform a TUG test is not trivial for patients who have limited mobility. This is a heavy burden for both patients and clinicians. Therefore, automatic TUG analysis methods, especially subtask segmentation methods [2,3,10–16], have recently attracted considerable research attention [4,17,18].

TUG tests have been widely studied not only for various patients but also in healthy young and older adults. For older adults, frailty is a common syndrome that embodies a high risk of critical declines, such as cognitive and functional decline, and falls with accelerated aging. Early diagnosis of a frailty syndrome can reduce harmful health outcomes and slow down the progression of frailty, for example, by prescribing suitable exercise programs [19–21]. This benefits individuals and relieves the burden on families and society. Healthy young people are also often selected as TUG subtask segmentation study subjects for several reasons: (1) to validate and evaluate the developed system [22–25]; (2) to adopt normal people as a criterion for comparison and perform an analysis on abnormal people (older adults, frailty syndrome, PD patients, and potential patients who do not yet show symptoms or those who already show symptoms) [24,25]; and (3) to obtain quick and meaningful data [25]. Therefore, this study collected data on both healthy young and older adults, and the performance of the proposed approach was then compared with those of the previous studies.

We think that there are two main challenges of TUG subtask segmentation studies: (1) accuracy improvement to provide more precise clinical information to patients and clinicians, and (2) automation to simplify TUG tests. If a TUG subtask is classified with higher accuracy, more precise clinical information, such as stride, gait speed, and turning speed, can be extracted. This may provide medical experts with a more elaborate diagnosis and therapy evaluation, e.g., after rehabilitation. In addition, automated TUG tests can reduce the test time and efforts of human operators, which may allow TUG tests at home for better accessibility to patients. For these benefits, the most advanced and readily available RGB-D cameras such as Azure Kinect must be used along with the current deep learning (DL) methods.

To the best of our knowledge, there are no previous studies on TUG subtask segmentation using a DL approach and an RGB-D camera. The RGB-D cameras have only been used in rule-based methods. The main contributions of this study are summarized as follows.

1. We propose a novel DL-based subtask segmentation of the TUG test using an RGB-D camera (Azure Kinect). In the proposed method, a dilated temporal convolutional network (TCN) is used to improve the accuracy and processing time of subtask segmentation compared to the existing Bi-LSTM.
2. We investigated several inputs to the dilated TCN model to determine the input(s) that is (are) better than the others in the TUG subtask segmentation. We showed that the input from the pelvis alone had the best accuracy among many inputs. This single feature point and the optimized dilated TCN architecture also reduce the processing time in the inference phase.
3. We evaluated the proposed method using the newly collected TUG data for three subject groups: (1) healthy young people, (2) older adults, and (3) stroke patients.

The test results showed significantly better accuracy and robustness than existing rule-based and artificial neural network (ANN)-based subtask segmentation methods.

The rest of the paper is organized as follows: Section 2 introduces the related work on the automation of TUG. Section 3 provides the applied methodology with details. Section 4 contains experiments and results, and discussion is presented in Section 5. Finally, Section 6 concludes the paper and suggests some future work.

2. Related Work

The automation of the TUG test can be divided into contact and non-contact methods. In contact methods, wearable devices, such as inertial measurement units (IMU) or infrared sensors, are usually used. For example, Hsieh et al. [4] segmented subtasks for a TKA patient using inertial sensors with machine learning (ML) techniques as the classifier. Ortega-Bastidas et al. [26] used a rule-based technique to segment subtasks for 25 healthy young and 12 elderly subjects using inertial sensors and an RGB camera for ground truth data. In contrast, with older adults, Hellmers et al. [27] used only a single IMU and applied four ML methods: boosted decision trees (BDT), boosted decision stump (BDS), multilayer perceptron (MLP), and adaptive multihyperplane machine (AMM). They classified the subtask into static (sit, stand), dynamic (walk, turn), and transition (sit-to-stand, stand-to-sit) using BDT and classified subtask activities using MLP. For more details on contact methods, refer to [2–4,14,23,28,29]. Although these wearable devices generally show high resolution and accuracy, they require complex and inconvenient setup and calibration processes, increasing the physical burden on medical professionals and patients.

Non-contact methods primarily use non-contact sensors, such as RGB or RGB-D cameras, because video data can be collected from almost anywhere with simple settings. Using a single RGB camera along with deep learning (DL) and postprocessing techniques, Savoie et al. [22] segmented subtasks for healthy young people. First, 2D keypoints were extracted from the RGB image through the mask R-CNN [30], and 3D local poses were extracted using a deep multitask architecture for human sensing (DMHS) [31]. Then, these 2D keypoints and 3D local poses were used together to extract the global 3D poses of the participants. However, for segmentation, a rule-based technique was used, considering the characteristics of the trajectories of each joint. Li et al. [18] segmented subtasks for 24 patients with PD using an RGB camera and ANN-based technique. Although these methods generate good results, they typically require several steps to extract global 3D poses [22], or they allow only frontal camera locations [18].

In contrast with the RGB cameras, the RGB-D cameras, especially Kinect, are popular because of their accurate built-in skeleton tracking application programming interface (API) and silhouette extraction capabilities. Moreover, Kinect has proven sufficiently accurate compared to golden standards such as marker-based systems in gait [32–34]. Using RGB-D cameras, Lohmann et al. [35] segmented subtasks for normal older people using two Kinect Xboxes (RGB-D sensors). They used a rule-based technique with a specific feature, such as maximum or minimum shoulder z-axis acceleration, among the skeleton data. For other purposes, Jannis et al. [36] segmented phases (the same as subtasks) for patients with PD using Kinect V2. They also used a rule-based technique that exploited periodically increasing distances between the feet. The segmented phases were then used to assess the PD disease scale clinically using the TUG. Kampel et al. [37] proposed automatic TUG subtask methods for a functional decline assessment of 11 older adults using the Kinect V2 and rule-based technique with a specific feature, such as the velocity of shoulder z-axis, and with other features. Note that all these methods using RGB-D cameras were rule-based and not DL-based.

The TUG subtask segmentation approaches are primarily classified into two types: (1) rule-based and (2) ANN-based. The rule-based methods are applicable to a small number of datasets, as described in studies [3,23,28,35,37]. However, rule-based methods generally use complex predetermined rules to detect the motion transition following physical attributes, such as the trunk bending forward, and moving the feet [38]. For example, by using two Kinects to find local maxima, the skeleton TUG (sTUG) detected ten events: start moving, end uprising, start walking, start rotating, start turning, max turn, end turning, end rotating, start lowering, and end moving [37]. As another example, Ortega-Bastidas et al. [26] used a single wireless IMU sensor to detect turning start/end events by determining the average maximum/minimum *Yaw* (rotation) signal of an inertial sensor placed on the back. With their increasing complexity, analyzing and adjusting rule-based systems is often cumbersome. Therefore, these approaches require the careful selection of strict rules by experts, which may not be objective. Moreover, rule-based techniques may not be robust because a high variance in movement patterns can occur, especially for transitions, in older adults.

In contrast, the ANN-based subtask segmentation approach builds a model using features from training data with ground truth data generated by experts. Hsieh et al. [4] studied five ML techniques as classifiers: support vector machine (SVM), K-nearest neighbor, naïve Bayesian, decision tree, and adaptive boosting (Adaboost); data were obtained using six wearable sensors containing a triaxial accelerometer, a gyroscope, and a magnetometer. As another example, Li et al. [18] explored two pose estimators (interactive error feedback (IEF) and OpenPose) and two classifiers (SVM and bidirectional long short-term memory (Bi-LSTM)) using a single RGB camera. However, learning an ANN model requires a considerable number of TUG test datasets. Unfortunately, an open dataset related to the TUG test [17] cannot be used for TUG subtask segmentation because there is no ground-truth label for subtask segmentation.

3. Methods

3.1. Data Acquisition System and Participants

Figure 1 shows the system configuration for the 3 m TUG tests, which is similar to that in [37]. One difference is that the Azure Kinect was installed at a height of 1.5 m rather than 0.7 m. The z-axis (viewing direction) was perpendicular to the walking direction of participants for capturing left and right lateral motions, the x-axis was perpendicular to the walking direction, and the y-axis was for up and down movement. A cone was placed at 3 m straight from a standard chair with armrests to indicate a turning point where the participant should perform a "turn." With this setting, participants sitting in a chair started to get up by the indicator's "start" signal, walked 3 m at a comfortable walking speed, rotated the turning point (cone), walked 3 m back, leaned back on the chair, and then finally waited for the "end" signal. At this time, additional data were collected for one second before and after the "start" and "end" signals to prevent data from becoming unstable and ensure equal data lengths between the start and end periods upon data entry. The skeleton data were collected at a 30 Hz rate, and the RGB image, RGB-D image, and skeleton data were separately stored for subsequent analyses.

We used three groups of subjects: (1) 50 older adult subjects (15 male, 34 female, 79.37 ± 5.08 years old), (2) 15 healthy young subjects (13 male, 2 female, 28.75 ± 2.05 years old) that were confirmed to have no known walking disorders, and (3) 23 stroke subjects (18 male, 5 female, 53.89 ± 6.12). All experiments were conducted in various environments, such as apartments, laboratories, and fitness rooms, because of the subject's availability in different locations. In this experiment, 15 trials of the 3 m TUG test for each participant were measured with a three-minute break after the first eight trials and one practice trial before the experiment. Some trials were excluded because of human error in measurement or missing skeletons. In this study, we obtained a TUG video dataset of 158 trials for healthy young subjects, 620 trials for older adults, and 268 trials for stroke patients. These datasets have an average video length of 11.34 s (healthy young), 15.33 s (older adults), and 43.11 s (stroke patients). The number of trials in this study is sufficient, considering a similar DL study with 24 PD patients and 127 trials using the Bi-LSTM model [18].

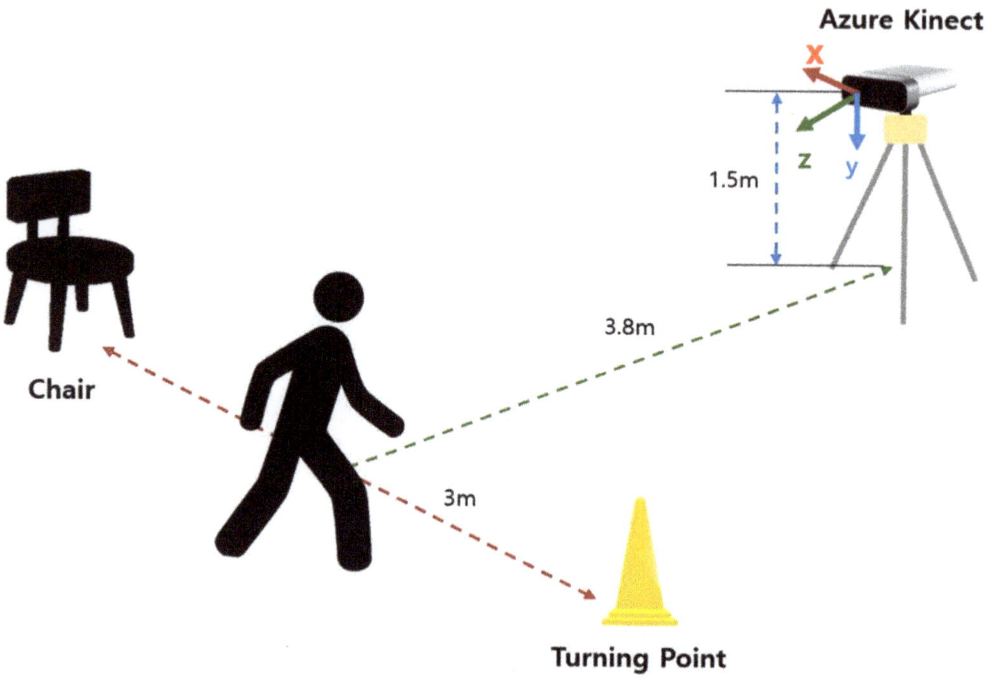

Figure 1. System configuration for the study. The participants performed a 3 m TUG test. A cone was placed at 3 m straight from a standard chair, and Azure Kinect was installed perpendicular to the walking direction at the height of 1.5 m.

3.2. Proposed Method Overview

Figure 2 shows the overall process of the proposed method with typical time plots in each process. The raw trajectories (e.g., pelvis x, y, and z) are the normalized preprocessor inputs, and a low-pass filter (LPF) was used for the deep model training with the dilated TCN model. The predicted frame-level action states were postprocessed by dynamic time warping (DTW) to modify the misclassified action states. Finally, many action capabilities were computed for the total time, subtask times, subtask speeds, etc.

3.2.1. Labeling Process

The proposed method detected six TUG events (StartMove, StartWalk, StartTurn, EndTurn, StartSit, and EndSit) and classified five subtasks (sit-to-stand, walk, turn, walkback, and stand-to-sit). The six TUG events were labeled for each TUG video to detect the six events. The labeling results of the two experts were averaged and used as the ground truth to avoid label biases by experts. We considered the labeling guidelines for six TUG events from [18], examples of which are summarized in Table 1. The reliability of the generated ground truth was measured using the intraclass correlation coefficient (ICC) [18]. The intra-rater reliability between the labeling results of the two experts showed high reliability with ICC = 0.99.

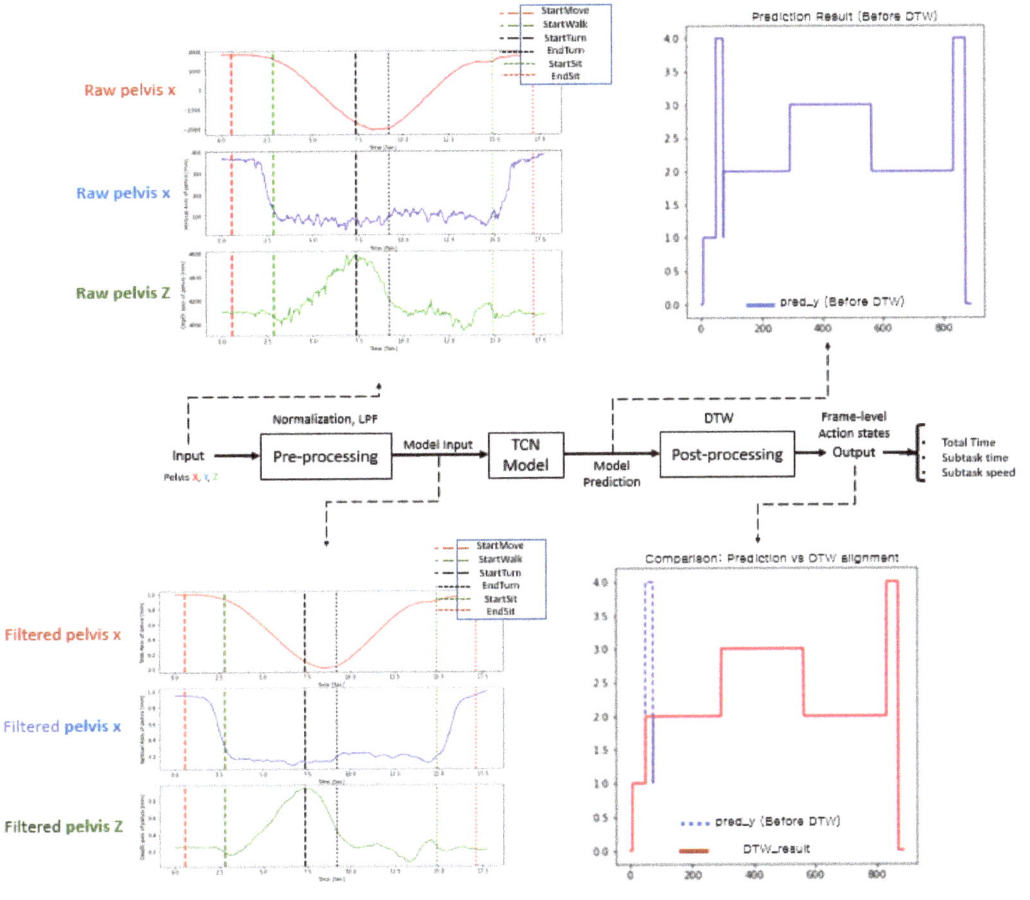

Figure 2. Overall flowchart of the proposed method.

Table 1. Labeling guidelines for six events.

TUG Events	Label	Criteria
StartMove	0	When body is tilted 45 degrees to get up from the chair
StartWalk	1	After getting up from the chair, when the first step is off the ground
StartTurn	2	When subject rotates the body to turn at the TUG marker
EndTurn	3	After turning at the TUG marker, when the body looks back at the chair
StartSit	4	When body stands against the chair after turning body to sit on the chair
EndSit	5	When body is titled 45 degrees to lean on the chair

3.2.2. Preprocessing

During preprocessing, an LPF was applied to remove noise generated during measurement, and min-max normalization was sequentially performed to normalize diverse ranges of subject data. Normalization of the trajectories is necessary because various height differences between subjects generate a diverse range of trajectories in the TUG data. In addition, to solve the problem of direction mismatch caused by camera coordinates, axis matching was performed to update the coordinates according to the subject's walking direction [37]. Figure 3 shows an example of preprocessing for the pelvis trajectories. The optimal cutoff frequency selected by trial and error was 0.3 Hz, and the min-max normalization technique was finally utilized together by applying the 4th Butterworth LPF.

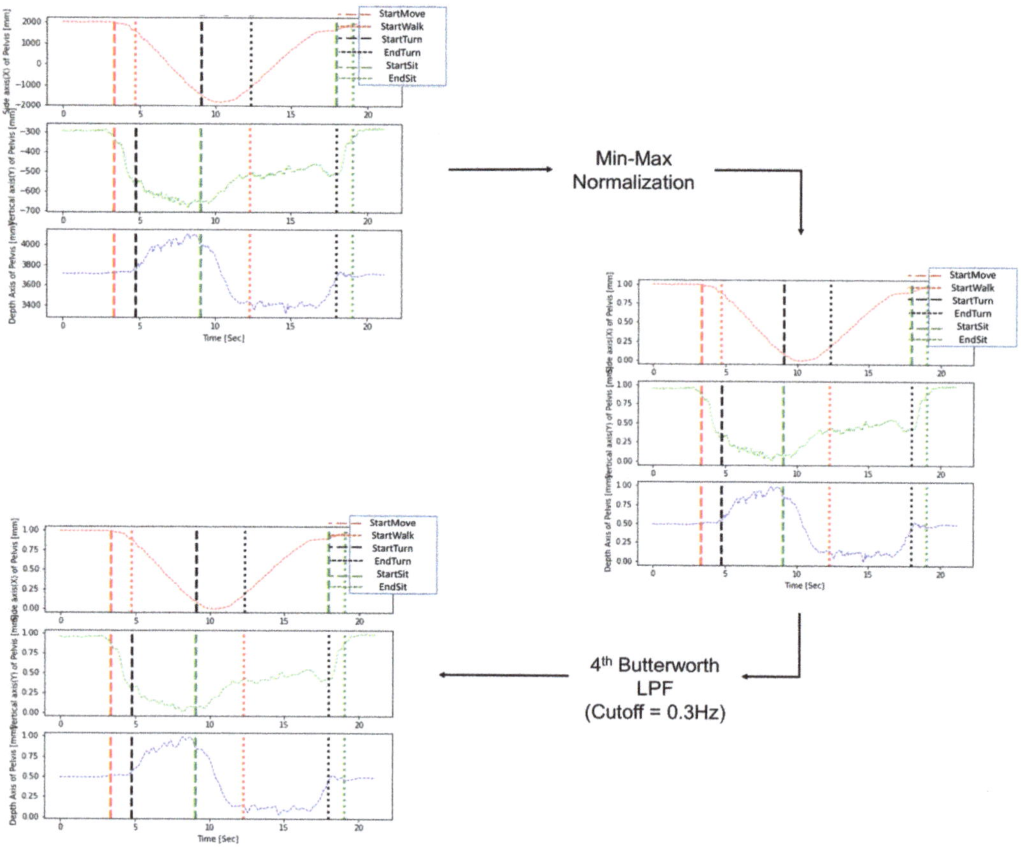

Figure 3. Preprocessing by low pass filter and normalization for a pelvis trajectory.

3.2.3. DL-Based TUG Subtask Segmentation Algorithm

Action segmentation, such as the TUG subtask segmentation, is crucial to analyzing the activities of daily living. TUG subtask segmentation aims to segment all subtasks in time for a given TUG video. For frame-level action state classification in TUG subtask segmentation, we investigated a class of time-series DL models, called temporal convolutional networks (TCNs), because they can capture features of action durations and pairwise transitions between segments and long-range temporal patterns more efficiently using a hierarchy of temporal convolutional filters [39]. This model is faster to train through parallel operations based on dilated convolution and tends to outperform canonical recurrent architectures, such as LSTMs and gated recurrent units (GRUs), across a broad range of sequence modeling tasks [40].

The dilated TCN uses a deep series of dilated convolutions instead of pooling and up-sampling, adding skip connections between layers. Each layer uses dilated filters that operate on only a small number of time steps. The dilated TCN has the following properties: (1) computations are performed layer-wise, which means that every time step is updated simultaneously; (2) convolutions are computed according to the time; and (3) predictions at each frame are a function of the receptive field. Figure 4 shows the architecture applied to the proposed TUG subtask segmentation. The normalized and filtered trajectories were input to the first temporal block. Three temporal blocks generated features, and the softmax activation function in the final layer output the frame-level action state classification results.

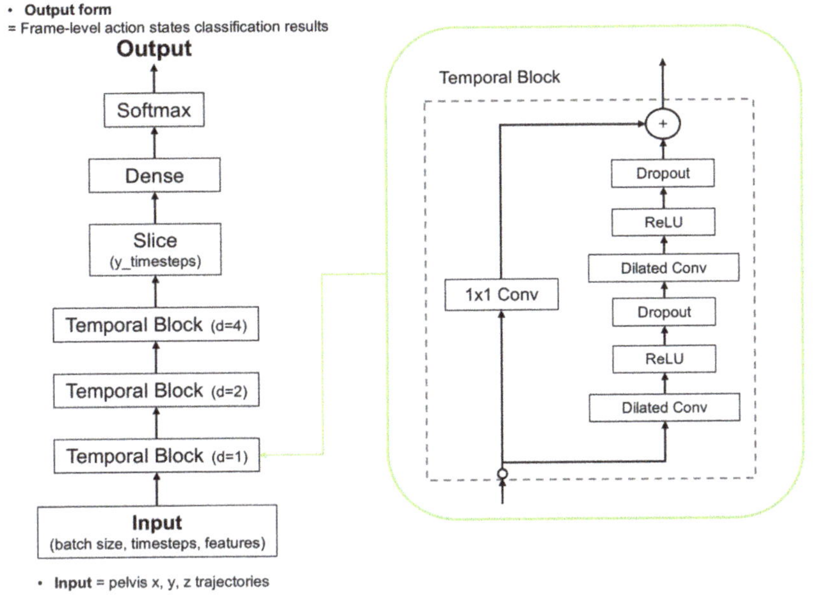

Figure 4. Architecture of dilated temporal convolutional network (TCN).

3.2.4. Postprocessing

The predicted frame-level action states from the dilated TCN can generate misclassified action states in each frame; this has already been discussed as a fragmentation error in [4,18]. Considering the strict time order of TUG subtasks (e.g., sit, sit-to-stand, walk, turn, walk-same as walk back, stand-to-sit, and sit), we also performed the DTW [18] algorithm to find the most appropriate subtask segmentation in the postprocessing stage. This was an application of the domain (field) knowledge in subtask classification to obtain the most appropriate subtask segmentation.

Figure 5 shows the example of the postprocessing for correcting the frame-level misclassifications. The green line represents the ground truth, the blue line represents the model classification result before postprocessing, and the red line represents the model classification result after DTW. In Figure 5a, the index of the y-axis represents each subtask, in which the walk and walk back subtasks existing between the turn subtasks are marked on the same index. Figure 5 shows the results before (Figure 5b) and after (Figure 5c) the DTW algorithm. Figure 5d shows the ground truth (green line) (Figure 5a) and the result (red line) (Figure 5c) from the DTW. Comparing Figure 5b,c illustrates that the prediction model incorrectly classified the "sit-to-stand" (index = 1) to "stand-to-sit" (index = 4) for the state transition from "sit-to-stand" to "walk" (index = 2). After the DTW was applied, this misclassification disappeared.

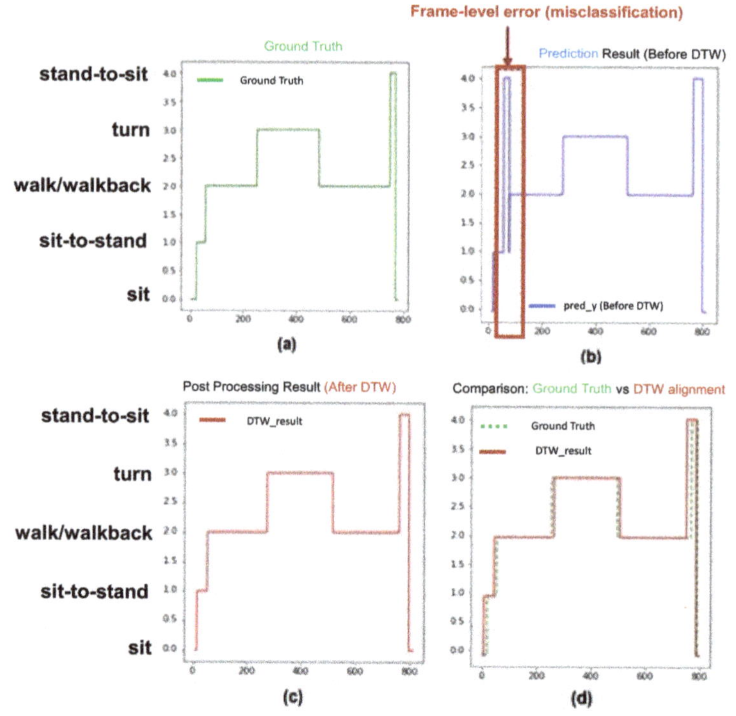

Figure 5. Sample postprocessing for correcting frame-level misclassification.

4. Experiments and Results

This section presents two studies for optimization of the input and DL models: (1) an input comparison study to find the most effective input for the proposed method, and (2) optimization of a DL model based on the results of the input comparison study. To demonstrate the superiority of the proposed method over the conventional methods, two methods were selected that are most similar to the proposed method regarding the experimental setup: (1) the rule-based TUG subtask segmentation method [37] and subject type (older adults); (2) the ANN-based TUG subtask segmentation method [22] and subject type (healthy young).

4.1. Metrics

The mean absolute error (MAE), standard deviation (STD), precision, recall, and F1 score were used to evaluate the TUG subtask segmentation classification performance for a fair and objective comparison with conventional methods. For the frame classification accuracy, we calculated the precision (prec.), recall (rec.), and the harmonic mean of the precision F1 score (F1) for each subtask, where TP, FP, and FN represent the number of true positive, false positive, and false negative frames for a given subtask. The total subtask classification accuracy (Acc.) was defined as the percentage of correctly predicted observations to the total number of observations in a trial. The equations below are for precision, recall, F1 score, and accuracy.

$$\text{Precision} = \frac{TP}{TP + FP} = \frac{TP}{All\ detections} \tag{1}$$

$$\text{Recall} = \frac{TP}{TP + FN} = \frac{TP}{All\ ground\ truths} \tag{2}$$

$$\text{F1 score} = 2 * \frac{prec * rec}{prec + rec} \quad (3)$$

$$\text{Accuracy} = \frac{TP + TN}{TP + TN + FP + FN} \quad (4)$$

Recall evaluated cases in which the correct answer was inferred from the standpoint of the ground truth, and precision evaluated from the perspective of the classification model. The best result was when both the recall and precision were high. However, because these two are conflicting concepts, it is generally difficult to judge because it is a trade-off relationship that decreases when one increases. The F1 score was defined as the harmonic mean between precision and recall and was used as a statistical measure to rate performance. The total TUG time was defined as the elapsed time between the first (StartMove) and last (EndSit) TUG events. The total TUG MAE was the mean absolute difference between its time duration estimated by the model and the ground truth.

4.2. Input Comparison Study

Human activity can be detected using various sensors (inertial, RGB camera, and RGB-D camera (Kinect)) and classified by suitable DL-based methods. However, using raw images or video frames is usually very time-consuming, even in ANN-based approaches. Moreover, it may require considerable training data from diverse viewing angles and involves significant time and effort. Because the TUG test consists of relatively simple activities, it can be easily distinguished using low-dimensional features, such as human posture detected by body keypoints [18]. Therefore, previous studies investigated various types of input data to obtain efficient results with subtask segmentation [3,4,18,22,23,28,35,37,41] or fall risk prediction [25,42]. We summarize these in Table A1 in the Appendix A.

Figure 6 shows the total 32 skeleton joints detected using Azure Kinect. Among these, previous automatic TUG research with wearable sensors, such as a single IMU, argued that the proximal location of sensors (e.g., at the lower back) could more accurately detect the main gait events and spatiotemporal and kinematic factors [2,43]. Additionally, according to Sprint's automatic TUG test research survey [2], 55.56% of wearable sensors, such as IMUs, are attached to the lower back. These findings motivated us to speculate that similar locations might be better than other locations in the TUG subtask segmentation task by using a lower resolution of RGB-D than the IMU. Previous studies using RGB or RGB-D cameras used complete skeleton data spread throughout the body as inputs to ANN models or rules [18,22,42]. We hypothesized that the input(s) only from the proximal joints, for example, center of gravity (COG) movement or head motion, and not with all 32 skeleton joints, should be sufficient for classifying TUG subtasks. Therefore, only five inputs (red box) were investigated for input comparison in this study because the accuracies of skeleton joint motion from the distal joints, such as ankles and hands, are usually very low due to fast movements and a wide range of motion. Using a smaller number of feature points can also significantly reduce the processing time in the inference phase, which is important for edge computing on mobile devices.

To determine which input(s) is (are) better than the others in TUG subtask segmentation, we investigated five different skeleton feature points from five groups of skeleton joints: (1) pelvis, (2) spine chest, (3) head, (4) a pair of left and right hands, and (5) a pair of left and right ankles (Figure 6). The pelvis and spine chest were selected because they are close to the COG inputs. The head, a pair of hands, and ankles were chosen to evaluate the effects of the inputs from the distal joints representing significant overall body motion.

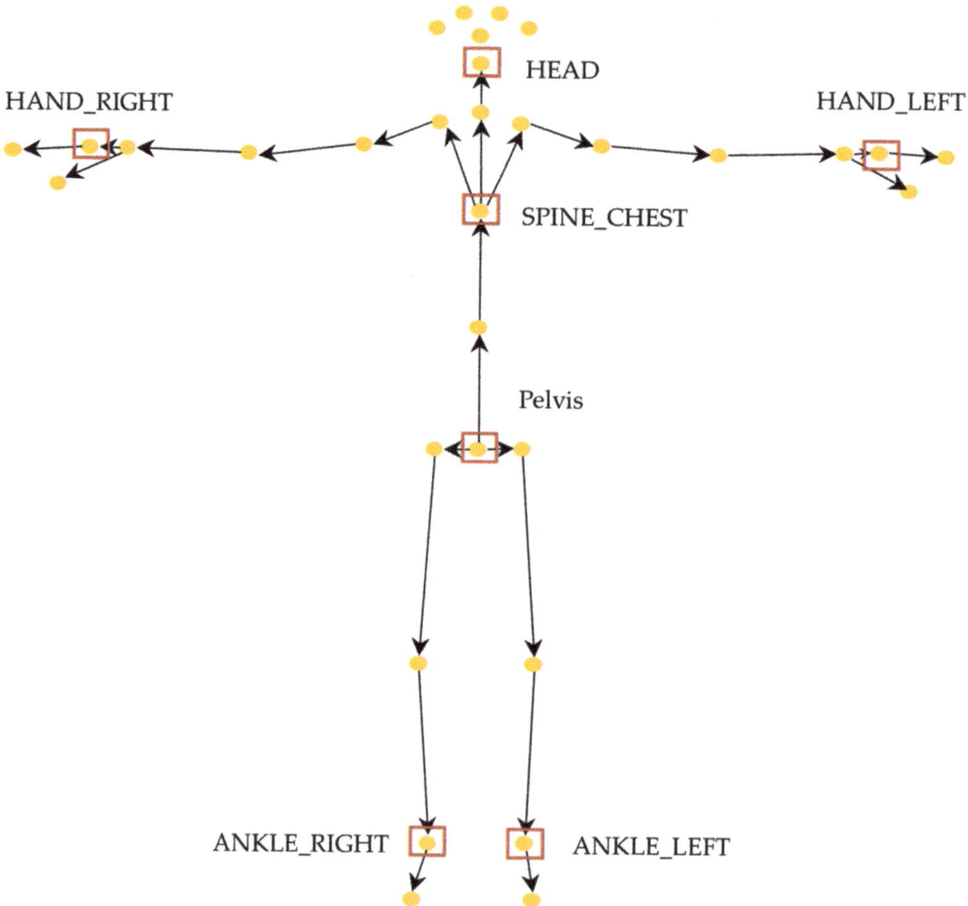

Figure 6. Skeleton joints tracked by the Azure Kinect and used inputs for comparison (red box).

Table 2 shows the results (5-fold average accuracy (%)) for the older adults, healthy young, and stroke patients with four different group combinations of skeleton feature points: (1) five single features (pelvis, spine chest, head, both hands, and both ankles), (2) four combinations of two features (pelvis + spine chest, pelvis + head, pelvis + both ankles, pelvis + both hands), (3) five combinations of three features (pelvis + spine chest + head, pelvis + head + both ankles, pelvis + both hands + both ankles, pelvis + head + both hands, and head + both hands + both ankles), and (4) a combination of all five features. Note that some feature combinations were excluded (e.g., spine chest + head, hands + ankles, head + spine chest + hands) because the distal joint inputs (hands and ankles) far from the COG have larger noise in the data as the accuracy was found to be worse. These results were obtained by training and testing each participant group separately.

Table 2. Comparison results of TUG subtask segmentation accuracy of joints closest to COG.

	Input		Healthy Young	Older Adults	Stroke Patients
	Joint	No.	Acc. [%]	Acc. [%]	Acc. [%]
Results	pelvis	input 1	95.46	94.53	93.58
	spine chest	input 2	94.29	94.25	92.83
	head	input 3	94.1	93.86	90.86
	hand (left/right)	input 4	92.24	91.32	79.46
	ankle(left/right)	input 5	89.89	86.53	80.58
	pelvis, head	input 6	94.4	94.22	91.59
	pelvis, spine chest	input 7	94.46	94.045	92.29
	pelvis, ankle	input 8	93.42	93.81	87.56
	pelvis, hand	input 9	93.46	93.68	87.07
	pelvis, head, spine chest	input 10	93.78	93.44	91.72
	pelvis, head, ankle	input 11	93.59	93.93	90.89
	pelvis, hand, ankle	input 12	93.15	93.84	91.944
	pelvis, head, hand	input 13	93.3	93.63	91.31
	head, hand, ankle	input 14	93.25	93.39	91.72
	pelvis, head spine chest, hand, ankle	input 15	93.62	93.95	92.4

Table 2 shows many interesting points: (1) a single feature of the pelvis had the highest accuracy of 95.46%, 94.53%, and 93.58% for healthy young, older adults, and stroke patients, respectively. This also shows that the accuracy decreased from the healthy young to older adults and stroke patients, as expected. In addition, the 5-fold average total TUG MAE (s) was 0.1449, 0.2213, and 0.5572 for the healthy young, older adult, and stroke patients, respectively, which indicates that the average total TUG MAE increases (almost doubled) from healthy young to older adult and stroke patients. (2) A single distal feature (head, hand, ankle) that is far from the COG showed relatively poor accuracy compared to the proximal features (pelvis, spine chest). This shows that the effects of the distal joint inputs are far from the COG. Note that the performance of the head input was better among these distal joint inputs (hand and ankle) because the head does not shake much, and skeleton tracking was easy. (3) The combination of more features on top of the pelvis showed no accuracy improvement for all subject groups. These results imply that the pelvis alone is sufficient for the subtask classification. Other previous approaches had also demonstrated the effectiveness of the pelvis point for other purposes for pathological gait classification with IMUs attached to the pelvis [43].

4.3. Optimization of Deep Learning Model

The proposed dilated TCN architecture, whose basic structure is shown in Figure 4, was optimized with pelvis input to determine the best kernel and window sizes. These two sizes are important hyperparameters in the dilated TCN architecture. We investigated three kernel sizes (3, 5, and 7) and four window sizes (4, 8, 16, and 32) because many other studies considered these ranges in the optimization process. Table 3 shows that a kernel size of three and a window size of eight showed the best performance for TUG subtask segmentation.

Table 3. Optimal values of kernel and window sizes.

Title 1	Kernel Size (Window Size = 8)			Window Size (Kernel Size = 3)			
	3	5	7	4	8	16	32
Accuracy	94.53	93.21	92.73	92.8	94.53	92.11	87.26
# of parameters	41,879	112,537	218,521	40,921	41,879	43,801	47,641

To find out the optimal hyperparameters, we used the grid search method: a heuristic approach. We considered the following hyperparameters: (1) learning rate in the interval of {0.00001, 0.00005, 0.00007, 0.0001, 0.0003, 0.0005}, (2) optimizer in the interval of {Adam, RMSprop}, (3) patience in the interval of {10, 20, 30, 50, 70, 100}, (4) batch size in the interval of {32, 128, 256, 512, 1024, 2048}. From this optimization process with the categorical cross-entropy loss in the proposed dilated TCN training, optimal parameters were determined as a learning rate of $7 \times 10 - 5$, optimizer of Adam, patience of 50, and batch size of 1024 to segment the TUG subtask (e.g., a multiclassification task). Figure 7 shows the behavior of loss and accuracy with the pelvis input. The best model was obtained using an early stopping criterion; the training will stop if the validation loss is not updated for a specific epoch (patience). Datasets were split into 6:2:2 (training: validation: test) to train the dilated TCN network for TUG subtask segmentation, e.g., (30: 10: 10) for 50 older adult datasets. To prevent overfitting, the model stopped training early when the validation loss did not improve during 50 epochs.

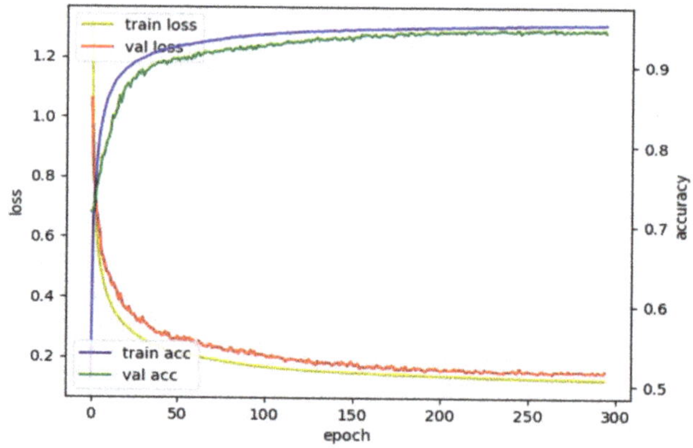

Figure 7. Loss and accuracy plot for model with pelvis input.

Based on the optimized parameters (learning rate of $7 \times 10 - 5$, optimizer of Adam, patience of 50, and batch size of 1024), we performed an ablation study to find out two optimal parameters of the dilated TCN model: (1) number of temporal blocks in the four ranges of {1, 2, 3, 4}, (2) number of convolutional layers in the three ranges of {1, 2, 3}. Five-fold cross-validation was used for the accurate evaluation and optimization of the model hyperparameters. The following results in Table 4 from the ablation study show the best accuracy for three temporal blocks with two convolutional layers for the TUG subtask segmentation task.

Table 4. Accuracy from ablation study of dilated TCN.

		Number of Temporal Blocks. Number of Conv. Layers	
	1.1	1.2	1.3
Acc [%]	92.3	93.53	92.54
	2.1	2.2	2.3
Acc [%]	93.94	90.42	92.03
	3.1	3.2	3.3
Acc [%]	92.34	94.53	93.44
	4.1	4.2	4.3
Acc [%]	92.75	92.31	91.86

We also compared the performance between the dilated TCN and Bi-LSTM architecture in [18] for encoding the input features. Bi-LSTM can combine temporal information in the positive and negative time directions. It also contained separate layers for batch normalization and dropout. The encoded features were then sent to a fully connected layer using the softmax function to predict the frame labels. The optimized Bi-LSTM architecture had 128 layers and a dense layer of 64 neurons. The 5-fold accuracy of the proposed method with TCN with 41,879 parameters was 94.53%, whereas that of the Bi-LSTM with 546,181 parameters was 94.23%. This comparison shows that the proposed method had slightly better accuracy and a significantly smaller number of model parameters.

4.4. Comparison with Rule-Based Method

Considering the objectives of the experiments, the experimental setup of the RGB-D camera for data acquisition, and subjects (older adults), we compared the TUG subtask segmentation classification results of the proposed method with those of the rule-based skeleton TUG in [37]. The proposed approach showed the best performance among the skeleton TUG [37], depth TUG [37], and sTUG [35]. We compare the TUG event detection and TUG subtask segmentation in the following subsections.

4.4.1. TUG Event Detection

To compare the results from the two methods, we first explain the TUG event detection (labeling) criteria because there are no standard criteria for segmenting subtask regions. Each subtask segmentation study uses its own criteria. The TUG event detection (labeling) criteria differ between the skeleton TUG and the proposed method. The skeleton TUG detected twelve events: the beginning and end of the six subtasks (ChairRise, FirstWalk, FirstTurn, SecondWalk, SecondTurn, SitDown). In this study, six TUG events (StartMove, StartWalk, StartTurn, EndTurn, StartSit, EndSit) were detected for the classification of five subtasks (sit-to-stand, walk, turn, walk back, and stand-to-sit). Note that the "walkback" of the proposed method can be represented with a combination of the "SecondWalk" and "SecondTurn" of the skeleton TUG. Furthermore, the "stand-to-sit" of the proposed method can be denoted as "SitDown" of skeleton TUG because "stand-to-sit" is the time elapsed from the moment the subject completes the turn around to sit back in the chair until the moment the subject's back lies against the backrest of the chair. Other studies have also considered five subtasks, using the IMU in smartphones [44] or UWB radar and insoles sensors [45].

Comparing the six events in both the studies, only StartMove and EndSit showed apparent differences in detection. Our study defined the StartMove event as when the body was tilted by 45°, whereas the ChairRise start event of the skeleton TUG detected the time when the body started to tilt. Our study defined the EndSit event as when the body was tilted by approximately 45° to sit in a chair. The SitDown end event of the skeleton TUG detected the time when the body was fully stretched with a slope of less than zero. These criteria for the ChairRise start and SitDown end in the skeleton TUG were made automatically based on predefined rules. In contrast, in our study, StartMove and EndSit events were detected by two experts, and averages were used as final detection, similar to previous studies [18]. The differences in these event detection criteria for the two events turned out to be insignificant, as manifested in a later result section.

Despite different definitions of subtasks, six events had similar geometrical meanings in both studies, as shown in Figure 8: (1) StartMove = ChairRise Start, (2) EndSit = SitDown End, (3) StartWalk = FirstWalk Start, (4) StartTurn = FirstTurn Start, (5) EndTurn = FirstTurn End, and (6) StartSit = SecondTurn End. Therefore, these six events can be used to compare the event detection performance.

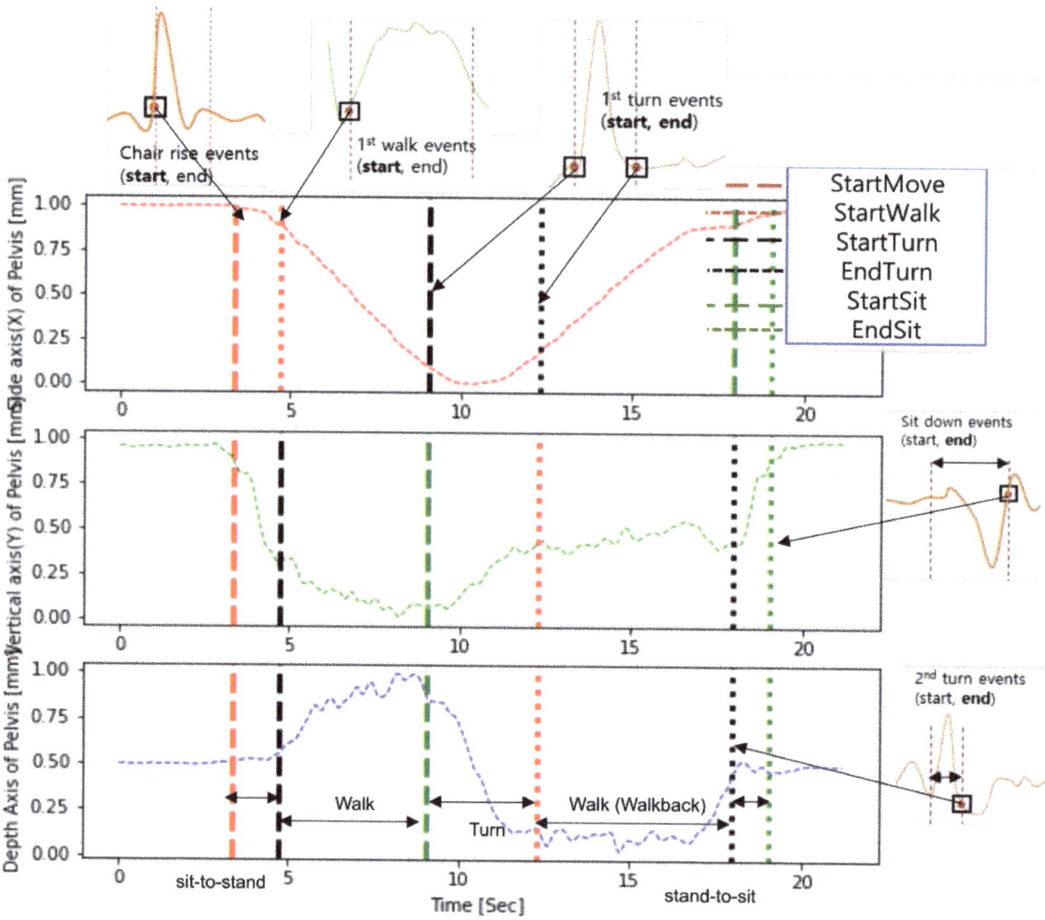

Figure 8. TUG events in Skeleton TUG and this study. Based on the dataset obtained in the study, it shows how the proposed method and the events of the skeleton TUG are matched, re-spectively. Each truncated plot is the result of detection of the event of the skeleton TUG.

Because the TUG subtask was segmented based on TUG event detection, Figure 9 compares TUG event detection errors for older adults [37] (MAE and STD) for six events in the bar chart for a more transparent comparison between the two methods. The MAE and STD for TUG event detection of the two methods did not show a significant difference between StartMove and EndSit. This indicates that the difference in the detection criteria for these two events did not cause significant differences in the MAE and STD. However, the other intermediate events (StartWalk, StartTurn, EndTurn, and StartSit) showed significant differences. The results show that the proposed DL-based method detected TUG events far more accurately (smaller MAE) and robustly (smaller STD) than rule-based methods.

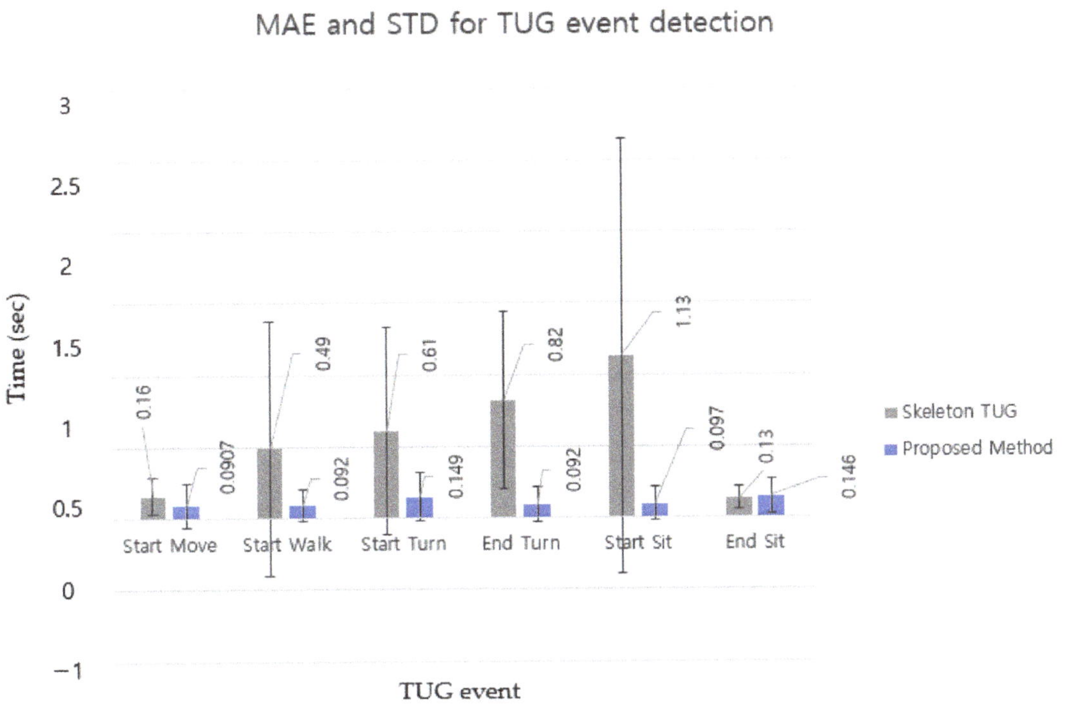

Figure 9. MAE and STD in seconds between skeleton TUG (gray bar) and the proposed method (blue bar) for each TUG event. Error bars are ± the STD of the values.

4.4.2. TUG Subtask Segmentation

Table 5 shows the subtask segmentation comparison regarding MAE, precision, recall, and F1 score for older adults. Because the F1 score for each TUG subtask of the skeleton TUG was not computed in the skeleton TUG, the F1 score was calculated here for a fair comparison. Note that the results in the skeleton TUG did not include the STD of MAE, whereas those of the proposed method included the STD of MAE to show the variability of the MAE.

Table 5. Comparison for TUG subtask segmentation for older adults.

Method	Metric	MAE, STD, Precision, Recall, and F1 Score of TUG Phases						
		Total TUG Time	Sit-to-Stand	Walk	Turn #1	Walk	Turn #2	Stand-to-Sit
Skeleton TUG [37]	MAE	0.227	1.024	0.903	1.061	1.224	2.182	1.570
	Prec.	0.997	0.647	0.961	0.793	0.831	0.832	0.593
	Recall	0.990	0.928	0.906	0.871	0.983	0.759	0.952
	F1 score	0.994	0.753	0.933	0.830	0.900	0.793	0.731
Proposed method	MAE	0.221	0.138	0.134	0.182	0.196	-	0.181
	STD	0.237	0.228	0.109	0.136	0.148	-	0.145
	Prec.	0.986	0.955	0.947	0.967	0.913	-	0.884
	Recall	0.990	0.973	0.966	0.96	0.932	-	0.818
	F1 score	0.988	0.964	0.957	0.963	0.923	-	0.849

Table 5 shows that for all five subtasks, the proposed method showed significantly lower timing errors (MAE) (9–10 times) and higher F1 scores. This means that the proposed method had a better classification performance for the multiclass classification task with imbalanced data, such as the TUG test, in which data were collected in an unbalanced form because the execution time of each activity was different for each subtask. However, in the total TUG time, precision, and F1 score were slightly better in the skeleton TUG than in the proposed approach. This may be because the skeleton TUG had a clearer criterion for sit-to-stand and stand-to-sit by their rules. In contrast, the proposed TUG in this study used the subjective criterion (45-degree inclination angle from the TUG video by two experts).

Figure 10 shows the results in a bar chart for a straightforward comparison between the two methods. The STD is also computed in the case of the proposed approach to demonstrate its robustness, whereas the STD is not calculated in [37]. The robustness of the proposed method may also be far better than that of the rule-based approach (despite no STD) if we consider the robustness of the TUG event detection results in Figure 9.

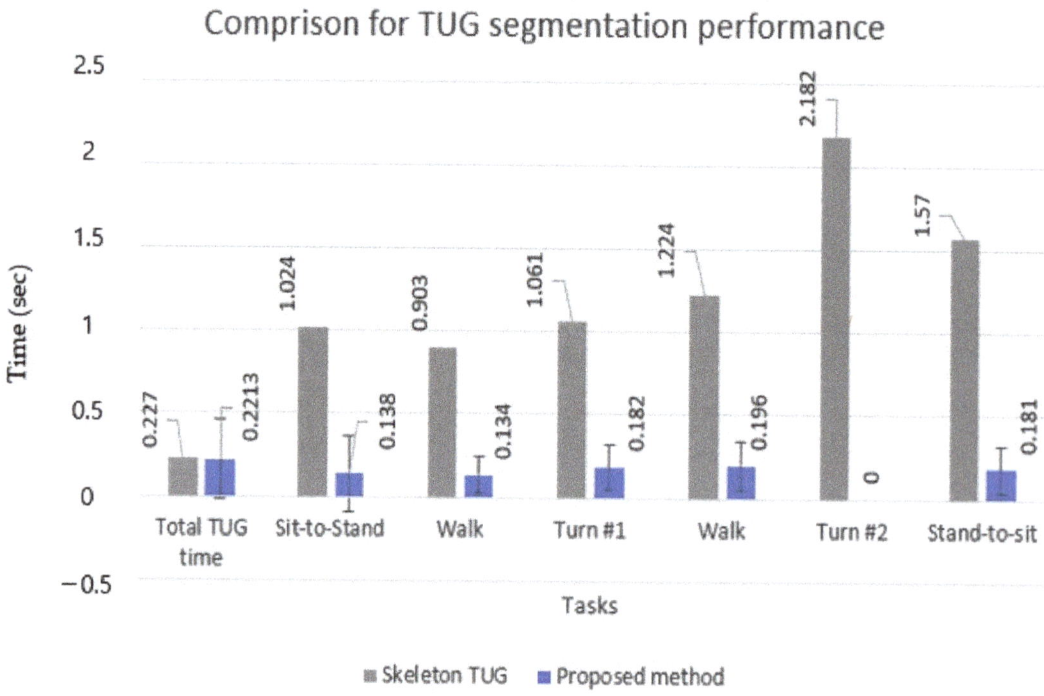

Figure 10. Comparison of MAE and STD for total TUG time and subtask segmentation.

4.5. Comparison with ANN-Based Method

We also compared the performance of the proposed method with an existing ANN-based vTUG technique [22]. This comparison should be fair in terms of 3 m TUG subtask segmentation on 30 healthy young people despite the differences in RGB installed in front of the subject [22] vs. the RGB-D installed at the subject's side in our approach. Note that vTUG used an ANN-based method only to obtain a 3D global pose. However, for segmentation, a rule-based technique was used, based on the characteristics of the trajectories of each joint. Comparisons with other ANN-based methods [4,18] were not performed because of subject differences (PD in [18], TKA [4]).

Figure 11 compares TUG event detection errors (MAE and STD) for six events in a bar chart for a more explicit comparison between the two methods. The MAE and STD for TUG event detection of the two methods showed significant differences. Overall results show that the proposed DL-based method detected TUG events far more accurately (smaller

MAE) and robustly (smaller STD) than the conventional ANN-based method [22]. A direct subtask segmentation comparison, however, cannot be conducted because vTUG only presented TUG events' performance.

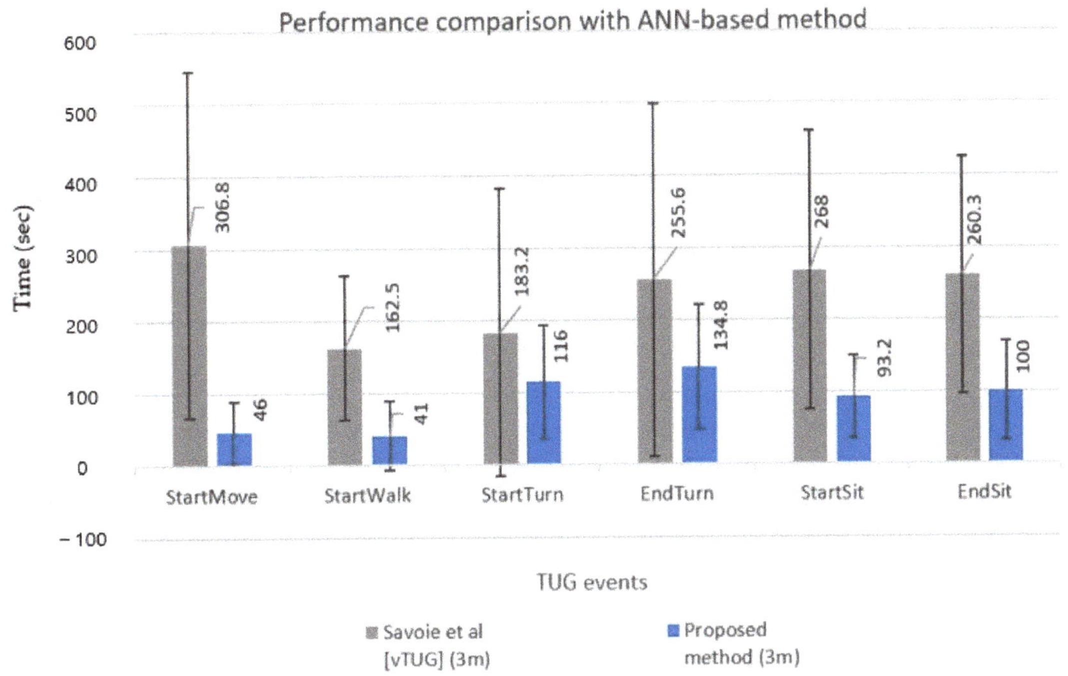

Figure 11. Comparison results with ANN-based method [22].

5. Discussion

We discussed many important points after presenting the results in Section 4; hence, we discuss only additional points here.

In some cases, DTW cannot correct other types of frame-level misclassifications, such as the ambiguity (or shift) error [4] caused by an ambiguity between successive subtasks, such as sit-to-stand, walk/turn, turn, and walk. In this case, the effect of DTW is small, resulting in a minor performance improvement. However, as shown in Figure 5, in most cases, it works well for frame-level misclassification.

The applied domain knowledge-based postprocessing step is a usual process in many ML/DL methods for action classification, such as subtask segmentation to obtain better accuracy, e.g., [4,18]. Unfortunately, fair comparisons with other ANN-based methods [4,18] cannot be performed because of subject differences (PD in [18], TKA [4]). Additionally, one should note that an existing ANN-based vTUG method [22] used an ANN only to obtain a 3D global pose. However, for segmentation, a rule-based technique was used, considering the characteristics of the trajectories of each joint. Therefore, we cannot compare our performance by using DTW with this method.

Furthermore, to fairly compare the performances using RGB-D data with those using RGB data in the proposed approach, the pelvis point (x,y,z coordinates) must be captured in the RGB images. However, the pelvis point (x,y,z coordinates) must be captured in the RGB images to compare the performance of RGB-D data with RGB data in the proposed approach. However, this requires a physical marker on the subject, the location of which may also be different from that of the pelvis point that is captured by an RGB-D camera. As an alternative to the pelvis point, the COM (Center of Mass) may be extracted from

RGB data. However, the COM is different from the pelvis point, which hinders a fair comparison between the performance with RGB-D data with that of RGB data. Moreover, COM computation requires many processes such as silhouette/background segmentation, background elimination, and COM computation from silhouette, which may accumulate errors in each process. Most seriously, the computed COM may not be robust because of pixel-based COM computation, as previously found in [22]. Although this study presented the results of the subtask segmentation for stroke patients, these results cannot be compared with other methods because of the unavailability of results for stroke patients from different approaches that use the same RGB-D cameras.

6. Conclusions

A novel DL-based subtask segmentation method was proposed for TUG tests using a single RGB-D camera and a dilated TCN. An evaluation of newly collected TUG data for three different subject groups (healthy young, elderly adults, and stroke patients) showed that the proposed method is more robust and accurate than the rule-based and ANN-based subtask segmentation methods. The evaluation results (healthy young = 95.458%, healthy adult = 94.525%, and stroke patients = 93.578%) demonstrated the generality and robustness of the proposed method. Moreover, an investigation of several inputs to the dilated TCN model showed that the input only from the pelvis is enough to achieve the best accuracy among many inputs. In addition, this single feature point significantly reduces the DL model's memory requirement and inference phase's processing time. If the TUG subtask is classified with higher accuracy, more precise clinical data (e.g., stride, gait speed, turning speed, etc.) can be extracted, which may provide medical experts with more elaborate diagnosis and therapy evaluation, e.g., for rehabilitation.

In the future, we will apply the proposed method to more diverse patient groups such as PD, MCI, AD, TKA, MS, lumbar degenerative disc disease, lower limb amputations, and COPD. In addition, we will implement the trained neural network model in the edge devices such as smart mobile phones for better portability. Finally, we will develop a method to correct shift errors for better accuracy.

Author Contributions: Conceptualization, Y.C.; Data curation, Y.C. and Y.B.; Formal analysis, Y.C., Y.B., B.C. and J.R.; Investigation, Y.C.; Methodology, Y.C.; Software, Y.C.; Validation, Y.C., Y.B., B.C. and J.R.; Writing—original draft, Y.C.; Writing—review and editing, Y.C., Y.B., B.C. and J.R. All authors have read and agreed to the published version of the manuscript.

Funding: This work was supported by an Institute of Information & Communications Technology Planning & Evaluation (IITP) grant funded by the Korean government (MSIT) (No. 2020-0-01129, Development of artificial intelligence technology for smart lower limb rehabilitation).

Institutional Review Board Statement: The study was conducted in accordance with the Declaration of Helsinki and was approved by the Institutional Review Board of the Gwangju Institute of Science and Technology (No. 20201124-HR-57-01-04).

Informed Consent Statement: Informed consent was obtained from all subjects involved in the study.

Data Availability Statement: Data sharing is not applicable to this article.

Acknowledgments: The authors would like to thank YoonWhan Kim and HeeJun Kim for their labelling contributions.

Conflicts of Interest: The authors declare no conflict of interest.

Appendix A

Table A1. Types of sensors and associated inputs.

Article	Purpose (Method)	Population	System (Sensor)	Inputs	Reason of Input Location
Hsieh et al. [23]	subtask segmentation (rule-based)	5 healthy people (25.6 ± 1.36 years old), 5 patients with severe knee osteoarthritis (67.4 ± 2.15 years old)	3 wearable sensors (3D-Acc/Gyro)	waist, R/L thigh	To acquire body acceleration and angular velocity while focusing on detecting changes of subtasks. (trunk bending, rotating, etc.)
Nguyen et al. [28]	subtask segmentation (rule-based)	16 healthy older adults (9 females; 68.7 ± 9.3 years, 7 males, 67.3 ± 5.8 years)	motion capture suit (17 IMU) (3D-Acc/Gyro), 3D-Magn	each body segment	To capture full-body 3D movement
Nguyen et al. [45]	subtask segmentation (rule-based)	12 older adults diagnosed with early PD (4 females; 67.8 ± 10.4 years; 8 males, 66.6 ± 3.6 years)	motion capture suit (17 IMU) (3D-Acc/Gyro), 3D-Magn	each body segment	To capture full-body 3D movement
Lohmann et al. [35]	subtask segmentation (rule-based)	5 older adults who suffered from age-related medical conditions, 4 healthy young	2 Kinect for Xbox 360	shoulder center (vel, acc), distance between R/L shoulder	To detect TUG events while focusing on detecting changes of subtasks
Kampel et al. [37] (SkeletonTUG)	subtask segmentation (rule-based)	11 older adults (89.3 ± 3.6 years)	1 Kinect v2	spine shoulder, distance between R/L shoulder	To detect TUG events while focusing on detecting changes of subtasks
Kampel et al. [37] (Detph TUG)	subtask segmentation (rule-based)	11 older adults (89.3 ± 3.6 years)	1 Kinect v2	Center of Mass	To acquire moving history data, COM is calculated using silhouette extraction method
Salarian et al. [3]	subtask segmentation (rule-based)	12 older adults in early stages of PD (60.4 ± 8.5 years), 12 age-matched control subjects (60.2 ± 8.2 years)	7 IMUs	forearms, shanks, thighs, trunk	To detect and analyze each subtask. For example, sensor attached to forearm used to detect walking
Hsieh et al. [4]	subtask segmentation (ANN-based)	26 patients with severe knee osteoarthritis (69.15 ± 6.71 years)	6 wearable sensors (3D-Acc/Gyro), 2 RGB cameras	chest, lower back, R/L thigh, R/L shank	To acquire body movement from various parts
Li et al. [18]	subtask segmentation (ANN-based)	24 PD patients (56.79± 9.48 years)	1 RGB camera	neck, R/L shoulder, R/L hip, R/L knee, R/L ankle	In total, 9 body keypoints were used to represent the human poses

Table A1. *Cont.*

Article	Purpose (Method)	Population	System (Sensor)	Inputs	Reason of Input Location
Savoie et al. [22]	subtask segmentation (ANN-based)	30 healthy young (26.1 ± 9.4 years)	1 Kinect V2, 1 RGB camera	center of shoulder, height between the center of feet and nose, right and left hip	To detect TUG events while focusing on detecting changes of the subtasks, e.g., center of shoulder for 'began to stand'
Ortega-Bastidas et al. [25]	fall risk prediction	25 healthy young (25–33 years), 12 older adults (59–93 years)	IMU sensor RGB video (GoPro hero 7 for answer)	back (acceleration, angles, and angular velocities)	To detect all gait, biomechanical elements of the pelvis, and other spatial and temporal kinematics factor
Jian et al. [42]	fall risk prediction	40 subjects (most tests were conducted on healthy subjects; few subjects were with high fall risk)	1 RGB camera, 1 RGBD camera	full joints (3D spatial)	To compute gait characteristics such as gait speed, step length, etc. Fall risk has been predicted based on gait characteristics using ML-method
Wang et al. [17]	abnormal gait classification	404 subjects (subjects are selected to fulfill the requirement of diversity)	1 RGB camera	vertical location sequence of R/L shoulder	Considering the visibility and stability of joint detection

References

1. Podsiadlo, D.; Richardson, S. The Timed "Up & Go": A Test of Basic Functional Mobility for Frail Elderly Persons. *J. Am. Geriatr. Soc.* **1991**, *39*, 142–148. [CrossRef]
2. Sprint, G.; Cook, D.J.; Weeks, D.L. Toward Automating Clinical Assessments: A Survey of the Timed Up and Go. *IEEE Rev. Biomed. Eng.* **2015**, *8*, 64–77. [CrossRef]
3. Salarian, A.; Horak, F.B.; Zampieri, C.; Carlson-Kuhta, P.; Nutt, J.G.; Aminian, K. iTUG, a Sensitive and Reliable Measure of Mobility. *IEEE Trans. Neural Syst. Rehabil. Eng.* **2010**, *18*, 303–310. [CrossRef]
4. Hsieh, C.-Y.; Huang, H.Y.; Liu, K.C.; Chen, K.H.; Hsu, S.J.; Chan, C.T. Subtask Segmentation of Timed up and Go Test for Mobility Assessment of Perioperative Total Knee Arthroplasty. *Sensors* **2020**, *20*, 6302. [CrossRef]
5. Timmermans, C.; Roerdink, M.; van Ooijen, M.W.; Meskers, C.G.; Janssen, T.W.; Beek, P.J. Walking Adaptability Therapy After Stroke: Study Protocol for a Randomized Controlled Trial. *Trials* **2016**, *17*, 425. [CrossRef]
6. Sebastião, E.; Sandroff, B.M.; Learmonth, Y.C.; Motl, R.W. Validity of the Timed up and Go Test as a Measure of Functional Mobility in Persons with Multiple Sclerosis. *Arch. Phys. Med. Rehabil.* **2016**, *97*, 1072–1077. [CrossRef]
7. Gautschi, O.P.; Corniola, M.V.; Joswig, H.; Smoll, N.R.; Chau, I.; Jucker, D.; Stienen, M.N. The Timed up and Go Test for Lumbar Degenerative Disc Disease. *J. Clin. Neurosci.* **2015**, *22*, 1943–1948. [CrossRef]
8. Clemens, S.M.; Gailey, R.S.; Bennett, C.L.; Pasquina, P.F.; Kirk-Sanchez, N.J.; Gaunaurd, I.A. The Component Timed-Up-and-Go Test: The Utility and Psychometric Properties of Using a Mobile Application to Determine Prosthetic Mobility in People with Lower Limb Amputations. *Clin. Rehabil.* **2018**, *32*, 388–397. [CrossRef]
9. Marques, A.; Cruz, J.; Quina, S.; Regêncio, M.; Jácome, C. Reliability, Agreement and Minimal Detectable Change of the Timed up & Go and the 10-Meter Walk Tests in Older Patients with COPD. *COPD J. Chronic Obstr. Pulm. Dis.* **2016**, *13*, 279–287. [CrossRef]
10. Greene, B.R.; Kenny, R.A. Assessment of Cognitive Decline through Quantitative Analysis of the Timed up and Go Test. *IEEE Trans. Biomed. Eng.* **2012**, *59*, 988–995. [CrossRef]
11. Wall, J.C.; Bell, C.; Campbell, S.; Davis, J. The Timed Get-Up-and-Go Test Revisited: Measurement of the Component Tasks. *J. Rehabil. Res. Dev.* **2000**, *37*, 109–113.

12. Shumway-Cook, A.; Brauer, S.; Woollacott, M. Predicting the Probability for Falls in Community-Dwelling Older Adults Using the Timed Up & Go Test. *Phys. Ther.* **2000**, *80*, 896–903.
13. Vernon, S.; Paterson, K.; Bower, K.; McGinley, J.; Miller, K.; Pua, Y.H.; Clark, R.A. Quantifying Individual Components of the Timed up and Go Using the Kinect in People Living with Stroke. *Neurorehabilit. Neural Repair* **2015**, *29*, 48–53. [CrossRef]
14. Aschneider, F.B.; Valencia, L.S.V.; Bastos-Filho, T.F.; Marques-Ciarelli, P.; Frizera-Neto, A. Automation and segmentation of timed up and go test with single redundant IMU. *J. Mech. Eng. Biomech.* **2019**, *4*, 45–51. [CrossRef]
15. Ponciano, V.; Pires, I.M.; Ribeiro, F.R.; Marques, G.; Garcia, N.M.; Pombo, N.; Spinsante, S.; Zdravevski, E. Is the Timed-Up and Go Test Feasible in Mobile Devices? A Systematic Review. *Electronics* **2020**, *9*, 528. [CrossRef]
16. Ansai, J.H.; De Andrade, L.P.; Nakagawa, T.H.; Rebelatto, J.R. Performances on the Timed up and Go Test and Subtasks Between Fallers and Non-fallers in Older Adults with Cognitive Impairment. *Arq. Neuropsiquiatr.* **2018**, *76*, 381–386. [CrossRef]
17. Wang, Y.; Zou, Q.; Tang, Y.; Wang, Q.; Ding, J.; Wang, X.; Shi, C.-J.R. SAIL: A Deep-Learning-Based System for Automatic Gait Assessment from TUG Videos. *IEEE Trans. Hum.-Mach. Syst.* **2021**, *52*, 110–122. [CrossRef]
18. Li, T.; Chen, J.; Hu, C.; Ma, Y.; Wu, Z.; Wan, W.; Huang, Y.; Jia, F.; Gong, C.; Wan, S.; et al. Automatic Timed up-and-Go Sub-task Segmentation for Parkinson's Disease Patients Using Video-Based Activity Classification. *IEEE Trans. Neural Syst. Rehabil. Eng.* **2018**, *26*, 2189–2199. [CrossRef]
19. Landi, F.; Abbatecola, A.M.; Provinciali, M.; Corsonello, A.; Bustacchini, S.; Manigrasso, L.; Cherubini, A.; Bernabei, R.; Lattanzio, F. Moving Against Frailty: Does Physical Activity Matter? *Biogerontology* **2010**, *11*, 537–545. [CrossRef]
20. Peterson, M.J.; Giuliani, C.; Morey, M.C.; Pieper, C.F.; Evenson, K.R.; Mercer, V.; Cohen, H.J.; Visser, M.; Brach, J.S.; Kritchevsky, S.B.; et al. Physical Activity as a Preventative Factor for Frailty: The Health, Aging, and Body Composition Study. *J. Gerontol. Ser. A Biol. Sci. Med. Sci.* **2009**, *64*, 61–68. [CrossRef]
21. Rubenstein, L.Z.; Josephson, K.R.; Trueblood, P.R.; Loy, S.; Harker, J.O.; Pietruszka, F.M.; Robbins, A.S. Effects of a Group Exercise Program on Strength, Mobility, and Falls Among Fall-Prone Elderly Men. *J. Gerontol. Ser. A Biol. Sci. Med. Sci.* **2000**, *55*, M317–M321. [CrossRef]
22. Savoie, P.; Cameron, J.A.D.; Kaye, M.E.; Scheme, E.J. Automation of the Timed-up-and-Go Test Using a Conventional Video Camera. *IEEE J. Biomed. Health Inform.* **2021**, *24*, 1196–1205. [CrossRef]
23. Hsieh, C.-Y.; Huang, H.-Y.; Liu, K.-C.; Chen, K.-H.; Hsu, S.J.; Chan, C.-T. Automatic Subtask Segmentation Approach of the Timed up and Go Test for Mobility Assessment System Using Wearable Sensors. In Proceedings of the 2019 IEEE EMBS International Conference on Biomedical & Health Informatics (BHI), Chicago, IL, USA, 19–22 May 2019. [CrossRef]
24. Hassani, A.; Kubicki, A.; Brost, V.; Yang, F. Preliminary Study on the Design of a Low-Cost Movement Analysis System Reliability Measurement of Timed up and Go Test. In Proceedings of the 2014 International Conference on Computer Vision Theory and Applications (VISAPP), Lisbon, Portugal, 5–8 January 2014.
25. Bergquist, R.; Nerz, C.; Taraldsen, K.; Mellone, S.; Ihlen, E.A.F.; Vereijken, B.; Helbostad, J.L.; Becker, C.; Mikolaizak, A.S. Predicting Advanced Balance Ability and Mobility with an Instrumented Timed up and Go Test. *Sensors* **2020**, *20*, 4987. [CrossRef]
26. Ortega-Bastidas, P.; Aqueveque, P.; Gómez, B.; Saavedra, F.; Cano-de-la-Cuerda, R. Use of a Single Wireless IMU for the Segmentation and Automatic Analysis of Activities Performed in the 3-m Timed up & Go Test. *Sensors* **2019**, *19*, 1647. [CrossRef]
27. Hellmers, S.; Izadpanah, B.; Dasenbrock, L.; Diekmann, R.; Bauer, J.M.; Hein, A.; Fudickar, S. Towards an Automated Unsupervised Mobility Assessment for Older People Based on Inertial TUG Measurements. *Sensors* **2018**, *18*, 3310. [CrossRef]
28. Nguyen, H.P.; Ayachi, F.; Lavigne-Pelletier, C.; Blamoutier, M.; Rahimi, F.; Boissy, P.; Jog, M.; Duval, C. Auto Detection and Segmentation of Physical Activities During a Timed-Up-and-Go (TUG) Task in Healthy Older Adults Using Multiple Inertial Sensors. *J. Neuroeng. Rehabil.* **2015**, *12*, 36. [CrossRef]
29. Reinfelder, S.; Hauer, R.; Barth, J.; Klucken, J.; Eskofier, B.M. Timed Up-and-Go Phase Segmentation in Parkinson's Disease Patients Using Unobtrusive Inertial Sensors. In Proceedings of the 37th Annual International Conference of the IEEE Engineering in Medicine and Biology Society (EMBC), Milan, Italy, 25–29 August 2015. [CrossRef]
30. He, K.; Gkioxari, G.; Dollar, P.; Girshick, R. Mask R-CNN. In Proceedings of the 2017 IEEE International Conference on Computer Vision (ICCV), Venice, Italy, 22–29 October 2017; pp. 2980–2988. [CrossRef]
31. Alin-Ionut, P.; Mihai, Z.; Sminchisescu, C. Deep Multitask Architecture for Integrated 2d and 3-d Human Sensing. In Proceedings of the IEEE Conference on Computer Vision and Pattern Recognition, Honolulu, HI, USA, 21–26 July 2017.
32. Pfister, A.; West, A.M.; Bronner, S.; Noah, J.A. Comparative Abilities of Microsoft Kinect and Vicon 3D Motion Capture for Gait Analysis. *J. Med. Eng. Technol.* **2014**, *38*, 274–280. [CrossRef]
33. Müller, B.; Ilg, W.; Giese, M.A.; Ludolph, N. Validation of Enhanced Kinect Sensor Based Motion Capturing for Gait Assessment. *PLoS ONE* **2017**, *12*, e0175813. [CrossRef]
34. Geerse, D.J.; Coolen, B.H.; Roerdink, M. Kinematic Validation of a Multi-Kinect v2 Instrumented 10-Meter Walkway for Quantitative Gait Assessments. *PLoS ONE* **2015**, *10*, e0139913. [CrossRef]
35. Lohmann, O.; Luhmann, T.; Hein, A. Skeleton Timed up and go. In Proceedings of the IEEE International Conference on Bioinformatics and Biomedicine, Philadelphia, PA, USA, 4–7 October 2012. [CrossRef]
36. Van Kersbergen, J.; Otte, K.; de Vries, N.M.; Bloem, B.R.; Röhling, H.M.; Mansow-Model, S.; van der Kolk, N.M.; Overeem, S.; Zinger, S.; van Gilst, M.M. Camera-Based Objective Measures of Parkinson's Disease Gait Features. *BMC Res. Notes* **2021**, *14*, 329. [CrossRef]

37. Kampel, M.; Doppelbauer, S.; Planinc, R. Automated Timed Up & Go Test for Functional Decline Assessment of Older Adults. In Proceedings of the 12th EAI International Conference on Pervasive Computing Technologies for Healthcare, New York, NY, USA, 21–24 May 2018; pp. 208–216. [CrossRef]
38. Beyea, J.; McGibbon, C.A.; Sexton, A.; Noble, J.; O'Connell, C. Convergent Validity of a Wearable Sensor System for Measuring Sub-task Performance During the Timed up-and-Go Test. *Sensors* **2017**, *17*, 934. [CrossRef]
39. Lea, C.; Flynn, M.D.; Vidal, R.; Reiter, A.; Hager, G.D. Temporal Convolutional Networks for Action Segmentation and Detection. In Proceedings of the IEEE Conference on Computer Vision and Pattern Recognition, Honolulu, HI, USA, 21–26 July 2017; pp. 156–165.
40. Bai, S.; Zico Kolter, J.; Koltun, V. An empirical evaluation of generic convolutional and recurrent networks for sequence modeling. *arXiv* **2018**, arXiv:1803.01271.
41. Nguyen, H.; Lebel, K.; Boissy, P.; Bogard, S.; Goubault, E.; Duval, C. Auto Detection and Segmentation of Daily Living Activities During a Timed Up and Go Task in People with Parkinson's Disease Using Multiple Inertial Sensors. *J. Neuroeng. Rehabil.* **2017**, *14*, 26. [CrossRef]
42. Jian, M.A. Predicting TUG Score from Gait Characteristics with Video Analysis and Machine Learning. *bioRxiv* **2020**. [CrossRef]
43. Nguyen, M.D.; Mun, K.R.; Jung, D.; Han, J.; Park, M.; Kim, J.; Kim, J. IMU-Based Spectrogram Approach with Deep Convolutional Neural Networks for Gait Classification. In Proceedings of the IEEE International Conference on Consumer Electronics (ICCE); IEEE Publications, Las Vegas, NV, USA, 4–6 January 2020. [CrossRef]
44. Galán-Mercant, A.; Cuesta-Vargas, A.I. Clinical Frailty Syndrome Assessment Using Inertial Sensors Embedded in Smartphones. *Physiol. Meas.* **2015**, *36*, 1929–1942. [CrossRef]
45. Ayena, J.C.; Chioukh, L.; Otis, M.J.; Deslandes, D. Risk of Falling in a Timed up and Go Test Using An UWB Radar and an Instrumented Insole. *Sensors* **2021**, *21*, 722. [CrossRef]

MDPI
St. Alban-Anlage 66
4052 Basel
Switzerland
Tel. +41 61 683 77 34
Fax +41 61 302 89 18
www.mdpi.com

Sensors Editorial Office
E-mail: sensors@mdpi.com
www.mdpi.com/journal/sensors

MDPI

www.ingramcontent.com/pod-product-compliance
Lightning Source LLC
Chambersburg PA
CBHW061452210926
51850CB00008B/1696
9783036581729